Accelerant

Accelerant

Energy Infrastructures and the Natural World in Making Modern Iran

Ciruce Movahedi-Lankarani

Stanford University Press

Stanford, California

Stanford University Press
Stanford, California

Library of Congress Cataloging-in-Publication Data available upon request.
ISBN 9781503643871 (case), 9781503646223 (paper), 9781503646230 (ebook)

Cover design: Michel Vrana
Cover art: schankz / istock
Typeset by Newgen in Adobe Garamond Pro 12/14

The authorized representative in the EU for product safety and compliance is: Mare Nostrum Group B.V. | Doelen 72 | 4831 GR Breda | The Netherlands | Email address: gpsr@mare- nostrum. co. uk | KVK chamber of commerce number: 96249943

To Cecilia
and the Triumvirate

Contents

Acknowledgements

As I write these final words in the early summer of 2025, Iran's energy infrastructures have come under sustained aerial assault for the first time since the Iran-Iraq War. Though details of what has happened are sparse, what emerges from the images of fire and smoke is the fact that the very foundations of Iranian society are being shaken. It is far too early to know what these events mean for Iran, the Middle East, and the world, but what is clear is that for all that modernity is seen as insulating us from nature's sharp claw, in many ways we have simply substituted one kind of vulnerability with another. In its sprawling filaments linking hydrocarbons and agricultural land and electricity generation and people across thousands of miles, industrialized life is fragile, and it does not take much to ravage it. May everyone, having peered over the ledge, step back from the brink, in this conflict and all others.

This project began in a simple conversation more than a decade ago. At the time, I was interested in how Iranians in the mid-twentieth century had encountered and incorporated new technologies into their culture and society. Toward that end, I asked my father, who grew up in Iran in the 1950s and 1960s, what device he remembered as having most changed his life during that time. His response surprised me. Where I had expected him to point to something like television or the automobile, he instead answered by saying: "I remember when your grandmother brought home a gas stove; she started cooking inside and it allowed me to use the yard to play." It was a modest response, one focused on his own experiences as a child, but also one that, as I considered all that is required to make a gas stove a useful appliance, progressively unfolded to encompass more and more of Iran's recent history. It has been a winding road since that early autumn afternoon, and while this project has long since come to focus on large-scale infrastructures and their entanglements with the natural world, it nonetheless still carries the imprint of people, their everyday lives, and the generosity with which so many share them. Without that spirit, this book would never have come to be.

While it has become *de rigueur* to note the community effort that goes into any single-author work, it makes the reality no less true, and I owe a great many people a very great deal for their support. At the University of Pennsylvania, I was profoundly fortunate to receive the guidance of Firoozeh Kashani-Sabet, whose probing questions and high standards consistently pushed me out of my intellectual comfort zone. In large part, I have become the scholar I am today because of her. I am also grateful to Eve M. Troutt Powell for her mentorship and the long conversations that helped me find my scholarly voice. I am further thankful to have been given the opportunity to work with a number of distinguished scholars across the country, including Nikhil Anand, Toby C. Jones, Vanessa Ogle, Adelheid Voskuhl, and John E. Woods, all of whom helped in honing my inchoate interests into a viable project. Valuable feedback and moral support were provided by my friends and colleagues: Nimrod Ben Zeev, Ayelet Brinn, Beeta Baghoolizadeh, Kelsey Rice, Josef Alaric Nothmann, and Lacy Feigh. Graduate school is transformative in ways both professional and personal, and if I have changed at all for the better, it is because of them.

The research and writing of this book received generous support from a number of institutions, including the National Endowment for the Humanities, the Social Science Research Council, the Farhang Foundation, and the University of Pennsylvania. Without their commitment to scholarship and the pursuit of understanding, projects like this are not possible. I also thank the University of Southern California and its Department of Middle East Studies, and particularly Ramzi Rouighi, for their continued support of me and my work. Historians are nothing without their sources, and I thank all the staff members at the National Library and Archives of the Islamic Republic of Iran; the Iran Petroleum Museum and Document Center; the Majlis Library, Museum, and Document Center; the Institute for Iranian Contemporary Historical Studies; the National Iranian Oil Company Information Center and Central Library; the Documents and Publications Center of the Plan and Budget Organization of the Islamic Republic of Iran; Princeton Special Collections; and the British Petroleum Archives at the University of Warwick. I am further indebted to the time and considered critiques of all the anonymous readers who have reviewed my work over the years; their feedback has pushed me and my scholarship in productive ways.

This book was strongly shaped by my time in Iran and all the friends and family I have there. Their company and assistance were instrumental in making my experience there enjoyable and productive. I give particular thanks to Dr. Hājj Seyyed Mahmud Marʿashi-Najafi for his advice and gracious assistance with archival access; Majid Zanjāni for offering priceless friendship and conversation, invaluable aid in navigating Iranian state bureaucracy, and helping open the doors to numerous ministerial document collections; Fuād Zanjāni for showing me Tehran as it lives and breathes today and making sure that I had fun during my time there; and my great-aunt Mansureh "Azam" Marʿashi-Najafi for her generosity in opening her home to me and safeguarding my health with a steady supply of delicious home cooking. I also thank Seyyed Muhammad Kāzem Marʿashi-Najafi for his recollections of working in the petroleum industry, his help in obtaining access to NIOC's central library, and for making me laugh. I thank Akbar Moqaddam, Mehdi Kamāli, and Abbās Khayāt for their warmth and hospitality, and Amir Abbāszādeh for both his friendship and his help in navigating the bureaucratic necessities of life in Tehran. I am also grateful to my grandfather Ali Lankarani and my great-uncle Jafar Lankarani for helping me settle into the old family neighborhood, for making sure shopkeepers did not charge me inflated prices, and for telling me stories of the old days. Beyond any shadow of doubt, this book would not exist without them.

No acknowledgements could be complete without special thanks to my friends and family, particularly my parents, Hossein (Professor of Mathematics at the Pennsylvania State University at Altoona) and Stephanie, my sisters, Sohayla and Maryam, and my grandmother, Cynthia. Special thanks are owed to my grandfather, John A. Jakle, Professor Emeritus of Geography and Landscape Architecture at the University of Illinois at Urbana-Champaign, for teaching me as a child to really see human landscapes. To my "cousins" and friends Alex, Robin, Kelly, Jamie, Catie, and Zack, I appreciate your efforts to make my life enjoyable despite my best efforts. Not all who provide comfort are human, and I am thankful to Esfandiyar, Rocky, and Homa for their companionship over the years; I wish you could be here to see the end of this journey. Finally, I owe a deep debt of gratitude to my spouse, Cecilia. Neither my academic career nor this project would have been possible without her exuberant cheer. I am deeply appreciative of all the late nights chatting; for the long days of hiking; for the companionship on interminable flights to far off locales; for the conversations and the debates; for all the moves (and for succeeding in a career that allows for that); for the proofreading; for the encouragement; and for the love.

No list of acknowledgements can ever be truly complete, and I owe far more to more people than is possible to name here. So, to all the brilliant, weird, wonderful, and courageous people I have met along the way: thank you.

Note on Conventions

All translations in this book are my own unless otherwise indicated. Transliteration has been undertaken with the guidelines established by *Iranian Studies*, including the use of ā for the long *alef* vowel and the representation of the *ezāfeh* with -e and -ye. Proper names with familiar English spellings have been rendered without diacritical marks (e.g., Reza Shah instead of Rezā Shāh). What counts as 'familiar' is an inexact science, and no significance is intended in the choice of which names are treated as such.

One of the core themes in this book's analysis is the historical importance of the overlapping nature of crude oil and natural gas. As such, when discussing the two fossil hydrocarbon resources together, the term 'petroleum' is used to refer to them in a combined manner (e.g., 'petroleum deposit' for hydrocarbon reservoirs containing both oil and gas).

List of Abbreviations

AIOC	Anglo-Iranian Oil Company
APOC	Anglo-Persian Oil Company
BP	British Petroleum
CNG	Compressed Natural Gas
ECAFE	Economic Commission for Asia and the Far East
IBRD	International Bank for Reconstruction and Development
IGAT-1	First Iran Gas Trunkline
IMEG	Iranian Management and Engineering Group
MEPL	Middle East Pipeline
MMSCFD	Million Standard Cubic Feet per Day
NIGC	National Iranian Gas Company
NIOC	National Iranian Oil Company
OPEC	Organization of the Petroleum Exporting Countries

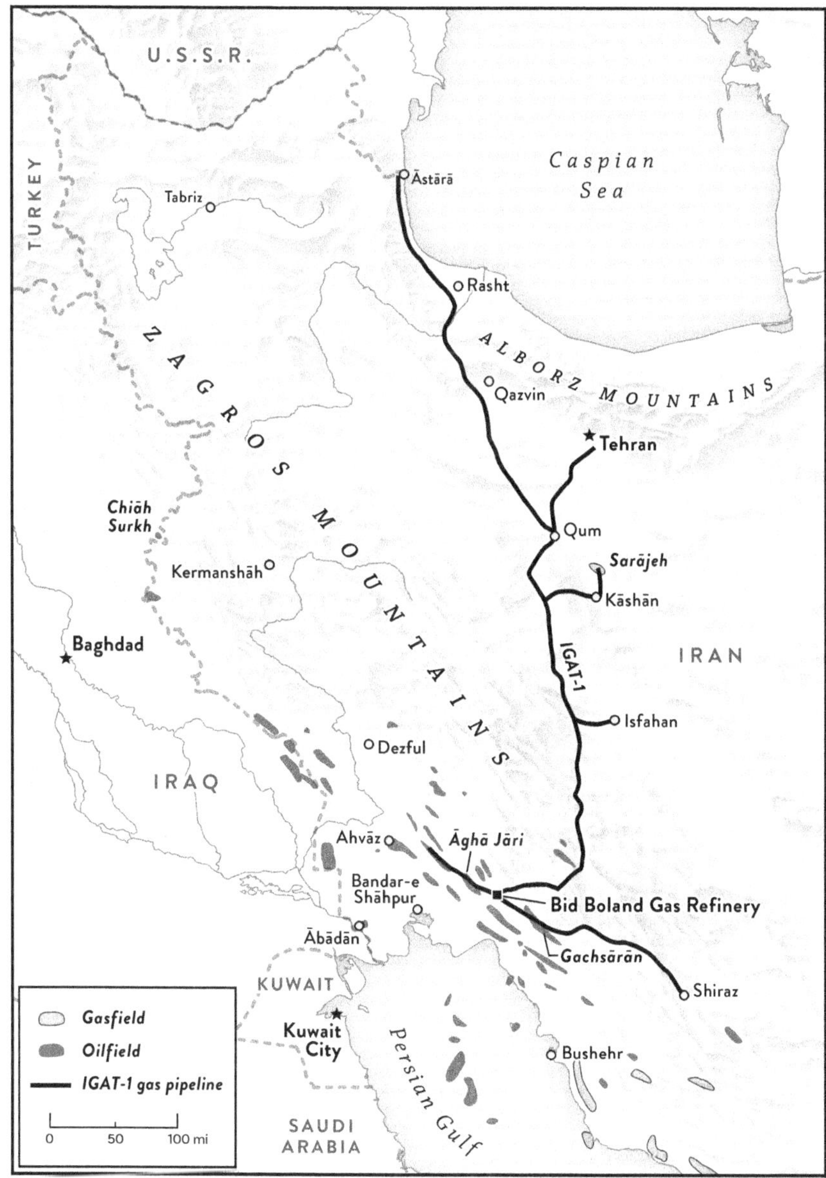

MAP. **Route of the First Iran Gas Trunkline (IGAT-1) and its Branch Lines Superimposed on a Map of Iran.**

Introduction

To walk through the streets of contemporary Tehran is to see the ubiquity of neighborhood restaurants, large and small, upscale and down, advertising *kabāb-e zoghāli*, or kabab cooked over charcoal. At most, large lunchtime crowds wait, watching while cooks, dressed in white and sweating over long coal pits topped with skewered meat, tend to their fires with the rhythmic waving of handheld fans, the wafting smoke marking their use of a traditional (and more flavorsome) fuel. To hail a cab after lunch is to listen to its driver bemoan the feebleness of compressed natural gas fuel in the city's hilly terrain. To exit the capital's subway at Haft-e Tir Square is to see an enormous and colorful banner celebrating the completion of yet another stage of the South Pars gas field's commercial development. To discuss politics is to hear persistent rumors that the people of Sistān-Baluchestān had supported the 2017 reelection of President Hassan Rouhani because their communities had finally been connected to the country's natural gas network. That the smoky aromas and glowing embers of neighborhood eateries have become notable to city residents, that most motor vehicles have been equipped to use both gasoline and CNG, that the Iranian state publicly celebrates seemingly obscure happenings in the country's gas industry, and that the residents of one of Iran's most restive peripheries would laud closer ties to the country's center are all a result of one of the most significant developments of Iran's recent history: the rise of natural gas as the country's primary source of energy.

With Iran sitting atop the world's second largest reserves of natural gas, this book argues that such petro-bounty is the spine around which Iranian society has become ordered, not only through the revenues that oil exports have brought, but in the energetic foundation that gas provided for the country's development. Powered by a shared commitment to industrialization and the intensifying energy consumption it would bring, the thorough integration of natural gas energy into Iranian life was the product of sustained attention by two ruling regimes over decades of time. It reflected the combined influence of development planners, oil firms, industrialists, engineers, consumers, mountain ranges, sedimentary rock, and natural gas itself. Beginning in the 1950s and accelerating thereafter, natural gas was deliberately woven into nearly every facet of Iranian life, changing how people heated their homes, fueled their vehicles, and cooked their food. During that time, factories and power plants in cities from Tehran to Shiraz began to use the new fuel too, a policy decision driven by a pursuit of rapid economic growth paired with deepening anxieties about the environmental violence that industrialization had begun to inflict. Cleaner burning than the charcoal and oil products that many Iranians relied on for fuel, natural gas seemingly promised a future that combined high energy consumption and unpolluted air—an improvement over what many saw as the environmental underside of the industrialized world's prosperity. Natural gas utilization became an emblematic expression of the Iranian state's developmentalist priorities both before and after the 1979 revolution, serving as an important tool in the legitimation strategies of the Pahlavi monarchy and the Islamic Republic alike. In that role, it became central to the imagined futures that guided Iran's developmentalist drive, embodying intertwined aspirations for modernizing technological sophistication, anticolonial resource nationalism, and environmental preservation. For many of Iran's modernizing officials, gas was a futuristic source of energy for a future world power, the primary means by which they hoped to build a prosperous modern society not just equal to those of the world's wealthiest regions, but superior.

Tracing the roughly half-century transformation of natural gas from a waste product burned in southwestern Iran's petroleum fields to a vital resource consumed across the country, this book studies the energy source as a crucial enabler of the country's developmentalist policies; a potent symbol within its highly charged politics of anticolonial modernization; and a failed technofix to the problem of urban air pollution. Beginning in the mid-1930s and accelerating thereafter, these three strands were tightly woven, linked by the industrialized futures that Iranians imagined as well as the geological, topographical, and chemical natures that shaped gas exploitation. Reflecting the intimate braiding of those thematic strands, this book is not a comprehensive history of Iran's natural gas industry in either the Pahlavi era or after. Nor is it an exhaustive account of Iran's accumulating environmental troubles in an age of rapid industrialization and global climate change. It is instead an effort to understand how natural gas, and through it the natural world, shaped the developmentalist trajectory and energy use of twentieth-century Iranian society. The choice to rely upon gas made its properties as both a natural substance and a refined product influential to the forms that modernizing industrialization took in the country. Iranians' mid-twentieth-century lionization of intense hydrocarbon consumption reflected an imagined future built on the comingling of resource nationalism and the materiality of natural gas itself—a materiality that was at once heterogeneous, shifting, and tied to the nature of mountain ranges and subterranean geographies. Utilizing gas meant overcoming the resistance of foreign oil companies like British Petroleum and their functional ownership of Iran's petroleum fields. That endeavor was built upon decades of negotiation and dispute as well as the ability of Iranian officials to create both gas markets and the enormous and technologically sophisticated infrastructures needed to serve them. But this was the great opportunity that gas presented Iranian modernizers: with it, they could build the energy-hungry society they desired without cannibalizing the petroleum exports that the oil majors (and thus their own developmentalist programs) depended upon. With natural gas, seemingly inexhaustible amounts of fuel would be made available for manufacturing, for electricity generation, for home heating, even for motor vehicles, all paid for by revenues generated from shipping oil abroad.

Joining Iran-as-state, Iran-as-society, and Iran-as-nature, over the course of the twentieth century, natural gas came to embody and empower profound imaginings of Iran's future, visions that expressed tangled understandings of industrialized prosperity, national independence, and environmental preservation. The story of natural gas in Iran is thus one that brings together the political, technological, social, and environmental histories of the country. It is each of those things, but it is also much more, for gas functioned not only as a substance of joining, linking together seemingly disparate strands of Iranian history, but also as a co-constitutive catalyst, influencing each thread in turn. Gas is thus a powerful and manipulable lens through which to examine Iran's recent past. Turn it one way to examine Iranian state building, making visible the extensive continuity between the social visions animating policies under the Pahlavi monarchy and the Islamic Republic. Turn it another way to see Iranian officials, planners, and engineers working to harness gas as a means of building an independent and sovereign modernity that rivaled the world's most prosperous. Turn it yet another direction to uncover how changing Iranian lifeways were shaped by more than ideology and religion, becoming reordered around new technologies of natural gas transmission and consumption. With a final twist, the boundaries between the human and natural worlds blur, revealing the extensive ways that modern Iranian society is as much a product of mountain ranges, ancient seas, and local climates as it is human hands. That these varied aspects of the story are overlapping and interlocking is an outcome of what natural gas energy has been for Iranians: a substance upon which they projected and attempted to build some of their most potent national aspirations; an ancient mixture of hydrocarbons and impurities lifted from deep beneath the earth; a source of fuel burned by energy-hungry consumers in ever-increasing quantities; a sprawling technological system of pipelines, refineries, and gas connections, one celebrated for its smashing of seemingly insurmountable natural barriers but nonetheless reflective of the country's geology and topography; and finally, a failed environmental savior, ultimately unable to fulfill aspirations for clean urban air.

Understanding such tangled threads is the aim of this book, not with the goal of picking them apart, but to see how their knotting fostered new political and social forms in which Iranians became voracious consumers of energy. From virtually nothing in the 1950s, gas use today accounts for approximately 70 percent of the country's total energy consumption, some 23.8 billion cubic feet per day in 2023.[1] Natural gas did not only or even primarily replace other fuels; rather, it enabled rapid and accelerating growth in Iranians' consumption of hydrocarbon energy across the second half of the twentieth century. For officials of both Iran's pre-revolutionary Pahlavi monarchy and post-revolutionary Islamic Republic, the country's enormous natural gas reserves were an opportunity to build an energy-intensive future with a fuel source under their own control. Iranians claimed the right and the ability to exploit their gas resources from foreign oil majors in the mid-1950s, the result of decades of pressure on quasi-colonial firms like British Petroleum to stop discarding the enormous quantities of natural gas they extracted as a byproduct of their oil operations. Natural gas and crude oil were commingled—in a material sense, as overlapping sections of a hydrocarbon continuum that accumulated together in underground deposits; in a political sense, as resources tied to state legitimation strategies; and in a socioeconomic sense, as the foundations of the country's developmentalist policies. Such significance was created and expressed through the infrastructures that gas use required. In the 1960s and 1970s, state officials oversaw the construction of massive refineries and country-spanning pipeline networks aimed at serving both domestic demand and export commitments to the Soviet Union. Official discourse used the activities of Iran's new National Iranian Gas Company to imbue gas utilization with notions of independence and self-modernization, championing them as embodiments of the country's developmentalist ambitions and the massive energy expenditures they would enable. Natural gas's calorific impetus was moreover paired with budding environmentalist hopes that its use would make for cleaner air in cities already seeing the smoggy effects of industrialization. But gas was a troublesome substance, one with widely diverging properties in its natural state and requiring careful management to transport hundreds of miles across mountainous terrain. Overcoming such challenges, to seemingly conquer nature itself, was offered as proof of an official commitment to bettering the lives of Iranians, a strategy of political legitimation that gave rise to the monumental pipelines and refineries that came to link southwest Iran's petroleum fields to urban areas across the country. More than a utilitarian technical system, in the heated dreams of state officials, natural gas and its systems of conveyance would be an enormous web drawing Iranians into a singular national whole. Later, in the 1970s, the anger of many rural inhabitants at being left out of that budding collective fused with the country's swirling revolutionary fervor to reinforce the sense that to be Iranian was to be a consumer of natural gas. It would be a notion that was cemented in the following decades as the new Islamic Republic prioritized the provision of gas to towns and villages.

Energetic Anticolonialism

Beginning in the 1960s and accelerating thereafter, Iran's natural gas resources would be the energetic foundation upon which many Iranians built their developmental aspirations. But natural gas as a potent sociotechnical force did not spring forth from the earth fully formed. Its roots lay instead in the great prize of the twentieth century: oil. Better understood as overlapping portions of a single hydrocarbon spectrum than as distinct substances, crude oil and natural gas frequently appear together. For eons they have silently filled petroleum reservoirs across a wide region now known as the Persian Gulf Basin, a place understood as being one of the richest hydrocarbon habitats in the world and encompassing much of southwest Iran. Industrial exploitation of Iranian petroleum dates to the turn of the twentieth century and the awarding of a concession for the prospecting, production, and marketing of petroleum—defined as oil, asphalt, ozokerite, and natural gas—to the British mining magnate William Knox D'Arcy. When his explorers struck (black) gold in May 1908 in the foothills of the Zagros Mountains, it marked the Middle East's first commercial-grade find and the beginning of a new petroleum-defined era for the region. Within a year, D'Arcy and his partners had created the Anglo-Persian Oil Company, and by 1912 oil was flowing out of the region through the company's new refinery on Ābādān Island. Boosted by the Royal Navy's 1914 decision to transition its fleet from coal to oil, APOC and its fields quickly became a critical strategic asset for the empire, prompting the government of the United Kingdom to take a controlling 51 percent stake in the company and turn it into a de facto arm of British imperial administration. Over the following two decades, production rose steadily, bringing new fields into use, making the Ābādān refinery into the world's largest, and generating substantial profit for the company's shareholders. At the same time, however, Iranians increasingly came to see Britain's oil concession as irrevocably unequal—a semi-colonial enterprise that extracted enormous benefit from Iran's natural resources but offered little in return. Such views extended to the highest levels of Iranian society, including its reigning monarchs, and prompted repeated attempts to win more influence over the industry and take a greater share of its revenues. First in the 1930s, again in the 1950s, and once more at the end of the 1970s, foreign domination of Iran's petroleum industry helped drive political turmoil, resulting in dramatic disputes in front of international courts, a coup d'état, and a revolution as three generations of Iranians sought to assert sovereignty over their country's natural resources. While British Petroleum, as APOC would eventually come to be renamed, roared about the considerable profits it stood to lose, for their part the British and American governments also resisted nationalization, worried not only about an independent (and, as they feared, potentially communist) Iran's ability to undercut international cartel arrangements and their stabilizing effect on global oil prices, but also the broader precedent Iranian control might set for anticolonial resource nationalism around the world.

The story of Iran's petroleum thus encompasses grand accounts of daring commercial ventures, high diplomacy, and Cold War geopolitics. For many years such were the histories written, whether focused specifically on Iran or more broadly on the region and the world. Foundational accounts like those of Daniel Yergin epitomize the prioritization of such themes in the history of oil, emphasizing the technical and capitalist ingenuity of early entrepreneurs and the world-shaking giants their firms would become.[2] The significance of oil in such narratives is the desire of people to possess it, whether to sell or to consume, and references to anything other than its status as a lucrative and strategic commodity are largely incidental. Focusing on the wealth that accrued to producer states and the political and economic developments they generated, other scholars developed ideas of the 'rentier state' and the closely related 'oil curse' in order to explain the ubiquity of autocratic governance among petroleum-rich nations.[3] Such viewpoints focus on the scale, source, stability, and secrecy of revenues generated from petroleum export, with their management and control by national elites driving oil's social and political significance in producer states. Indeed, with the world's petroleum concessions reflecting a geography of imperial and commercial interests, control over oil and its enormous revenues became a critical political issue across the Global South. In Iran, such pressures galvanized the rise of the National Front in the late 1940s, a nationalist umbrella party led by the parliamentarian Mohammad Mosaddeq that coalesced around the issue of oil and its control. Propelled in 1952 into the position of prime minister by the party's electoral victory, Mosaddeq promptly nationalized Iran's oil industry and expelled the Anglo-Iranian Oil Company. What followed was a year of international boycotts and political and economic turmoil that ended in August 1953 when a combined Anglo-American coup deposed the prime minister, reinvigorated the shah's authority, and put Iranian oil firmly back under foreign control. With new waves of scholarship attending the periodic declassification of British and American documents, Mosaddeq's rise and fall is seen as a seminal moment in Iran's recent history.[4] While much of that work has focused on the relative responsibility of the two powers for the coup and its significance for the reinstated shah's political fortunes, the events of 1953 were in other ways the epitome of what has come to be understood as a longer quasi-colonial history.[5]

Iran was never formally colonized by any of Europe's great powers, but in the nineteenth and early twentieth centuries, Russia and Great Britain exerted considerable and competing sway over the country. Using Iran as a buffer state between their imperial holdings, the two agreed to an unofficial partitioning that placed the north of the country under Russian domination and the south under British. Often directly involved in the affairs of Iran's weak Qājār state, the two empires jockeyed for primacy in Tehran while expanding their influence over the country's economy through numerous concessions. Many Iranians—few explicitly nationalist in the early years but increasingly so as time went on—chafed at the situation, angry at both the predations of the two European powers and the seeming inability of the Qājārs to resist them. In response, in the final decades of the nineteenth century, a current of economic nationalism began to coalesce among Iranians, first breaking into the open with the 1890 Tobacco Protest and its rejection of a monopoly concession for the production and sale of tobacco in the country. While the uprising was successful in forcing the cancellation of the concession, albeit at the cost of significant financial penalties that were paid through new state debts, it did little to meaningfully prevent further encroachment by foreign interests or assuage the political resentments that they inspired. Indeed, in the middle of the next decade, discontent over concessionary grants helped spark the onset of the bloody and tumultuous 1905–11 Constitutional Revolution. This uprising also failed to meaningfully diminish British and Russian involvement in Iran's affairs, though it did result in the creation of a new parliamentary body with significant nationalist representation. With many Iranians linking the idea of national progress to control over the country's natural resources, it was in this revolutionary moment that a developmentalist politics of resource nationalism began to coalesce in Iran.[6] Of more immediate significance, however, was the weakening of the Qājār grip on power and, amidst persistent rumors of covert British machinations, the 1925 rise of Reza Khan as the first Pahlavi monarch.[7] It would be under this new dynasty that Iran was set on a course of modernizing development, one that sought greater claim over the country's fossil hydrocarbons as abundant sources of both revenue and energy.

While European influence in Iran was in large part exercised indirectly through trading partners and the royal court in the nineteenth century, the discovery of oil prompted the British to assume increasingly direct control over Iran's southwestern reaches. Over the previous three decades, Khuzestān—before the time of oil, a peripheral backwater in the eyes of the shahs—had been "opened up" to international trade by the British. Generally following the course of the Kārun River, new roadways and other infrastructural improvements oriented the region toward the British-dominated Persian Gulf and away from Iran's central government in Tehran.[8] The discovery of commercial-grade reserves of petroleum in 1908 put the region further beyond Tehran's reach as British imperial officials dealt directly with local Bakhtiyāri khans for access and protection and incorporated many tribesmen into their workforce.[9] At the same time, D'Arcy and his financial backers schemed to systematically exclude Iranians from their promised ownership stakes in the newly formed Anglo-Persian Oil Company, even those that had been critical in securing the concession from Qājār Iran's notoriously cryptic and personality-driven power structure.[10] Within its territories, APOC thus ruled as a quasi-state, largely free of interference from Iran's central government, drawing in workers from across Iran and British India, and managing its labor force through a harsh and racialized caste system that rigidly capped the rise of Iranian employees.[11] APOC's imperious hand and poor treatment of Iranian workers became a major impetus for the politics of resource nationalism in the country, one that persisted in the wake of the oil nationalization crisis as the British company maintained its dominant position even as de jure ownership was transferred to the new National Iranian Oil Company. It was within this long history of quasi-colonial domination that Iran's natural gas was understood, and it gave the resource's eventual exploitation and widespread utilization a nationalist edge crucial to its political significance within the country.

While Iran's efforts at oil nationalization largely failed, other producer states in the Global South had more success, casting off the colonial and quasi-colonial arrangements that structured their petroleum concessions. In many cases, those efforts resulted in the formation of opaque state-owned companies ripe with inefficiencies and mismanagement, conditions that often reflected the salience of oil revenues within national politics and the competing demands that domestic constituencies placed upon the new firms.[12] But nationalization and state ownership were not the end of oil's anticolonial significance, as the resource became in the postwar decades a crucial site where battles over the broader issues of sovereign rights and economic justice were fought. For many of the world's postcolonial elites, political independence meant relatively little to their countries without a concomitant ability to claw back economic power from the developed world. With economies dependent on the export of a lucrative but nonetheless raw commodity, petrostates like Iran thus found common cause with a wide swath of the developing world. The 1960 formation of OPEC and its success at pushing up oil prices in the early 1970s was an expression of that desire to retain a larger share of commodity revenues, one that had much in common with the hopes of non-oil commodity exporters. But the centrality of oil to the global economy meant that rewards flowed disproportionately to petroleum producer states, ultimately undermining the nascent solidarity of the Global South on the issue of economic decolonization. Its close geopolitical ties with the United States and monarchical conservatism notwithstanding, Pahlavi Iran's petroleum policies therefore exhibited important revisionist tendencies that were championed at the highest levels of government.[13] But while scholars have increasingly recognized the anticolonial tenor of Iran's petroleum related activities on the world stage, this book argues that it extended to the domestic arena as well, and that a crucial aspect of natural gas's significance in the country, in terms of both energy and politics, was the fact that it had in large part escaped foreign control. This had both legal and operational consequences. Iranian disputes with British Petroleum over questions of the ownership and utilization of natural gas were not nearly as fraught as those fought over oil, defined less by the highly charged politics of nationalization and more by questions of technical and economic viability. But the fact that it was Iranian officials who drove the transformation of gas from a byproduct into a useful commodity carried significant political meaning within the country, quickly becoming a critical plank in state legitimation strategies centered on developmentalist ambition and the politics of resource nationalism that accompanied it.

Petro-Materiality

Natural gas did not evade the control of British Petroleum because of an oversight on the company's part or because Iranian negotiators possessed uncommon prescience about its future utility. It instead rested on the materiality of gas itself, the available technologies of the day, and the political and commercial consequences that flowed from that combination. In this, natural gas was not unique, for the physical properties of fossilized hydrocarbons have had significant repercussions for the kinds of social arrangements that have been built upon their use. Going beyond accounts of revenue flows and geopolitics, for many decades the bread and butter of energy scholars, in recent years, greater attention has been given to the materialities of fossil fuels and their shaping of labor relations, international commerce, and the world's geographies of democratic and authoritarian governance.[14] As one of the twentieth century's most critical resources, petroleum was moreover as directly constitutive of the period's social and cultural transformations as it was the economic and political.[15] In this light, gas has largely been seen through its creation of geopolitical geographies of energy dependence.[16] But as a substance with its own material and technical characteristics that overlap with, but are nonetheless distinct from, those of oil, natural gas has been largely overlooked in analyses of petroleum's historical influence. Such differences were instrumental to gas's role in Iran's twentieth-century developmentalist policies. Large volumes of natural gas are commonly found in oil deposits—a reality reflective of their shared processes of formation and connected hydrocarbon essences—and are necessarily lifted as a byproduct of oil operations. But natural gas was much harder to confine, control, and transport than liquid oil. Not only was methane, its primary constituent hydrocarbon, lighter than air and troublesome to corral, but the ubiquity of the impurity hydrogen sulfide could corrode equipment and threaten the lives of workers. Moreover, unlike crude oil, which could be piped to Iran's Persian Gulf ports and shipped to markets across the world (or, in the very early years, pumped into barrels and carried by animal to waiting cargo ships), in the early and mid-twentieth century, gas required enormous and costly pipelines directly connecting producer fields and consumer markets. As it was, there were no markets for gas in Iran, the Middle East, or any neighboring region at the time, leaving the country's sizeable reserves functionally stranded.

With seemingly no practicable outlet within reach, British Petroleum treated the gas it extracted as a non-economical waste product, either using it freely in its own operations or, as was true for the vast majority, burning it day and night in bright flares above the oil fields. For their part, Iranian officials saw such disposal practices as needlessly squandering a portion of their country's hydrocarbon wealth, and over the years, they insisted in increasingly strident terms that a method be found for utilizing or preserving something they saw as having considerable (if latent) value. In this way, Iran's natural gas was a liminal substance, straddling the line between resource and waste, its categorization dependent on the political, technical, and commercial contexts within which it was considered.[17] In contrast to framings of 'waste' as something akin to dirt, or polluting "matter out of place," both Iranian and British officials viewed natural gas through the lens of utility and the degree to which they could extract economic value from it.[18] These perspectives had a long antecedent in the histories of capitalism and development, for just as 'wasteland' had been understood as possessing the wrong characteristics for human utilization—too much or too little water, poor soils, uncontrollably wild flora and fauna, inhospitable climates—so too was natural gas seen as exhibiting the wrong material properties for profitable exploitation.[19] Overcoming the challenges posed by gas's volatile materiality was a complicated proposition, and, simply put, with the primary aim of exporting oil, British Petroleum was not inclined to accept the technical risks and financial costs of doing so. The company's categorization of natural gas as 'waste' thus primarily reflected its own commercial priorities, and it was exactly that dismissal that opened the door for Iranians to claim real ownership of their gas resources. Iranian gas was thus at once waste and not, and much like development has often meant the transformation of 'wasteland' into spaces more amenable to the extraction of value, Iranian industrialization would in large part be built upon the reconfiguration of the country's natural gas reserves into something useful. Building gas markets and the sprawling infrastructures needed to feed them would be an expensive and decades-long undertaking, but it was nonetheless one through which Iranian officials both expressed a politics of resource nationalism and asserted the utility of the country's gas reserves to their vision for energy-intensive development.

Even beyond the difficulty of controlling and moving natural gas, however, planners and engineers struggled to manage heterogeneous and variable gas compositions and pressures, working to transform what was naturally unstable into a carefully standardized product. From extraction to transmission to consumption, such material considerations shaped gas infrastructures at every point and helped determine the choice of source fields, the routes of pipelines, and, ultimately, the dynamics of distribution and the political consequences that flowed from them. In recent years, a growing body of scholarship has explored the close connections between infrastructures—"built networks that facilitate the flow of goods, people, or ideas and allow for their exchange over space"—and the construction of social and political worlds.[20] In addition to a foundational understanding of politics as being embedded in the design and function of technological systems, the study of infrastructure has been further characterized by a specific awareness of the technical and natural materialities shaping the construction, operation, and sociopolitical significance of such systems, an approach that has increasingly found its way into scholarship on the making of modern societies in the Middle East.[21] This perspective is built most directly on the work of Bruno Latour and Michel Callon and their contention of a generalizable and irreducible causal equivalence between humans and nonhumans in the production of social phenomena.[22] Better understood as an intellectual orientation or strategy of analysis than a rigorously defined theory, Latour and Callon's rejection of human agency's presumed centrality has joined with others to inspire a broad wave of scholarship that sees social systems as co-constituted by human and material factors.[23] This approach has been controversial, with some criticizing it as an abandonment of the moral and political stakes of social analysis in favor of pure descriptivism and unsupportable attributions of intentionality to nonhumans.[24] Others like Jane Bennett have softened the strict symmetry of Latour and Callon's theory, arguing instead for the notion of the more-than-human assemblage: "a collectivity whose origins are historical and circumstantial," one marked by an "uneven topography" where "power is not evenly distributed," is "not governed by a central [authority]," and is "made up of many types of actants: humans and nonhumans; animals, vegetables, and minerals; nature, culture, and technology."[25] In contrast to a presumed equivalence of human and nonhuman factors and the stability of their connections in space and time, Bennett's approach leads to a sprawling and open-ended view of causality that embraces the significance of nonhumans without seeking to rigidly determine their relative importance in advance. Her notion of "distributive agency" thus decenters people without requiring the ascription of equal agency to nonhumans, an idea useful for tracing the causal influence of not only living things like mosquitoes and microbes, but also nonliving matter like rocky strata, mountain chains, and the compositions of petroleum deposits.[26] Nor is the significance of nonhumans confined to mere responsiveness to human action in this perspective, for in their diverse natures as both stable and unstable objects they are capable of limiting, shaping, and catalyzing the creation of new social forms.[27] Notions of distributive agency underlie the analysis that this book undertakes, and the history presented traces the thread of causality not just in the minds and actions of government officials, engineers, and consumers, but also through the materiality of technological systems, natural gas, and the land of Iran itself. It was their physical properties that opened and closed possibilities for the utilization of Iran's natural gas resources, not as passive objects waiting in anticipation of human intervention, but actively and continually as part of the co-constitution of a newly industrialized and energy-intensive society. Developmentalist politics and the anticolonial resource nationalism with which it was intertwined may have been the spur for Iran's embrace of gas consumption, but it was also true, as has been written in other circumstances, that "engineers did not have infinite choices."[28] In large part, industrializing Iran's growing hunger for energy was met with cheap natural gas *because* it was a resource that was hard to control, making the country a place built not in defiance of material limits, but in concert with them.

The Infrastructure of Development

The story of Iranian natural gas was one written in the language of infrastructure. It was written with steel and telex machines, inside the corporate boardrooms of London and the lungs of Tehran's residents, amidst the high passes of mountain ranges and the alleys of urban neighborhoods. It combined webs of pipe, expert networks, financial flows, and the fractured matrices of limestone strata. Far from neutral technological systems inertly shuffling gas from place to place, these amalgamations of steel, hydrocarbons, expertise, rock, and money were co-constituted with a rapidly changing country. Such systems were built upon the "organizational work" of oil—the "variety of technical, legal, scientific, and administrative networks for its production and sale on multiple temporal-spatial scales."[29] Such work depended on the emerging petroleum sciences, systems of knowledge that were closely connected to the business of oil and were aimed at producing generalizable schema from the variable and often hyperlocal natural realities of petroleum reservoirs.[30] But these infrastructures of representation and calculation reflected political and commercial imperatives for control and productivity as much as they did the material realities of oil deposits. In Iran, that twin nature was exploited by British Petroleum to minimize royalty payments and ward off escalating demands by government officials to oversee an industry that Iranian state coffers had come to rely upon. This careful production and management of information supported BP's dominant position within Iran's oil industry, but the marriage of materiality and expertise that marked the company's operations was one that would be mirrored throughout all of Iran's oil and gas infrastructures. That included not just systems of petroleum production, but also transportation. Even more than wells and refineries, it was the gas pipelines and distribution networks that worked to produce new sociopolitical forms in Iran. Systems of energy transmission like canals, pipelines, and wires have been as crucial to the creation of industrialized and energy-intensive societies as those of production and consumption, a reality that has often been driven by changes in top-down policies for supply rather than bottom-up policies for demand.[31] Indeed, transportation infrastructures have often proved to be amongst the most potent arenas for the extension of modernist (and often authoritarian) state power, seen by their advocates as tools useful for enacting the substitution of the particular with the general, the disordered with the ordered, and the local with the national.[32]

Such was the case in Iran, where a flurry of road and rail construction in the 1920s and 1930s helped forge a new national market out of what had once been a number of smaller regional ones segregated by the high cost of overland transport.[33] New transportation networks further aided the country's political centralization as the state's reformed army used its newfound mobility to break the power of the country's tribal confederations.[34] More broadly, infrastructural expansion was part of the Pahlavi shahs' programs of modernizing social change, heavy-handed efforts aimed at remaking Iran through policies of secularization, military reform, educational expansion, industrialization, and the promotion of consumer culture.[35] Largely inchoate under Reza Shah (r. 1921–41) but increasingly organized under his son and heir Muhammad Reza Shah (r. 1941–79), the Pahlavi state's broader modernizing ambitions propelled its efforts to exploit the country's natural gas resources. While 'modernity' is an unstable and deeply fraught concept—frequently overdetermined by later cultural and political proclivities and doing more to obscure than illuminate historical relationships—Iran's encounter with expansionist European empires and the modernities they claimed to embody proved to be deeply influential on the country. In the late nineteenth and early twentieth centuries, Iranian intellectuals, many of them seeking to understand how their society had become so comparatively weak, began to engage extensively with European philosophy and political thought. Out of their efforts came not only Iran's constitutional movement, but also the foundations of Iranian nationalism and national identity.[36] Represented most strongly among relatively small classes of educated urban professionals, the efforts of both Iran's intelligentsia and a growing cadre of state officials to reshape their society along 'modern' lines found enduring purchase in the early and middle decades of the twentieth century.[37] Many changes, particularly those dealing with personal status and family law, were, however, far from uncontroversial, and topics like unveiling and women's ability to obtain an education and work outside the home became flashpoints for highly charged debates about the country's future.[38] But while shifting social norms often grabbed the lion's share of attention and pitted more conservative members of society against Reza Shah's secularizing policies, no less important were his efforts to foster new industries through infrastructural expansion, tariff barriers, and financial support. By 1941, the year of Reza Shah's abdication, Iran boasted 346 industrial facilities, up from twenty in 1925, while over the same period the Trans-Iranian Railway had been completed and the highway network expanded from 2,000 miles to 14,000.[39] Under the second Pahlavi monarch, Muhammad Reza Shah, such programs were not only accelerated but also, with the 1948 creation of the governmental Plan Organization, increasingly organized through extensive five- and seven-year development plans. Their sprawling scope meant that in the postwar decades nearly all of Iranian society would become subject to technocratic intervention aimed at catapulting the country into the company of the world's most industrialized and energy-hungry nations.

With the rise of centralized economic planning in the late 1940s, Iran's developmentalist agenda began to increasingly reflect the global postwar ascendency of theories of modernization and the growing influence of the United States in the country.[40] Mirroring the importance of macroeconomic analysis to Pahlavi Iran's modernizing initiatives, scholars have frequently studied the country's developmental plans and their effects from a technical perspective.[41] For these scholars, development is something assumed, and they have given comparatively little attention to how its aims were formulated. Others have more closely examined the planning process in Iran, often with an eye to its interactions with the country's broader political currents and the influence of prominent personages.[42] Largely neglected in these bodies of literature was the depth of Iranian officials' preoccupation with industrialization and the provision of infrastructural services. It was a fixation that linked Pahlavi modernizers to both their counterparts across the Global South as well as to a group of influential theorists then shaping U.S. foreign policy. Such experts, many economists by training and speaking to the American Cold War fears, offered prescriptions for transforming what they deemed to be stagnant agrarian societies into dynamic industrialized cultures marked by rapid economic growth and inoculated against the seductions of communism.[43] Educated in the same elite institutions from which those American theorists hailed, many Iranian officials likewise emphasized industrialization, new infrastructural services, high energy use, and urban development as necessary for the rapid and sustained economic expansion they hoped to achieve.[44] Indeed, industrialization was the foundation upon which many of their developmental initiatives were conceived, representing not just prosperity but also the kinds of technoscientific mastery they saw as "the essence of modernity."[45] This was the key intellectual commitment around which Iran's developmental policies were built. For the country's modernizing officials, as was true for many leaders across the Global South, development was synonymous with industrialization or, at the very least, predicated upon it. That, in turn, meant access to cheap and bountiful quantities of energy, a need that Iranian officials saw as best met through their country's vast reserves of natural gas.

Underpinning Iranian thinking on industrialized modernity was a fundamentally futurist orientation that produced development in the often fraught negotiations between idealized transformative aspirations and the mundane material, political, and economic circumstances within which people lived and worked.[46] Even highly rationalized variants like modernization theory and economic planning expressed the "inclination to utopias, to images of future perfection" of their practitioners, making development as much an imaginative project as it was a political, economic, and technical one.[47] But the future was not an empty category waiting to be filled; rather, it was overdetermined by the attachments and preoccupations of those who conceived it, laden with powerful and crosscutting notions of civilization, advancement, and prosperity.[48] With development's rooting in ideas of progress and linear temporality, the historical experiences of countries like the United States thus came to serve as templates for developmentalist programs around the world. In practical terms, postwar modernizers often drew upon the lessons of the New Deal-era Tennessee Valley Authority and its totalizing model of infrastructural improvement, industrialization, and poverty abatement. Animated by deeper assumptions about the inherent validity of mid-century American lifeways and drawing upon the marriage of "salesmanship and statesmanship" that had made the United States the source of modern consumer practices in Europe, such experts further encouraged the spread of consumer culture to developing societies like Iran.[49] That twin lineage has prompted historians to argue for understanding America's postwar promotion of modernization as an effort to export its own political and economic systems, build new markets for American manufacturers, and showcase the benefits of liberal governance to societies it deemed in danger of falling to communism.[50] Iran was firmly understood through that paradigm, and under multiple postwar administrations official American interest in Iran expanded to include military aid, economic planning assistance, rural development, cultural exchange, and the promotion of new consumer lifestyles.[51] Given the country's hydrocarbon wealth, however, little direct economic assistance was forthcoming, and most developmental programs in Iran were funded by revenues generated from the country's oil exports. Iranian development was for this reason closely tied to petroleum's importance on the world stage, a connection central to the mutually reinforcing positions of Muhammad Reza Shah, the oil majors operating in the country, and the United States. It was these conditions that led successive American administrations to accept the shah's turn toward authoritarian modernization despite worries that his aggressive development policies and lack of political reform would lead to economic ruin and social unrest.[52] Intertwined flows of oil, money, and modernizing expertise thus structured much of Pahlavi Iran's relationship with the outside world. It was a potent combination, one that allowed Iranian officials to draw upon an enormous pool of wealth to quickly implement sprawling programs of economic and social change.

Charged with providing both the energy and money needed for the country's developmental programs, NIOC and its subsidiary NIGC were in effect fully integrated arms of the Pahlavi state. Muhammad Reza Shah, seeing Iran's petroleum as the key to his ambitions, maintained very close oversight of the firms and was often personally involved in negotiations with foreign oil companies and their national governments. On the other hand, the strong link between Iran's oil revenues and modernizing policies left the country, like many others, in what some saw as a quasi-colonial position of exporting cheap raw commodities in exchange for expensive goods and services. Responding to this dynamic, a loose group of largely Marxian scholars has advanced ideas of dependency theory and world-systems analysis, arguing that the societies of the Global South have been deliberately peripheralized in the international economy as the rewards of trade disproportionally flow to the world's industrialized core.[53] Moreover, as some have argued was true for Pahlavi Iran, such economic arrangements have undermined the sovereignty of developing states and left them as little more than clients to stronger powers.[54] But while dependency theorists have successfully brought attention to the structural inequalities of global trade, their tendency toward a totalizing view of capitalism often causes them to treat developing societies as passive fields where foreign capital and expertise dominate. Such analyses leave little room for either cooperation or resistance, struggling to explain wide historical variations in developmental experiences and their divergent trajectories. Missing in such accounts is the question of how developmental goals were imagined and set, a process involving complex and historically contingent interactions between transnational expert networks, state officials, and everyday citizens both within the Global South and outside it. Rarely was there any real unanimity among Iranians, even amongst those who accepted the general terms of modernizing industrialization, and fierce disputes would at times erupt as people debated the practicability and desirability of various projects. The rivalry between different perspectives and the conflicting, if often overlapping, narratives of their advocates were important factors in the creation not just of Iran's natural gas infrastructure but also the broader story of industrializing development of which it was part.[55] But what was rarely in question was the idea of (technological) progress itself or its centrality to Iran's future, and the notion became a discursive vehicle for linking particular engineering projects to more general beliefs about Iran's present and future conditions.[56] More than just expressions of methods and means, Iranian modernizer's descriptions of their projects—articulated both publicly and behind closed doors—were statements about their futurist imaginings for their country.

Strongly shaping Iran's postwar developmentalist orientation was Muhammad Reza Shah's own visions of the future. Projecting that within a few decades Iran would possess one of the world's largest economies, the shah imagined a revitalized nation of productive industry, cutting edge technology, glittering infrastructure, and universal well-being—a utopian "Great Civilization" (*tamaddon-e bozorg*) that promised Iranians not just the material benefits of economic prosperity but also the pride of returning to the cultural grandeur of their country's ancient past.[57] While the contours of that vision would not be fully articulated until the 1970s, beginning in the 1960s monarchical legitimacy began to be explicitly banked on modernizing development. Epitomized by the White Revolution, a set of six (eventually nineteen) developmentalist reforms introduced in 1963 and named in contradistinction to the red of a feared communist revolt, the shah sought to use improved living standards to bind Iranians to his rule and erase the specter of his return to power on the back of a foreign coup.[58] His was a top-down linking of state legitimacy and industrializing developmentalism that treated Iranian society as a passive object waiting to be modernized—a perspective that has in turn shaped much of the scholarship on the period.[59] But the wide-ranging program also cemented the importance of a "politics of material promise" that had first emerged a decade earlier amidst vocal consumer demands for expanded electricity service in Tehran.[60] Driven by middle- and upper-class Iranians' growing cognizance of prosperous consumer lifestyles around the world, during that time, Pahlavi officials committed the state to high-profile infrastructure projects like hydroelectric dams and electricity grids. In this way, Iranians often understood infrastructural systems as embodiments of the state, enacting a spatialized politics of practice that molded contours of citizenship and national belonging.[61] Rather than uniting Iranians via a shared national experience, however, such interactions often created "differentiated citizens" as access to infrastructural services remained uneven.[62]

In Pahlavi Iran, natural gas infrastructures acted as an expression of the modernizing ambitions of state officials, becoming a crucial site in which Iranians as national subjects would eventually, it was hoped, become folded into their aspirations. But gas could drive apart too. "Infrastructures were intended to be a universal public good," Brian Larkin has noted, one "in which every member of society was presumed to be equal to everyone else."[63] In practice, however, Iranian technocrats working in state planning institutions prioritized the construction of piped gas distribution networks for industrial consumers and newly built urban areas, particularly in Tehran. That decision left consumers in older neighborhoods and rural regions to be served by what was understood to be an inferior system of bottled propane and butane. Such Iranians, in the unequal spread of piped gas across the country, seemingly saw their value as national subjects reflected in their exclusion from Iran's most modern and celebrated of energy systems. It was a sight that displeased many of them. As a potent symbol of Pahlavi neglect, in the 1970s, that inequality of service quickly became part of the country's growing revolutionary movement. In the 1980s, those same political currents would strongly influence the gas policies of the new Islamic Republic as it sought to differentiate itself from its monarchical predecessors through the symbolically charged provision of gas to all citizens. These expanded infrastructures served to bring a new source of energy to people previously left out, in the process refiguring but nonetheless reproducing the legitimating power of natural gas for a new regime. Rather than volumes of gas produced, transported, and consumed, scale was now measured in the number of households and businesses connected to the network. But this rhetorical shift away from large cities and industrial consumers to towns and individual families did not upend Pahlavi-era understandings of gas exploitation as something embodying technological sophistication, national benefit, and the conquest of nature. The persistence of these themes underscores the fact that fossil hydrocarbon-based Iranian developmentalism was not purely the top-down imposition of a modernizing autocrat, but instead something formulated in the interactions between official ambitions, the demands of consumers, the politics of resource nationalism, and the varying materialities of technology and nature.

That both the shah and the Islamic Republic chose legitimation strategies based in part on the provision of infrastructural services demonstrates just how influential developmentalist ideas had become in twentieth-century Iran. With the notable exception of the monarchy's secularizing tendencies, even the revolutionary movement's overt anticolonialism and rejection of foreign influence did not cause Iranians to fundamentally question whether their society could benefit from industrialized modernity. In this they were not unique, for as some postcolonial scholars have argued, the potency of modernization lay in the readiness of people to adopt its perspectives and the ubiquity with which they did so. In this telling, development is best understood as a discursive practice, one in which practitioners, both outside the Global South and within it, positioned the world of industrialized capitalism (and, to a lesser extent, the communist) as an aspirational goal, decided which societies did and did not exhibit its signifiers of modernity, and then defined those that failed as impoverished and in need of intervention. Crucially, in this understanding, development was not only an explicit imposition of outside empire, even if it was largely a continuation of earlier Euro-American notions of civilizing mission and colonization; instead, it was something adopted and fiercely embraced by state leaders and national elites in much of what came to be known as the Third World. This "historically singular experience" enfolded both the proponents and critics of development through the demarcation of debate in social scientific, particularly economic, terms: overpopulation, income and poverty, GDP and economic growth, deficits of capital and technological know-how, and sclerotic political and cultural institutions.[64] For their part, leaders across the Global South embraced the seeming ability of development experts to understand their diverse societies through 'scientific' abstractions, using them to implement top-down technocratic policies that subsumed fraught questions about the distribution of resources into ostensibly neutral technocratic plans that promised a general improvement in living standards.[65]

State modernizers across a broad swath of the developing world, including Iran, thus embraced the transformational goal of "un-underdeveloping" their societies, an ambition of such sweeping scope and urgency that it seemingly justified their use of violent dispossession and authoritarian control.[66] Understanding development as discursive is important for tracing how knowledge and violence were coproduced as people and ideas moved through the transnational networks that connected governments, businesses, NGOs, and academic institutions. It moreover helps explain why so many promising interventions have proven underwhelming or even backfired.[67] But treating modernizing development as a singular, colonizing force facilitated by local compradors undercuts the agency of those in the Global South who chose to welcome it. It moreover risks reproducing the universalizing pretensions of modernization's original theorists by flattening the process of gaining local support, romanticizes those who resisted, and overlooks the (very often flawed) work of local officials to navigate developmental networks and translate their ideas into tangible projects. This is not to argue that the Cold War period did not see a global wave of development that was closely connected to geopolitical power and exhibited a preoccupation with industrialization and economic growth. Rather, it is to say that development was multiple, a varied outcome of diverse encounters between societies' different circumstances and the transnational currents of modernization.[68] Attending to such granularity is critical for understanding why development was instantiated in unique ways around the world. It is further necessary to see how developmentalist projects were influenced not just by politics and culture, but also the natural and material characteristics of the world's numerous regions.

Smoke and Modernity

This book aims to complicate top-down accounts of Iranian development and highlight the influence of Iranian consumers and the politically charged relationships they had with their country's natural gas infrastructures. It further argues that Iran's state-driven developmentalist projects were shaped by the materiality of technological systems and the natural world, making them not just expressions of Iranian officials' modernizing ambitions, but more-than-human assemblages reflecting the tangled connections between human endeavors and the physical properties of topography and geology. Indeed, as has been true in many times and places around the Middle East and the world, Iranian modernization was predicated on the domination of nature, as much an environmental project as it was an economic and political one.[69] Monumental infrastructures like hydroelectric dams and natural gas pipelines asserted that dominance in a hypervisible form, tangible proof of the pre- and postrevolutionary Iranian state's ability to fulfill its promises of development. All the crucial differences between the Pahlavi monarchy and the Islamic Republic notwithstanding, the developmentalist policies that natural gas energy embodied—the comingling of anticolonial politics, political legitimation, and energy-intensive industrialization—remained remarkably similar across the violent revolutionary divide. Far from a network of pipes silently and invisibly operating in the background, gas energy was made deliberately visible through celebrations of its gleaming, sophisticated technology and their enabling of 'modern' lifeways through the provision of clean and convenient fuel. Such depictions were deeply reflective of official imaginings of Iran's future, and by drawing upon such depictions this book treats the Ministry of Petroleum, the National Iranian Oil Company, and the National Iranian Gas Company as not only institutions responsible for the construction of Iran's natural gas infrastructure, but also as cultural producers that were instrumental in developing and sustaining important political narratives of gas. Both before and after the revolution, the public relations arms of these organizations produced large volumes of printed media for circulation among their employees and beyond. Through their textual and photographic depictions of natural gas infrastructure, a sophisticated idiom of technological modernity was created to 'prove' the desirability of gas-based development, the possibility of an energy-intensive society in Iran, and the ability of the Iranian state officials to create it.

In treating these depictions as media and cultural artifacts, and attending to how they did and did not change over time, it is possible to gain insight into the political orientations that underlay them. More than public relations fluff or dry institutional accounts, these publications offer a window into what Sheila Jasanoff and Sang-Hyun Kim have termed sociotechnical imaginaries: "collectively held, institutionally stabilized, and publicly performed visions of desirable futures" that are rooted in scientific and technological advancement.[70] The multiple and intermeshing currents of social and political thought that comprised these textual and pictorial utterances, some overt and some less so, are revealed by the contextual framings in which they are placed. Read simultaneously as a single interlocking corpus and as a diachronic series of utterances in dialogue with each other, it becomes possible to see that the sociotechnical imaginary of natural gas was one that operated across different institutions and forms of media.[71] Each article or image can be understood as polyphonic, simultaneously communicating (often intentionally) both the specifics of any particular gas project as well as an "undertow of seduction" toward the broader developmentalist orientation that underpinned it.[72] In the late Pahlavi era, industry publications celebrated rising household gas use in Iran, tying it to the rise of a new consumer culture among the country's urban middle and upper classes. But in these publications, Iran's natural gas infrastructure was also significant for its symbolic spectacle. Official discourse frequently engaged in almost formulaic recitations of the size and scale of Iran's natural gas infrastructure, hailing it as a bridge to a sovereign and technologically sophisticated future. The immensity of those systems—the millions of cubic feet of gas produced per day, the hundreds of miles traversed, the thousands of feet ascended, the measurements of pressure and the resilience of steel—was harnessed as proof of the state's ability to build a new Iran. This discourse was backed by the disciplinary output of Iranian researchers working within government ministries and national petroleum companies, publications that reflected not only their expert findings but also the modernizing political currents within which they undertook their work.[73] This broader discourse was moreover one that was reproduced and performed by everyday Iranians in the form of petitions that requested, at times demanded, the inclusion of their communities in the growing national gas network. That meeting of top-down and bottom-up impulses gave the Pahlavi-era sociotechnical imaginary of gas resilience across the 1979 revolution, and it in many ways formed the basis of the Islamic Republic's own articulations of the significance of the resource. In this way, through a triumvirate of image, word, and statistic, natural gas energy was built to be a fundamental legitimating factor for the state, proof of its ability to construct a brighter future through the hard stuff of technology. It was a heroic vision of massive steel pipes traversing mountain ranges and rivers, of gleaming purification towers rising into the sky, of deep petroleum wells and the convenient twist of a stove's knob—in other words, a rooting of Iran's state-directed modernization program with all its economic and social heft within the land of Iran.

Even before the advent of modern nationalism, Iranians had long seen themselves and their civilization through their relationships with place.[74] In the nineteenth century, suffering defeat and territorial loss at the hands of expanding British and Russian empires, Iranians' coalescing sense of national identity was built on a foundation of land as they mourned the diminution of their country and sought to defend it against further encroachment.[75] It was a connection that would be maintained in the decades after the Second World War, including as part of the sociotechnical imaginary of natural gas that emerged during the period. Defined by the conquest of nature instead of the contests of empires, the operational success of Iran's gas infrastructure was publicly celebrated as proof of the state's ability to tame the land. Articulated was a highly aestheticized technological sublime built on the physical heft of pipelines and refineries, a visual grammar that borrowed heavily from globalized depictions of petroleum infrastructure and one that emphasized the technological sophistication, efficiency, and modernity of systems that seemingly traversed landscapes with ease.[76] In these publications, images of rent earth and steel pipe were married to descriptions of miles traversed and mountain ranges subdued, an assertion of modernizing triumph that rested on the assumption of a passive earth. The reality, however, was much more muddled. Those same systems that were celebrated for demonstrating the developmentalist state's ability to impose its will were also shaped by the specific natural environments within which they operated. Channeling Pahlavi gas policies were topographic, geological, climatic, chemical, and technological realities, all-enveloping and interpenetrating contexts that pushed Iranian policymakers toward identifying natural gas as the fuel around which to build the energy-intensive society they desired. They moreover shaped the design and technical function of Iran's new energy systems. The natural compositions of petroleum reservoirs influenced the qualities of gas and its availability; pipeline routes were dictated not just by existing agglomerations of urban consumers and their potential demand, but also by the terrain that such machinery could feasibly traverse; even the pressures at which pipelines operated were shaped by topography and climate. Iran's natural gas infrastructures were thus hybrid assemblages, bringing together numerous human and nonhuman forces to create systems that connected consumers to the hydrocarbon pools of southwest Iran. In this way, rather than distancing Iranians from the natural world, the widespread use of gas energy brought them closer, sinking the foundations of Iran's budding industrialized modernity deep into the earth.

In the 1960s, that material reality had moreover met a nostalgic longing among Iranians for a past of unpolluted skies and environmental cleanliness. Owing to a complex assemblage of factors both human and nonhuman—mountainous topography, the chemical compositions of hydrocarbons, a semi-arid climate, industrialization, and waves of rural-urban migration—in the decades after the Second World War Iranian cities saw a precipitous decline in air quality. Many urban Iranians had a particular understanding of air pollution, one centered on the tangible experiences of living amidst skies dense with smoke and smog. Spurred by the blighting of beautiful vistas and the burning of their throats, residents of Tehran, amongst the hardest hit by this environmental development, expressed their laments through photographs, air quality measurements, penned reminiscences, and the formation of official and unofficial organizations to fight it. Iranian officials were an integral part of that movement, with some expressing fears that their cities were becoming like the polluted urban areas of the industrialized world—a specter that they understood as threatening the fundamental validity of Iran's entire modernizing project. In contrast to accounts that have largely treated Iranian environmentalism as a post-revolutionary phenomenon, this book argues that natural gas energy and its infrastructures were important avenues by which environmental anxieties influenced the country's developmentalist priorities in the late Pahlavi era.[77] Rather than have their relationships with nature progressively disrupted by urbanization and the spread of infrastructural services, Tehran's residents instead found them refigured as social and natural phenomena interacted in unexpected ways.[78] Tehran, like any city, was a hybrid environment, a space where complex and contingent historical encounters created more-than-human assemblages that render impossible sharp analytical distinctions between people and the natural world.[79] Iranian officials implicitly saw their cities as hybrid, even if their analyses did not employ such language, and they paid heed to the significant role of topography and climate in both creating the problem of air pollution and in limiting their own choices for fighting it. While they could not move mountains in either metaphorical or literal ways, they could and did seek to change the primary fuel source of city residents, embracing natural gas as a way to build a clean future that would not necessitate a retreat from the developmentalist choices they had already made. In this way, Iranian officials combined their sociotechnical imaginary of natural gas with a spatial imaginary of urban areas as sites marked by their air quality, an explicit rejection of what they saw as a polluted global norm and an articulation of developmentalist ambitions combining both the prosperity and convenience of modern technology with the clear skies of yesteryear.[80]

For decades, Iranian cities had been sites where petroleum's influence within the country had been amongst the most pronounced.[81] Some like Ābādān had long borne that imprint directly, treated as an environmental sacrifice zone where high levels of pollution were deliberately ignored so as to not disrupt oil operations and the economic benefit they brought.[82] Outside of such areas, in the era of natural gas, many Pahlavi officials saw in the new energy source the possibility of a straightforward technofix to the problem of air pollution. They expressed that viewpoint through fuel conversion programs for both industry and motor vehicles. Such ideas of "ecological modernization"—a notion that locates both the causes of and solutions to environmental troubles within existing institutions—were aimed at reconciling Iran's politically charged push for industrializing development with growing worries about the environmental violence it wrought.[83] It was a position intimately tied to the Pahlavi shah's aspirations to transform Iran into an exemplar of a new modernity "free from the blemishes of . . . [Western] society."[84] Development and notions of environmental stewardship were thus intertwined in Iran, and the return of clear skies was a sought-after feature of the modernity many Iranians longed to build. Active participants in the world's budding environmentalist movement, they moreover perceived looming "limits to growth" and an impending inability to continue industrialized consumption without ecological exhaustion and collapse.[85] For Pahlavi officials, natural gas was a way to escape that trap, a potential new source of energy that was not only voluminous and almost wholly untapped, but one that burned more cleanly than the oil products then in widespread use. But foundering on the inherent insufficiency and brittleness of technical solutions to the problem, their bet on gas as a clean energy source failed, and in subsequent decades, Iranian cities remained as polluted as ever. Such ambitions nonetheless reflected a powerful and persistent vision for their country, one that promised a prosperous, sovereign, and modern society built on intensive energy use but without the poisonous clouds and hacking coughs that marked the dark environmental underbelly of the industrialized world.

***

What follows is a history narrated in roughly chronological fashion, beginning in the early twentieth century with the search for petroleum in Iran and concluding in the early 1990s and the start of the reconstruction efforts that followed the Iran-Iraq War. Its primary emphasis is on the quarter century between Muhammad Reza Shah's 1953 return to power and the end of his dynasty in 1979, a period in which official ambitions for natural gas exploitation were translated first into detailed plans and then into the steel and concrete of new infrastructures. It was at this time that the foundations of Iran's natural gas system were laid, and though the majority of the country's gas network would be constructed in the late 1990s and thereafter, such efforts built upon and extended plans and frameworks that had been first conceived in the final decades of Pahlavi rule. Under the Islamic Republic, natural gas moreover maintained much of its potency as a tool of political legitimation, even if the weighting of its rhetorical emphases in some cases shifted. Employing much of the same visual grammar and developmental idioms that had been evident under the shah, the Islamic Republic's official lionization of gas exploitation largely reproduced the preoccupation with scale, technical sophistication, and domination of the Pahlavi era, even as it drew contrasts with the previous regime by celebrating a new rapidity with which rural areas were connected to the national gas network. In light of such continuity, this book largely concludes its historical account in the 1990s, a choice that reflects not the end of natural gas's salience in Iranian society but its transition to a period more strongly emphasizing expansion and implementation rather than conception. The specificities of that period, and there is much to study in it, more than can be reasonably addressed here, are left to another work.

Given the prominent place that government officials took in setting the terms of both modernizing development and gas utilization in Iran, the archived correspondence of employees of Iranian state ministries and the national petroleum companies—primarily the National Iranian Oil Company and its subsidiary the National Iranian Gas Company—comprises the core of this book's source base. The bulk of that material was found and collected in Iran's centralized national archive. Reports produced within ministerial and company offices, some released publicly but many not, are likewise important sources for the analysis undertaken here. Many of these were found in the document collections of the Plan and Budget Organization, NIOC, and the Ministry of Energy. Official attention was largely dedicated to managerial and financial considerations; bottom-up factors like consumer demands and the limits of technology and the natural world were more seldom given direct attention. In some cases, the debates of local councils and consumer petitions have been preserved in provincial archives, and these have been used as an important additional source base for understanding how national gas policies were received. On the other hand, officials actively worked to shape public perceptions of natural gas utilization, and it was both a frequent topic of speeches by government ministers and a consistent presence in public relations materials. Both are useful for studying the sociotechnical imaginary of natural gas that operated in state agencies, particularly its overarching political import, how it fit into the country's broader developmentalist agenda, and how it did and did not change over time.[86]

Because Iranian records dating to the Pahlavi era have gaps, sometimes remain classified, or are still uncatalogued and unavailable, published oral histories with principal figures in Iran's petroleum industry are also employed. While the challenges of using oral histories are well known—time and distance, the influence of later events on people's recollections, a possible eye toward posterity and self-promotion on the part of interviewees, potential guidance (subtle or overt) from interviewers—their careful use can open doors to accessing information not well represented in the archival records. This was particularly relevant for studying the construction of urban distribution networks in 1970s and the high-level technical choices that Iranian planners faced in their designs. Indeed, technical considerations are not well-represented in the ministerial archives, in most cases making it difficult to decipher how different technologies worked to shape the decisions of Iranian officials. To some extent, this gap has been offset by using the archives of British Petroleum and the Development and Resources Corporation. While strongly defined by the commercial interests of the two firms, these collections contain information more closely connected to the technological, geological, chemical, and topographic forces at work, and they form a crucial third evidentiary pillar for the analysis undertaken in this book. Conversely, because Iranians themselves oversaw the creation of their country's natural gas infrastructures, these records contain relatively little direct information on how those systems were designed. Rather than a detailed account of the co-constitution of natural gas politics and technology in Iran, what emerges from the broader source base was the strength with which the politico-commercial concerns of natural gas exploitation interacted with the demands and limits of the natural world. Technological considerations nonetheless maintain an important presence in this book's analysis, at times on the back of against-the-grain readings, but often in closer connection to natural materiality than had been anticipated. In a few instances, most notably in discussions of air quality, contemporaneous scientific papers published by Iranian researchers are available to provide further insight on those connections. Where possible, they have been used to understand both the natural factors influencing Iran's encounter with gas energy and people's perceptions thereof.

This book's focus on natural gas's multivalent nature is an effort to go beyond the questions of religion, ideological contestation, and Great Power meddling that have shaped much of the scholarship on modern Iran. It aims to understand how such histories have been co-constituted with technology and the environment, highlighting the tangled knot of physical infrastructure, anticolonial sentiment, and developmentalist ambition that gave natural gas its political significance. It further examines the interactions between the natural world and gas utilization as a modernizing project, revealing how the fracture systems of limestone rock, the composition of underground petroleum deposits, and the rugged mountain chains of Iran opened and foreclosed developmental possibilities. Through these connections, Iranians sought to build an industrialized society predicated on intense energy consumption, one that expressed politically potent aspirations for their nation's future. Natural gas was the great enabler of Iranians' developmental ambitions, an energy source that not only did not compete with the need to export oil but was in fact built upon it. It made energy consumption a practical expression of Iranians' resource nationalism, made energy infrastructures a tool of political legitimation, and made fossil hydrocarbons a seeming source of clean air. Over the past century, natural gas has become the foundation upon which contemporary Iranian society now rests; why it was made to be so is the history presented here.

One
Petroleum Lands

Before the place known as Iran became a nation of widespread gas consumption, of pipelines and refineries, of cities and towns and agriculture, before even humanity existed as such, there was a landscape of snowy mountains and arid basins, sitting at the crossroads of what we now call the Middle East and Asia. Buffeted by cold, dry air from the north and warm, humid winds from the south, given shape by two mountain chains—the Zagros and the Alborz—and encompassing dark alluvial soils and thin dusty ground alike, the land is as ecologically and climatically diverse as it is topographically dramatic. For millennia, people have lived in those mountains and basins, their states rising and empires falling, all the while shepherding their flocks and tending the pistachio and pomegranate trees so central to their, and the land's, identity. On its southwestern edge, where the high peaks of the Zagros give way to the rolling foothills and marshy plains of Mesopotamia, lies a region long distinct for its agricultural productivity and more recently for its enormous hydrocarbon deposits. A center of civilization since at least the fourth millennium BCE, southwestern Iran's protracted human history has for more than a century been overtly intertwined with a pre-human one that is hundreds of millions of years older. It is a history given shape by the movement of continents, the eons-long advance and retreat of water, and the tectonic rise of the Iranian highlands. Three times warm, shallow seas opened and closed in the area, each home to countless cycles of microscopic life and death, minuscule remains accumulating in deep drifts on ocean floors. Buried, compressed, and heated by millions of years of the earth's movement, that biological debris would become the marl and limestone layers of southwest Iran's very bedrock. But such organic-rich stone would mold the region further still, for transformed by heat and pressure, it would in time produce petroleum, the great natural resource of the twentieth century.

As one of the world's most oil- and natural gas-rich nations, Iran has long borne the physical imprint of fossil hydrocarbons. The country's southern and western reaches, a region stretching a thousand miles from the waters of the Iraqi border through the high peaks of the Zagros Mountains to the arid coastal plain of Makrān and the boundary with Pakistan, are marked by petroleum's escape from beneath the earth. In southeastern Baluchestān, large mounds of hardened white earth, dubbed mud volcanoes, are produced where pressurized natural gas erupts through saltwater-bearing clay to form dark bubbling craters. In Fārs, the ancient land of wine and poetry, the sulfurous smell of natural gas seepages wafts through the air as fetid pools of black water foam and *gach-e torsh*, powdery deposits of dark brown calcium carbonate and crystalline sulfur, grow where natural gas reacts with exposed outcroppings of limestone gypsum. Such silent accretions, muddy pools of oil, and naturally burning vents of gas litter the ground in southwestern Iran—the heartland of Iran's petroleum industry—and hint at the vast quantities of wealth and energy lying beneath the surface.[1]

Premodern peoples made full use of such natural seepages. Ancient and medieval Iranians gathered bitumen, pitch, and naphtha to variously caulk watercraft, produce building materials, and wage war.[2] In the second half of the nineteenth century, the significance of such shows changed, becoming important less as direct sources of petroleum and more as potential signposts for underground deposits of oil and gas. Indeed, in the 1850s, it was in areas near such noticeable seepages that the commercial exploitation of underground petroleum arose. Edwin

Drake's drilling of the first true oil well in Oil Creek, Pennsylvania in 1859 has become the most famous, but the era saw a veritable gold rush as prospectors sunk hundreds and thousands of wells near surface shows all over the world. Quickly, however, aspiring oilmen discovered that petroleum seepages did not guarantee the presence of meaningful reserves at all, let alone in their immediate vicinity. As would be found to be true in southwest Iran, oil and gas could travel long distances beneath the earth, following fault lines and other fractures in the bedrock, leaving surface eruptions only weakly correlated with meaningful underground accumulations.[3] Far more important for determining the location of a petroleum deposit was the geological structure of an area, and in the final decades of the nineteenth century it was increasingly understood that significant finds were associated with the crests of anticlinal formations and their long ridge-shaped arches of stratified rock that slope downward and trap accumulations of oil and gas.[4] But for all their ambiguity as signposts, seepages remained useful indicators of regions likely to possess large deposits, and within Iran, European oil prospectors used them as guides in their hunt for the country's suspected petroleum bounty.

European travelers to Iran had long noted an abundance of surface shows near the city of Baku, but the 1813 cession of the area to Russia also meant the loss of its budding oil industry. When a report on the geology of Iran's border region was published in 1855, however, written by a member of a diplomatic mission sent to settle the country's boundary with the Ottoman Empire, the numerous seepages in the country's south began to gain wider notice. Tempted by the possible riches such indications hinted at, foreign concessionaires began to aggressively seek out Iran's petroleum resources, in the process becoming part of the long series of interventions European powers were making in the country's economic, administrative, and political affairs. In the nineteenth century, Iran was a weak state, one ruled by the shahs of the Qājār dynasty through an inefficient and corrupt system of royal patronage. Beset by competing centers of power that resided in nomadic tribal confederations and the Shi'a religious establishment, the Qājār shahs' influence outside the gates of Tehran was limited, and they largely found themselves unable to fend off the predations of expansionist

European empires. By midcentury, alarmed by the Russian Empire's steady southward march and the threat that it posed to its Indian possessions, Great Britain committed itself to preserving Qājār rule as a buffer between their imperial holdings. In the following decades, the two powers thus jockeyed for influence over Iran's government and economy, in the process gaining considerable commercial privileges and incorporating Iran into global markets as a commodity producer. It was in this context that Nāsir al-Din Shah Qājār (r. 1848–96), needing to fund the royal treasury and in any case possessing little power to otherwise resist their demands, awarded increasingly extensive concessions to the two Great Powers. As a consequence, in the late nineteenth century, monopoly concessions were awarded to European businessmen for everything from laying and operating telegraph lines to banking to the production and sale of tobacco, often for scandalously small sums. By the 1870s, those concessions had begun to include a right for the production and sale of petroleum.

The first to do so was the Reuter Concession of 1872, a wide-ranging agreement that granted unparalleled control over Iran's transportation, mining, and agricultural sectors to a British businessman by the name of Baron Julius de Reuter. Deeply unpopular among Iranians and facing intense opposition from Russia, within a year the concession was canceled by Nāsir al-Din Shah. In 1885, however, he again granted rights to Iran's petroleum to Reuter as part of a narrower concession for banking and mining in the country. Though Reuter would duly found the Persian Bank Mining Rights Corporation and hire experienced oilmen to lead the hunt, little came of his investments as the search was repeatedly stymied by Iran's weak and chaotic governance, its limited transportation infrastructure, and its mountainous terrain.[5] Interest in Iran's as yet unproven oil reserves nonetheless remained high. In 1901, William Knox D'Arcy, another British businessman, received a sixty-year concession for the exploration, production, and sale of petroleum in all of Iran save the five northern provinces that bordered Russia. Given his concession by Muzaffar al-Din Shah (r. 1896–1907), the fifth Qājār monarch and successor to Nāsir al-Din Shah, D'Arcy in return promised a one-time payment of £20,000, £20,000 worth of company shares, and a 16 percent cut of any annual profits. It was, all told, a comparative pittance for the enormous amount of wealth that would soon be extracted from beneath Iran's soil. With the right to search for oil firmly in hand, D'Arcy's prospectors quickly turned to Iran's numerous surface shows to guide their work. In this they were in good company, for until the 1950s, and despite their notorious unreliability, most petroleum fields discovered in the region were found through their proximity to such natural seepages, the most productive of which led prospectors to the foothills of the Zagros Mountains. Forming the boundary between the Mesopotamian plains of Iraq and the highlands of the Iranian plateau, the Zagros are a rugged range running from southeastern Anatolia to the Gulf of Oman. Created by the collision of the Arabian and Eurasian plates, the ongoing tectonic advance has deformed deep sedimentary layers of Eocene, Cretaceous, and Jurassic stone into high, jagged peaks while also exposing many ancient petroleum deposits to the sky. While many of these surface shows, distributed in a wide belt between Mosul and Bandar Abbās, represent the final gasps of reservoirs exhausted by a steady loss of hydrocarbons to the hot sun and dry air, their widespread proliferation nonetheless marked the area as one hospitable to the formation and accumulation of oil and gas.[6] All that was needed was for D'Arcy's oilmen to find those hidden underground deposits, a task that in the early twentieth century depended as much on luck and perseverance as it did study and geological knowledge.

In 1901, the best available data on the region was that of M. Jacque de Morgan. Traveling to Iran as part of a scientific mission sent by the French government, Morgan had roamed western Iran between 1889 and 1891, collecting information on the culture, archaeology, and geology of the area. It was his 1892 article describing the Kurdish region's oil seepages and his 1895 follow-up confirming the territory's petroleum potential that became the basis for D'Arcy's initial explorations. Following Morgan's lead, by 1902 company prospectors were drilling in Chiāh Surkh, a nearly inaccessible plateau in the mountainous region bordering Ottoman Iraq.[7] Lying on the far side of the Zagros from Iran's capital and some three hundred miles north of the Persian Gulf, the site was remote, rugged, barren, and almost wholly ungoverned. Nearer to Baghdad than Tehran, work at Chiāh Surkh depended entirely upon a transportation route that ran through Ottoman territory, one that saw supplies disembarked in the port city of Basra, loaded onto river steamers for the journey to Baghdad, and then transferred to wagons for the remaining one-hundred-mile trip to the drill site. When oil was thus struck in October 1903, the low volumes produced—twenty-five barrels per day—proved to be hardly worth the effort, once again proving that using petroleum shows for guidance was a financially dangerous game.[8] Indeed, while they hunted for the land's buried hydrocarbon riches, D'Arcy's prospectors struggled to stay ahead of the commercial clock, seeking to find and bring to market economically viable quantities of oil before investors chose to cut their losses. It was not an idle fear, for with costs mounting and the concession in danger of being lost should exploration cease, D'Arcy had been forced earlier that year to seek outside investment. Concerned that the venture might be sold to foreign interests and aiming to secure stable oil supplies for the Royal Navy, the British government engineered a takeover by the Burmah Oil Company. Established in 1886 and headquartered in Glasgow, Burmah had been tapped to supply the navy with fuel despite long concentrating its operations within Britain's Indian colonies. Concerned that the Iranian concession might threaten their dominant position, in 1905 the company agreed take on D'Arcy's operations as a subsidiary while leaving the concessionaire to direct it. His endeavor thereby saved from immediate dissolution, D'Arcy transferred prospecting work from Chiāh Surkh southward to the province of Khuzestān, once again relying on Morgan's accounts of petroleum seepages in the area. Coming up empty in their first attempt near Shārdin, in 1907 operations were again moved, this time to a site some fifty miles to the northwest known as the "Plain of Oil" (*Maydān-e Naftān*).[9]

Located in the foothills of the Zagros near the ruins of an ancient Zoroastrian temple named Masjid-e Sulaymān, the "Plain of Oil" was remote and uncultivated, used only for winter grazing by the region's nomadic Bakhtiyāri tribes. Moving annually between summer territories in the mountains and winter lands in the lowland plains, with their ability to muster thousands of armed horsemen the Bakhtiyāri enjoyed, and did not hesitate to assert, considerable autonomy. With the mountain chain forming a major barrier to communication with Tehran,

Muzaffar al-Din Shah's grant of concessionary rights to D'Arcy carried little weight in a peripheral territory like Khuzestān, a reality that forced the British oilmen to accommodate the demands of the Bakhtiyāri khans. Though often portrayed as backward and thieving, the khans understood that D'Arcy and his prospectors sought something of value in the areas they controlled, and they shrewdly deployed their ability to disrupt company operations in order to extract compensation for the use of tribal lands. There was thus a great deal of grumbling among company officials about unending negotiations with the khans and the payments they demanded, but in the longer term the disruption of traditional grazing patterns and the use of oil rents to elevate tribal leadership into a distinct economic class transformed the Bakhtiyāri into a pool of cheap labor. For that reason, the company's early operations came to quickly depend upon local inhabitants, many of whom worked to build roads in the rough terrain and haul supplies to the drill site.[10] But as challenging as their thirty-five-mile overland journey was, D'Arcy's new ground was far more accessible than Chiāh Surkh had been, a beneficiary of the nearby Kārun River and more than two decades of British effort to open the waterway to commercial navigation.[11] When a great gusher of oil thus rose into the late afternoon sky on May 25, 1908, even as a letter ordering the cessation of drilling operations was en route from the notoriously tightfisted Burmah company directors, there was little doubt that Iran's petroleum wealth had not only been proved but also made accessible to the world. Few imagined just how enormous a find was made that day, however, for the new field, named Masjid-e Sulaymān after the nearby temple, was what would later come to be known as a supergiant. Eventually shown to encompass a surface area of seventy-five square miles and inhabiting a column of rock one thousand feet thick, over the next century some 1.5 billion barrels of oil would be produced from the field.[12] It was off this gigantic find that the Anglo-Persian Oil Company was established the following year, and the foundations of what would eventually become one of the world's largest oil companies were thus laid in the bedrock of southwest Iran.

The way forward would nonetheless be long and difficult, for bringing Iranian oil to market would require a railroad, a pipeline, a port, and, most importantly of all, a refinery to be built. Moreover, the high sulfur content of Iranian crude oil was something British engineers were not accustomed to handling, and it would take decades for the problems it caused to be entirely resolved. In 1908, however, all that lay in the future. What was true then was that the landscape of southwestern Iran had begun, after years of dedicated searching, to finally reveal the secrets of a hydrocarbon habitat that was second to none. But for all the sense that D'Arcy and his oilmen had at last conquered the harsh landscape, in reality their work had been all along responsive to its demands. Guided by the presence of natural seepages, prospectors had drilled where the earth suggested, seeking oil in regions shaped by the oceanic accumulations and tectonic collisions of the last 500 million years. Needing supplies, they had left the difficult passages of Chiāh Surkh for the much more accessible Masjid-e Sulaymān, modern industry giving way to the simple need to traverse the earth. Their communication with Tehran impeded by the high peaks of the Zagros, they rooted their work in the social geographies of Khuzestān, both contending with and coming to depend upon the long presence of the Bakhtiyāri people in the region. In this close connection to land, the story of D'Arcy's search for oil, as briefly recounted as it is here, is a microcosm of the history that will follow. Though at times eclipsed by the sheer humanness of events, always at the heart of Iranians' encounter with natural gas and its calorific potential was the earth, the imprint of surface topography and underground geology a sometimes hidden but nonetheless powerful presence in one of modern Iran's most consequential developments. Uncovering that influence is at the core of this book, a story of how an industrializing society's search for cheap energy would alter but nonetheless deepen its embedding in the natural world. But before any pipelines were built or any gas stoves lit, before there were any complicated debates about air quality or heated demands for service, a decision needed to be made that natural gas mattered at all.

No Gas, No Oil

Between the discovery of oil in 1908 and the mid-1930s when the utility of natural gas first began to be seriously considered, the Anglo-Persian Oil Company was built into a thriving commercial enterprise and a key component in British imperial power. Central to company operations was the Ābādān oil refinery. Constructed between 1909 and 1912 on a narrow island in the Shatt al-Arab waterway and built to take crude from the field at Masjid-e Sulaymān, the facility would eventually grow to become, for a time, the world's largest oil processing station. Though it refined a range of products like benzene and kerosene for markets in Asia and Europe, the refinery's most notable output was fuel oil, a heavy distillation for which Iran's petroleum deposits were particularly well suited. This development dovetailed with the changing requirements of the British armed forces, for having begun experiments with it in 1905, by 1914 the Royal Navy had deemed oil to be superior to coal and was well on its way toward adopting the new fuel. Amidst mounting fears of war with Germany, that decision made petroleum into a resource of great strategic importance and prompted the British government to take a controlling 51 percent stake in Anglo-Persian. With the company now a de facto arm of British colonial administration, the importance of Iran's oil exports induced changes in the country's border with Ottoman Iraq. Responding to British pressure, Iran traded the very area around Chiāh Surkh that had been explored by D'Arcy's prospectors for improved access to the Shatt al-Arab, thereby allowing the new stream of ships bringing supplies and exporting oil to avoid anchoring in foreign waters.[13]

For all the newfound attention given to petroleum, for the next quarter century, natural gas would not be a significant concern for APOC officials, its presence in southwest Iran largely treated as an adjunct to the extraction and refining of oil. Their view was not wholly without merit, for given the centrality of oil to British warfighting, and the fact that APOC had become by the 1920s an important purveyor of petroleum products across the empire, almost all company efforts had been channeled toward developing efficient systems for producing, refining, and marketing Iranian crude oil.[14] From its outset, however, Anglo-Persian had been just as much a producer of natural gas as it had been of oil. Indeed, when D'Arcy's prospectors watched that great geyser of oil reach high into the afternoon light, they were simultaneously engulfed by a volume of natural gas so overwhelming as to threaten them with suffocation.[15] Such venting of gas into the atmosphere, inadvertent at first but later done intentionally as part of oil production, was standard operating procedure in Anglo-Persian's fields, and considerable volumes of natural gas were discarded in this way. Though the company did not begin recording the amount of natural gas it was producing until 1933, retroactive assessments indicated steadily rising amounts over the previous quarter century. In 1911, APOC produced an estimated fifty million cubic feet of gas at the Masjid-e Sulaymān field; by 1937, that number had risen to more than twenty-seven billion cubic feet, primarily from Masjid-e Sulaymān and Haft Kel, a field some thirty-five miles to the southeast that had been opened in 1927. Total production varied over time in accordance with oil production, but by the late 1940s, a period when the economic potential of Iran's natural gas reserves would begin to attract real attention, more gas was being produced from more fields than ever before. By 1947, more than eighty billion cubic feet of gas had been extracted from five fields: Masjid-e Sulaymān, Haft Kel, Naft Sefid, Gachsārān, and Āghā Jāri.[16]

Driving such large totals was the natural materiality of petroleum itself and the circumstances of its accumulation in the bedrocks of southwest Iran. Prior to the 1979 revolution, nearly all gas produced in the country was found in close association with oil and extracted alongside it.[17] Dissolved within the liquid crude or separated out and sitting atop it as a gas cap (or both), large volumes of such associated gas are an integral part of most oil deposits. Though their different physical states have tremendous implications for their utilization, oil and gas are closely related, better understood as constituent aspects of a broader petroleum whole than as distinct substances. Accounting for different, if partially overlapping, sections of a single continuous hydrocarbon spectrum, oil and gas are differentiated by the weights of the various molecules that predominate within them. Discounting the impurities that are commonly present in petroleum reservoirs, natural gas consists of the four lightest hydrocarbons—methane (CH_4), ethane (C_2H_6), propane (C_3H_8), and butane (C_4H_{10})—whereas crude oil contains a far more complicated mix of compounds ranging from pentane (C_5H_{12}) at the lightest to very heavy asphaltics of more than sixty carbon atoms. Under the temperature and pressure conditions that are prevalent on the earth's surface, the four lightest hydrocarbons exist as gases, while pentane and heavier hydrocarbons appear as liquids, semisolids, and solids. It is this contrast between their states of matter that primarily differentiates the two petroleum resources.

But that above-ground—and decidedly anthropocentric—distinction is more imposition than true reflection of petroleum's wide subterranean variability and mutability. In the early twentieth century, scientific understandings of the resource were formulated in close relation to the business of selling oil. With commercial concerns paramount, significant effort was put toward producing knowledge that could subject oil to human control through generalizable and transportable understandings of petroleum.[18] There are limits to the utility of such schema, however, for oil and gas deposits are the result of hundreds of millions of years of natural history, and each reservoir has its own story of creation and accumulation that is embodied in its particular reservoir characteristics. In southwest Iran, such histories began when the detritus of ancient planktonic and algal life accumulated in deep beds on ocean floors. Buried and compressed by time and overlying sediment, such microscopic remains became the sedimentary layers that constitute much of the region's geology. They would also eventually come to serve as both source and reservoir rocks for the region's oil and natural gas accumulations. First, however, before petroleum would be generated within such organic-rich stone, long epochs and the heat of the deep earth would be needed. While the precise mix of organic remains in the rock will influence the character of the resulting petroleum, the ratio between crude oil and natural gas is primarily driven by the depth to which sediments are buried. Below 7,000 feet, oil and gas are generated concurrently, while the higher temperatures and pressures at 18,000 feet and lower result in deposits composed entirely of natural gas. Reflecting the influence of such local conditions, petroleum deposits are thus differentiated by their precise hydrocarbon mixtures as well as in the impurities they contain. Indeed, in southwest Iran, the extensive activity of ancient sulfate reducing bacteria in the region's deep biosphere is today manifested in the presence of notably sulfur-rich oil and gas.[19] But generation is only part of the story, for without specific conditions that allow petroleum to accumulate, the resource would be too widely dispersed for human use. Exploitable deposits are thus formed when oil and gas, both lighter than groundwater, migrate toward the surface, passing through permeable rock riven by countless pores, fissures, and fractures. That movement is at times blocked by an impermeable layer, and if the right kind of geological formation exists—a dome of sedimentary rock, an offset layer, a salt plug, and others—then oil and gas may be trapped in commercially viable quantities. Such is the case in southwest Iran, where the Zagros foothills are famous for broad anticlinal ridges that curve like the spine of a diving whale. It is a landscape given shape by a thick and porous layer of fractured limestone that has been dubbed the Asmari Formation. Capped by an impermeable seal of mixed salt and crystalline anhydrite, within the whaleback folds of the Asmari enormous quantities of oil and gas have over time accumulated. It was these deep pools that were tapped by D'Arcy's prospectors, first at Masjid-e Sulaymān and then elsewhere in the region. Created through extended and overlapping processes of deposition, deformation, generation, and accumulation, southwest Iran, and the entire Persian Gulf region, has become one of the richest hydrocarbon habitats in the world. But every deposit in the region is also in many ways unique, their characteristics produced by long and contingent natural histories that were not uniform across the area. For that reason, despite their treatment by scholars as functionally homogenous substances, oil and gas are in fact multiple and heterogeneous, and each reservoir's complicated individual history is reflected in its mix of hydrocarbons, the presence and concentration of various impurities, and the pressure under which petroleum is held.[20]

With oil and gas created in similar circumstances and migrating along similar pathways, oil deposits inevitably contain gas as well (the reverse is not true, as very deep gas deposits will not contain oil). At times, when the contours of the ground's rocky layers allow for it, a portion of that gas sits atop the oil, a dome of gas trapped between the reservoir seal and the heavier oil beneath. More often, under the intense pressures of the deep earth, the light hydrocarbons of natural gas remain dissolved among the heavier liquid and semisolid hydrocarbons of crude oil. Under such conditions, the two petroleum resources are, in effect, one. It is only upon being brought to the earth's surface that they take on their distinct identities, for it is at that point that the light hydrocarbons evaporate out of solution, becoming a gas and leaving the liquid oil behind. Inevitably, to produce oil is thereby to produce gas as well. As an Iranian delegation to the United Nations wrote in 1965,

> since the solution gas is present and since it is an integral part of the oil, intimately mixed and co-mingled with the oil, it must be produced if the oil is to be produced. In effect, 'no gas, no oil.'[21]

Subsequent debates regarding the utilization of Iran's natural gas reserves therefore revolved around a resource that was not only found in abundance but was also already being extracted from the earth in great quantities. For that reason, from the moment Iran became an oil-producing state in 1908, it became a gas-producing one as well, the extraction of the two petroleum resources as "intimately mixed and co-mingled" as their presence in the underground reservoirs of the Asmari. The question of natural gas in Iran was one that centered not on searching to meet a preexisting need, but on what was to be done with the overwhelming supply of something that no one as yet knew how to usefully employ.

A Low Convenience Value

By the mid-1930s, officials within the Anglo-Iranian Oil Company, as Anglo-Persian had been renamed in 1935, had begun to recognize the potential of Iran's large gas reserves. At the time, George Martin Lees, AIOC's chief geologist, reported that "in addition to gas produced with our oil which is surplus to requirements we have, in the ground, proved gas fields which are among the biggest of their kind in the world."[22] Contemporary estimates put the total availability of that natural gas at some 4.4 trillion cubic feet, approximately half of which was contained in the company's two operational fields at Masjid-e Sulaymān and Haft Kel.[23] It was a notable figure, for in a decade that saw AIOC produce six million barrels of oil in 1930 and ten million barrels in 1939, the total energy content of those proven gas reserves (equivalent to roughly 105 million barrels) was seen as very significant.[24] Limited usage of that associated gas in the 1920s had moreover shown the resource's potential value, with small amounts serving refinery operations as boiler fuel and as a source of sulfur. On the other hand, despite high hopes, experiments conducted between 1926 and 1928 that aimed to produce benzol from the gas had been a failure.[25] Given such challenges, even as finding outlets for the "colossal and increasing" supplies of natural gas in Iran had come to be seen as a "problem of the very highest importance," as the chairman of AIOC explained at a 1937 meeting held to discuss the issue, it was far from clear what commercial ventures might prove viable.[26] Standing in the way was what A.C.G. Egerton, Chair of Chemical Technology at the Imperial College of Science and adviser to Anglo-Iranian, deemed natural gas's "low convenience value." Unlike liquid crude oil that could be easily transported in barrels and ships, gas was much harder to corral and control, a trait rooted in its gaseous nature and, particularly, methane's lighter-than-air volatility. To overcome this challenge, he argued, the conversion of gas into products more easily transported to distant markets was needed.[27] Indeed, by the late 1920s, AIOC was already siphoning off the heavier hydrocarbons of the associated gas it was producing, largely butane and pentane, in an effort to spike its oil and boost Ābādān's gasoline production, the most in-demand and lucrative of the company's products.[28] There was no obvious solution for using the remainder, however, and attendees at the 1937 meeting on the subject heard proposals that ranged from the mining of magnesium and calcium carbide to cement production to the manufacture of petrochemical precursors to conversion to other hydrocarbons, all of which faced significant technical and commercial headwinds.[29]

The need for basic research was not alien to Anglo-Iranian, and over the years a great deal of scientific and geological knowledge was developed by company employees.[30] But while the manufacture of products like acetylene and aviation fuel from Iran's gas reserves were real technical possibilities, in practice the pursuit of such research was governed by AIOC's identity as a profit seeking enterprise.[31] Indeed, it was not clear that any such effort would be entirely in the company's interests, and it was reported that Baron John Cadman, the firm's chairman, had

> stressed the opinion, both in relation to the utilisation of Fields gas and refinery gas, that in conducting such research work the aspect of ultimate monetary value to the Company must not be lost sight of, nor should 'long distance research' be allowed to prejudice the prosecution of researches of more immediate commercial utility.[32]

For that reason, with seventy-six million cubic feet of gas being flared or vented each day, and a great many technical unknowns surrounding petrochemical and refining projects, AIOC officials turned instead to the idea of mining magnesium from southwest Iran's dolomite hills.[33] Used in numerous industrial processes and metal alloys, magnesium is one of the most abundant elements on earth; what made the possibility of production in southwest Iran attractive was the availability of cheap natural gas energy for the refining process.[34] "More [natural gas] calories could be exported in the form of magnesium than in any other way," company officials noted, and with demand for the metal outstripping available supplies in the 1930s, they argued that Iranian magnesium could compete "very favourably" with British suppliers dependent on more expensive coal power.[35] Nonetheless, despite the promise of the idea, the transformation of unexportable natural gas into exportable magnesium was not pursued, foundering in the face of concessionary hurdles—Anglo-Iranian did not have rights to the country's non-petroleum resources—and the outbreak of the Second World War.[36]

In the postwar period, magnesium was no longer of particular interest to Anglo-Iranian officials, as by then new and potentially more lucrative proposals for utilizing southwest Iran's gas reserves were being put forward. Foremost among them was the production of carbon black, an ultrafine powder used for applications like pigmentation and the strengthening of rubber tires. An idea first proposed in the late 1930s, carbon black's low market value in the prewar period had made it an unworkable outlet for Iran's gas resources. By 1944, however, impending postwar shortages and the inquiries of an American chemical firm, the Cabot Company, spurred renewed interest.[37] Southwest Iran, with its inexpensive natural gas and relative proximity to Indian Ocean markets, was an attractive locale for a carbon black plant, enough that Anglo-Iranian quickly decided to engage Cabot as a technical adviser and explore the idea.[38] There was a great deal of uncertainty in the proposed project, however, including from which field the facility might draw natural gas. In 1944, the two most promising possibilities were Pāzanān and Āghā Jāri, neither of which were in active operation.[39] Though Āghā Jāri was only a year away from entering commercial oil production, Pāzanān was the more straightforward option, as its gas was already being vented and could instead be captured and piped directly to any new plant. AIOC officials nonetheless opted for Āghā Jāri, choosing to prioritize the company's efforts to manage the compositional variability of southwest Iran's petroleum deposits. Their primary concern was hydrogen sulfide gas (H_2S), a common petroleum impurity but one that varied widely in its concentrations both within and across the region's petroleum deposits.[40] Toxic, corrosive, and flammable, hydrogen sulfide is a significant menace to oil and gas operations as well as an impediment to the manufacture of carbon black. In ideal circumstances, any gas used for carbon black would contain no more than 0.1 percent hydrogen sulfide by volume, a concentration that very few deposits in southwest Iran met. At Masjid-e Sulaymān, for example, dissolved gas was 11 percent hydrogen sulfide and dome gas 4 percent. At White Oil Springs, both sources contained almost no hydrogen sulfide at all.[41] Pāzanān's levels, measured as 0.024 and 0.046 percent at two separate well heads, were well within the requirements.[42] Āghā Jāri's 0.29 percent, in contrast, was far too high.[43]

Any gas produced at Āghā Jāri would therefore need to be desulfurized, a theoretical disadvantage but one that in practice might "facilitate selection of the most suitable gas" for the carbon black project.[44] More broadly, however, the variability in hydrogen sulfide content that was demonstrated at Pāzanān meant that even low-sulfur "sweet" gas would need continual monitoring and control, a reality that made Āghā Jāri's shortcomings less stark than they had at first appeared.[45] In the end, in light of such technical challenges and despite a continuing need to find outlets for the "disposal of Fields gas," it was decided in the summer of 1947 that the idea of a carbon black plant in Iran was impractical.[46] On a more fundamental level, however, the project had also come to be seen as superfluous to Anglo-Iranian's primary task of producing and exporting oil.[47] That prioritization of Anglo-Iranian's commercial interests, expressed in the 1930s as a desire to "maintain a balance between immediate and long-distance research," was a theme that consistently defined the company's relationship to Iran and its natural gas resources.[48] When it was determined that significant investments would be required to overcome the challenge that hydrogen sulfide presented to the carbon black project, the whole idea was shelved.[49] Seen differently, however, with hydrogen sulfide a nearly inescapable fact of oil and gas production, the project's cancellation was driven as much by the natural history of Iran's petroleum deposits as it was by Anglo-Iranian's pursuit of profit. In the decades that followed, AIOC officials would continually resist the Iranian government's insistence that something be done with the vast quantities of natural gas the company was discarding. Structuring that resistance was the simple hierarchy of concerns that Baron Cadman had articulated in 1938: that whatever good might come of increased natural gas utilization in Iran, Anglo-Iranian would pursue only those ventures that were directly beneficial to oil operations. But the company's ladder of economic interests was also a material one, for it was an arrangement that privileged hydrocarbons of high molecular weight over low. For decades, the conceptual disentangling of oil and gas allowed the company to treat the light, and inconveniently volatile, hydrocarbons of Iran's petroleum deposits as waste. AIOC officials thus saw and understood the geology of southwest Iran through the lens of liquid oil, choosing to deliberately unsee the ubiquitous presence of natural gas and leaving it to others to claim its productive potential.

Staking Claim

In February 1946, prompted by a request from an irrigation company working near Shushtar in central Khuzestān, officials from Iran's Ministry of Finance wrote to Anglo-Iranian to ask that "oil gases" be piped to the city for use as "fuel for the inhabitants instead of being wasted." Taking on a sharp tone, the letter continued, stating that

> now that the war has been finished, the utilization of oil gases should, in principle, be exposed to quick and fundamental investigations and measures. These gases may be conveyed by pipes to neighbouring towns to be used as fuel by the inhabitants, or they may be used in the

> localities of origin for the generation of electric power, which could be conveyed to the neighbouring cities and boroughs. Finally and alternatively, the gases may be returned to the oil wells.[50]

While the ministry's proposals were much easier said than done, they were not idle, and they prefigured many years of debate over Iran's natural gas reserves and the sprawling infrastructures that would be needed to exploit them. Animating the ministry's letter was an enormous sense of waste and the belief that AIOC was needlessly discarding a valuable resource that could be used to build a modern and prosperous society. Indeed, in the late 1940s, Pahlavi officials began to imagine a network of pipelines and refineries that would provide cheap energy to homes and factories across the country, a petrochemical industry that would be built on the use of gas as a feedstock, and a new source of export revenues that could further support their modernizing ambitions. In the following decades, such aspirations would govern a long and arduous effort to make natural gas the energetic foundation of Iranian society. First, however, new plans for an extensive and state-directed program of industrialization would need to be formulated.

Official developmental programs first took root in Iran in the mid-nineteenth century, chiefly in reaction to the devastating territorial losses the country had experienced in wars with Russia and Great Britain. Largely piecemeal and focused on the military and elite education, it was with the rise of Reza Shah Pahlavi to the throne in 1925 that such policies were broadened with an explicit focus on industrial expansion and economic growth. Between 1925 and 1941, the year he was deposed in a combined Anglo-Soviet invasion, Reza Shah sought to use his almost total control over the state to reform a society he deemed to be backwards and hidebound. Most visible (and controversial) were his secularizing policies, including those that mandated Western-style clothing for men, banned the *chador*, encouraged the education of women and girls, forbade some public religious observances, and undercut the authority of the Shi'a clerical class. While such decrees would become the subject of fierce political and ideological debates in the ensuing decades, less controversial were programs like the construction of the Trans-Iranian Railway; the expansion of the country's road network; the fostering of hundreds of new industrial firms; and the rapid growth of Iran's educational system, including the creation of the University of Tehran.[51] Financing for such initiatives came in large part from the country's growing oil revenues, and over the course of Reza Shah's reign, the Iranian state became increasingly dependent on its share of Anglo-Persian's profits. Relations between the Iranian government and the oil company were nonetheless fraught, a result of increasingly heated conflicts over revenue sharing, production rates, and the poor treatment of Iranian workers. Taking on nationalist dimensions that recalled the popular backlash to the 1872 Reuter Concession and the 1890 Tobacco Protest, in November 1932, the Iranian government unilaterally canceled the concession that underpinned APOC's operations in the country. In doing so, Reza Shah accused Anglo-Persian of a litany of offences, including the manipulation of financial data to reduce royalty payments, the exclusion of Iranian workers from skilled positions, and the exorbitant pricing of oil products sold in the country. For five months, Anglo-Persian and the Iranian government fought, their dispute going as far as the League of Nations, before a deal was struck in April 1933. The new terms stipulated that Iran would receive 20 percent of annual profits, up from 16 percent; a reduction in the size of the concession area by some 400,000 square miles; and a promise to integrate Iranians into administrative roles in the company. In return, Anglo-Persian had its concessionary period extended by a further thirty-two years, from 1961 to 1993. It also retained sole access to the company's technical and financial data, a privileged control of information that it would go on to use to minimize royalty payments and gain advantage in negotiations with the Iranian government. It was an outcome that was arguably worse for Iran than what had existed prior.[52]

When Muhammad Reza Shah Pahlavi (r. 1941–79) ascended to the throne in 1941, he inherited a country that was both under foreign occupation and rapidly descending into economic turmoil and famine. Seen as a puppet of the Allied forces that had deposed his father, the young king possessed little authority and even less legitimacy. The 1940s were, for that reason, a period of great political ferment in Iran, one that saw the rise of new political parties, the dominance of landed notables, and, late in the decade, the first coordinated efforts to modernize the country. In the postwar period, Muhammad Reza Shah sought to extend and accelerate his father's ambitions, and over the course of his reign, he created an overarching framework for governance that was rooted in a mix of royal patronage, industrialization, and technocratic oversight. It was a drive that would be in large part articulated through formal five- and seven-year development plans, and though they were less aggressively secularizing than the policies of Reza Shah's era, they nonetheless continued to embody a social vision that looked to the North Atlantic world for inspiration. That tendency was reinforced by the often extensive involvement of foreign consultants in setting the terms of Iranian development. Indeed, Iran's First Seven-Year Development Plan, ratified by parliament in February 1949, was based entirely on a sprawling program for the "improvement" of Iran that had been created by the Morrison-Knudsen International Company.[53] An American civil engineering and construction firm with experience working on infrastructure projects like the Hoover Dam and the San Francisco-Oakland Bay Bridge, Morrison-Knudsen had been hired by the Iranian government to create a modernizing blueprint for the country that could attract funding from the International Bank for Reconstruction and Development.[54] In its 1947 report, the company recommended $1.25 billion worth of work in sectors ranging from agriculture, industry, and mining to media, education, and public health. With strong emphases on industrialization and agricultural modernization, Morrison-Knudsen's plan turned on the availability of "more and cheaper" energy, something it argued would "stimulate . . . industries and promote the comfort and health" of all Iranians. At the time, Iranian industry largely relied upon scarce coal resources and expensive oil fuels, while the country's predominantly rural population was left to scavenge for "straw, weeds, or animal dung." Though it was believed that increased oil use could meet Iranians' growing need for fuel, Morrison-Knudsen instead identified natural gas as Iran's most promising option, arguing that the country's enormous reserves were an untapped and nearly limitless supply of cheap energy.[55]

In the late 1940s, as was already well understood by the Iranian government, some 120 million cubic feet of associated gas was being extracted and discarded every day in the country. Though Anglo-Iranian had long portrayed that loss to be of little concern, Morrison-Knudsen did not agree, arguing that because "natural gas produced in Iran must be used in Iran or wasted," the true price of oil fuels was thus "the cost of the oil plus the value of the equivalent amount of natural gas wasted."[56] It was an interpretation that Pahlavi officials welcomed, and they quickly accepted the company's recommendation that a pipeline network be built to make use of the country's gas resources. Though more than two decades would elapse before such a system would become a reality, for Iranian officials its mere possibility became a useful tool for pressing Anglo-Iranian into doing more with the associated gas it was lifting. For its part, AIOC strongly resisted, arguing that Morrison-Knudsen's proposed pipeline would never be commercially viable and should not be considered a realistic proposal for the country. At the same time, however, company officials began to fret that the tide was turning in favor of natural gas utilization, and that in the coming years they would find it increasingly difficult to avoid such demands. As they noted in September 1947,

> demand for a supply of gas for industrial and domestic uses in various towns in Iran will obviously become more insistent as time goes on. Although we may hold the view that practical issues concerning the distribution of gas in the towns and its safe use by domestic consumers render the use of gas on a scale which would justify the piped supply extremely doubtful, it will be difficult to continue to ignore such demands on these grounds without practical experience to back up our arguments.[57]

Without the "assurance of a substantial gas market" that could make gas utilization commercially viable, Anglo-Iranian officials feared being roped into an expensive and open-ended scheme for gas distribution that would serve only the Iranian government's developmental goals.[58] It was a prospect so worrisome that company leaders considered launching their own pilot program in Ahvāz solely for the purpose of demonstrating the difficulty of building a functional and profitable gas network.[59] They moreover sought to obscure just how much associated gas the company was producing, choosing in the summer of 1948 to provide Pahlavi officials with only "generalised" figures rather than the detailed technical data that they had requested and was readily available.[60] While such sleight of hand was not an unusual move for AIOC, as company officials had long perfected the art of manipulating data when dealing with Iranian requests for information, it did mark a new level of significance for the issue of natural gas.[61]

Between Morrison-Knudsen's recommendations and the growing insistence by Iranian officials that the country's gas resources be put to productive use, by the late 1940s, Anglo-Iranian was no longer able to evaluate potential gas projects solely in light of the company's own interests. Indeed, despite enjoying the benefits of a concession that continued to privilege their own commercial priorities, company officials increasingly found themselves evaluating proposals in terms that were set by the Iranian government. In August 1948, Iran's minister of finance informed AIOC that an investigation into the question of gas utilization in Khuzestān had found that Āghā Jāri's gas composition and reservoir pressures made it a viable source of energy for the city of Ābādān. The minister estimated that each day some forty-eight million cubic feet of gas, roughly equivalent to a yearly total of 475,000 tons of oil products, could be supplied to the city via pipeline. Going further, he referred to the 1933 concession agreement and cited its requirement that Anglo-Iranian "preserve the deposits of petroleum and . . . exploit its Concession by methods in accordance with the latest scientific progress" before declaring that the proposed plan "seems practicable and useful . . . [and] it is appropriate that you [AIOC] should study it as soon as possible."[62] As it happened, the company had already considered using gas from Āghā Jāri in its refinery operations at Ābādān, as in October 1946 it had been reported that more gas was needed than could be separated from the refinery's supply of crude oil. It was energy deficit that had been met via increasing quantities of fuel oil, and the refinery's consumption had quadrupled between 1938 and 1947.[63] Attracted to the prospect of selling that product rather than burning it—"the oil is wanted," they wrote—Anglo-Iranian officials estimated that the refinery's projected demand for fuel could be met with some sixty million cubic feet of natural gas per day. With the problem only growing, in the spring of 1948, approval was given for experimental infrastructure to be built in order to test methods of supplying the refinery with gas. When hurdles inevitably arose, however, the project was quickly set aside. In their meetings with Pahlavi officials, who continued to advocate for supply to Ābādān and other nearby cities, AIOC managers blamed the delay (and eventual cancellation) of the project on a shortage of steel and the need to preserve what was available for oil operations. They did, however, allow that the refinery represented "perhaps the best outlet" for Āghā Jāri's associated gas, and they promised the project's resumption "in due time."[64]

By the early 1950s, all mention of the Ābādān refinery's gas project had ceased and would not return. The episode was nonetheless illustrative of the fundamental orientations that the Iranian government and Anglo-Iranian had toward the question of southwest Iran's natural gas resources. For Pahlavi officials, gas had intrinsic value that was being wasted, and they sought to pressure AIOC into finding productive outlets for what they considered to be an important national resource. Anglo-Iranian, on the other hand, while never overtly disputing the idea that Iran's natural gas could be valuable, continued to prioritize the production and sale of oil over all else. While these positions reflected the different commercial and developmental interests of the two parties, the entire issue was structured by the natural conditions of southwest Iran's petroleum deposits. Gas utilization was a disputed topic only because AIOC's oil production inevitably lifted large quantities of natural gas as well. With nowhere for that gas to go, it had long been discarded in the country's oil fields, a literal burning away of Iran's fossil hydrocarbon resources. With the British company reluctant to mitigate that ongoing waste, it thus fell to Iranians themselves to find ways to either harness gas for the country's industrialization or conserve it for future use. For that reason, as the country's industrialization accelerated and enormous natural gas infrastructures rose to feed it, the willingness of Iranian officials to stake claim to the country's gas resources would come to imbue their use with highly charged notions of resource nationalism and sovereign development.

Surplus Production

In the 1950s, with vast quantities of associated natural gas being discarded and any sizeable program for domestic consumption years away from becoming a reality, Pahlavi officials working for the National Iranian Oil Company began pursuing the idea of gas recycling. Founded in 1951, NIOC was a direct outgrowth of the 1951–54 oil nationalization crisis, a period of profound upheaval that refigured political power within the country and altered Iran's relationship with both the Anglo-Iranian Oil Company and its own petroleum resources. It was a watershed moment decades in the making, but one that had accelerated during the turmoil that accompanied the Allied occupation of the country during the Second World War. With Reza Shah forced to abdicate in 1941 and the onset of a period of weak monarchical rule under his son, the 1940s had seen the rise of new mass urban movements. From them emerged the country's first true political parties, most notably the communist (and Soviet-aligned) *Tudeh* party and the National Front, a nationalist umbrella group that brought together a number of smaller parties rooted in the country's growing urban middle classes, its merchant guilds, and its religious associations. Led by Mohammad Mosaddeq, a longtime member of parliament with a reputation for incorruptibility and a commitment to constitutionalist principles, the National Front was strongly opposed to both foreign influence in Iran and the control that Anglo-Iranian exercised over the country's petroleum resources. Going well beyond the desire for greater revenue sharing that had animated Reza Shah's renegotiation of the original D'Arcy Concession in 1932, Mosaddeq and the National Front instead advocated for the expulsion of AIOC from the country and the total nationalization of Iran's oil industry.

By 1951, the National Front had become the majority party in parliament, and it used its newfound influence to both elevate Mosaddeq to the position of prime minister and nationalize the country's oil industry. Overwhelmingly supported by Iranians from all walks of life, oil nationalization prompted the creation of NIOC and propelled the National Front to new heights of popularity. Anglo-Iranian, on the other hand, backed by the British government, staunchly rebuffed the legitimacy of the decree and refused to cooperate with the prime minister's demand that it facilitate a smooth transition. With seemingly no negotiated resolution possible, the clash put Iran at the center of a major international crisis that saw heated debates in the halls of the United Nations and a worldwide boycott of Iranian oil. The United States, though not unsympathetic to Iran's position, chose to support AIOC and the British government, worried about the strength of the communist *Tudeh* party and not wishing to encourage similar anticolonial opposition towards its own business interests around the world. But even as Iran's economy suffered under the effects of the boycott, Mosaddeq chose to push controversial reforms that aimed to vest political power with parliament and render the shah as little more than a figurehead. Mosaddeq and the National Front, with their popularity eroding in the face of these twin crises, thus became vulnerable, and it was resolved in London to overthrow the prime minister and restore control of Iran's oil to AIOC. It would be an effort backed by American money and clandestine expertise, and one that found a base of support among conservative factions in the army, bazaar, and clerical establishment that mistrusted Mosaddeq's constitutionalist reforms. Carried out over four tumultuous days in August 1953, the coup succeeded in toppling Mosaddeq and restoring the shah's authority. In its immediate aftermath, Muhammad Reza Shah took full control of the state and crushed the National Front and *Tudeh* parties while also clamping down on the open political culture that had fostered their rise. Despite his return to unquestioned power, however, the shah found his legitimacy to have been greatly undermined, a victim to what many Iranians saw as the obvious meddling in their domestic affairs by foreign powers. In contrast, Mosaddeq, though tried and sentenced to lifelong house arrest, would be remembered as a national hero.

While the nationalization drive of Mosaddeq and the National Front ultimately failed, it nonetheless forced changes to the terms under which Anglo-Iranian operated in the country. Though AIOC had expected to recover its pre-crisis position as the only foreign oil firm working in Iran, staunch public opposition, and an American desire to receive compensation for its role in the coup, made that position untenable. In 1954, a new twenty-five-year Consortium Oil Agreement was signed that divided control of Iran's petroleum industry between NIOC and a group of eight oil firms that included AIOC—renamed British Petroleum—Royal Dutch Shell, five American companies, and a French firm. With a 40 percent ownership stake, British Petroleum remained the dominant player and, despite NIOC's nominal ownership of all fields and equipment, continued to operate with minimal oversight and largely at its own discretion. More important for Iran was the inclusion of a new 50-50 profit sharing agreement. Pioneered in the Middle East by Saudi Arabia's Aramco in 1950, the new system of revenue sharing, while still not permitting the Iranian government to examine the consortium's financial ledgers, replaced the easily manipulated per-barrel payments of the 1933 revised concession. As a result of that change, Iran's oil revenue jumped more than fivefold, going from $34 million in 1954–55 to $181 million in the following year, and increasing yearly thereafter.[65] But even as the 1954 agreement left Iran's oil under the control of foreign companies, it also included other terms that would prove to be of equal, if not greater, importance for Iran: the rights to all associated and nonassociated natural gas in the country.[66] Within a decade, Pahlavi officials, aiming to make gas into a source of cheap energy for their industrializing society, would set about creating new markets for that gas and building new infrastructure to serve them. In the mid-1950s, however, all that was yet to come, and Pahlavi officials instead pursued the idea of gas recycling, hoping to preserve for future use much of the associated gas that was then being discarded.

A theoretically straightforward idea, recycling was in practice a complicated and potentially expensive endeavor. Under such schemes, extracted gas could be collected and then reinjected into underground petroleum reservoirs, a practice that could be used to prevent loss or to repressurize oil deposits that were otherwise approaching the end of their productive lives. For Pahlavi officials, either use was better than flaring or venting gas for no gain, but, as had long been the case, their ability to pursue any such plan was sharply curtailed by the limited influence they exercised over oil operations. For their part, despite recognizing that reinjection was the "only practical means of avoiding burning the greater proportion of the gas," companies like British Petroleum had long viewed the idea as uneconomical.[67] As early as the

1920s, company employees had explored the idea as part of broader initiatives to recycle oil during times of elevated supply. Indeed, despite widespread notions of oil scarcity, for most of the twentieth century, global production had far outstripped demand. Cartel arrangements and organizations like the Iranian consortium were therefore used to control oil supplies and prevent gluts that could depress prices and hurt profits. Oil recycling could aid that effort, but in order to prevent the reinjected oil's "total loss to humanity," enough natural gas would also need to be recycled to keep reservoir pressures high enough for continued production.[68] Still, in the prewar era, despite such potential benefits, company officials had resisted the possibility of recycling gas as commercially unviable. In 1938, for example, they rejected an Iranian proposal that leftover refinery gas from Ābādān be reinjected into the Masjid-e Sulaymān and Haft Kel oil fields, arguing that while the plan was feasible, "there could be no possible economic justification of re-pressuring for gas storage purposes."[69]

Shelved during the Second World War, in the late 1940s the prospect of recycling was revived by Iranian officials beginning to more seriously consider the prospects of natural gas utilization. Concern for immediate commercial feasibility rather than long-term economic potential nonetheless continued to govern British Petroleum's approach. Indeed, one 1945 study found that oil production "would either remain the same or would be somewhat reduced by gas recycling," and it thus argued that

> [a] policy of gas recycling would therefore be entirely for the purpose of conservation. The justification of a heavy expenditure on this account may be questioned on the grounds that gas vented to the atmosphere should be regarded as legitimate losses in the process of production and is in fact an exceedingly small proportion of the total losses when irrecoverable crude in the reservoir is included.[70]

As with all prospects for gas utilization in Iran, company officials focused on short-term "expenditures"—the costs of building, operating, and maintaining a system of gas injection—and gave little weight to either long-term availability or the potential value of natural gas. In February 1946, responding to repeated inquiries from the Iranian government on the issue, one exasperated director asked, "are we or are we not entitled . . . to burn or vent off gas . . . or are we bound to return the gas at substantial cost to the underground reservoir in order to conserve it for possible use at some certain date in the future?"[71] So strong was the opposition to the idea of recycling that company officials were willing to risk permanent declines in oil production to avoid it, going so far as the hide unexpected changes in reservoir pressures for fear that the Iranian government would use such opportunities to press for a costly gas recycling program.[72] By 1948, the company had gone further, building small experimental systems at Haft Kel and Masjid-e Sulaymān in order to demonstrate their difficulty and expense.[73]

But even more important for many company officials was the fact that their existing production and reservoir management practices already conserved much higher volumes of associated natural gas than was the global norm. As one wrote,

> [It] should be appreciated that the method of multistage gas separation and crude stabilisation employed in our high pressure fields substantially reduced the amount of gas actually produced in these fields and thus constitutes indirectly a utilisation of the gas contained in the crude.[74]

Indeed, British Petroleum had pioneered the technique of multistage gas separation, and while it had been originally conceived to boost volumes of recoverable natural gasoline, it also enabled the careful control of associated gas.[75] Particularly in high-pressure fields like Haft Kel where the method was developed, aggressive rates of oil production could also draw dome gas, an outcome that would not only result in more wasted gas but also "a loss of natural energy and possibly of ultimate recovery" of oil.[76] With multistage separation reducing the scope of that problem, company officials thus argued that their careful production practices constituted a meaningful system of natural gas conservation. "This in itself may be claimed to be a measure of gas conservation," they wrote, as it "follows from this that when the bulk of the recoverable crude reserves have been produced, the bulk of the gas reservoirs will remain."[77] It was not an unreasonable claim. Company officials reported that in 1947 the multistage separation process had resulted in some 5.0 billion cubic feet of associated gas to remain within the reservoirs at Masjid-e Sulaymān and Haft Kel rather than be discarded. But it was also not the full truth, for the total lift of associated natural gas from the two fields had been 35.7 billion cubic feet, of which 12.7 billion was used in oil operations and 18.0 billion was flared. Moreover, Masjid-e Sulaymān and Haft Kel were not the only fields in operation, and in the same year a further 49.4 billion cubic feet of gas had been flared at Naft Sefid, Gachsārān, and Āghā Jāri. All told, approximately 67.3 billion cubic feet of natural gas, or 84 percent of the total produced in southwest Iran, had gone unused. In their reporting to the Iranian government, company officials sought to obscure the fact that so much gas had been lost, writing that since "Agha Jari, Gach Saran, and Naft Safid oilfields are still in the course of development, figures given for utilisation of gas would only be misleading."[78] They were also well aware of the productive potential that they had let evaporate. "Surplus gas production . . . [at] Agha Jari burnt to waste during 1947 could have replaced Abadan's liquid fuel consumption more than twice over," one engineer wrote, before adding that a nearby irrigation project could have had its water level "lifted 50 feet by utilising about half a million tons/year . . . of surplus gas production, now burnt to waste."[79]

Fractured Stone

In the years after the 1951–54 oil nationalization crisis, the possibility of gas recycling was again sidelined as Iran's new oil regime took shape. By the 1960s, however, the topic had once again become a significant point of contention between the Iranian government and British Petroleum. It was a complicated debate, one that involved commercial imperatives and production practices while also turning on minute geological features. In the late 1940s, while compiling estimates on the price of recycling gas in southwest Iran's various petroleum fields, company officials had found that gas volume and reservoir pressure, the primary determinants of reinjection costs, could mix in unexpected ways.[80] As a result, during any gas reinjection program, no two wells or reservoirs would need the same injection pressure or would react in the same way, and it thus required "very much more intimate knowledge of the conditions in the limestone" than the company possessed to make accurate predictions.[81] Indeed, as would become apparent in the 1960s, the region's geological characteristics were among the most important factors shaping the debate over gas reinjection. At issue was the region's "porous limestone"—the Asmari Formation—and its influence on the movement of oil and gas within the ground.[82] A long belt of sedimentary reef rock that runs along the southwestern front of the Zagros range, the Asmari is "by far the most important proved reservoir in Iran" and one of the richest in the world.[83] Making it so uniquely productive is an extensive system of vertical fissures and fractures that had been created by the region's slow tectonic deformation. Though the primary permeability of the Asmari limestone itself is poor, the fracture system, "confirmed in numerous localities in Khuzestān," created an "extremely responsive flow system" during oil production that enabled the "rapid segregation of oil and gas to take place in the fissures through which the gas migrates readily upward into the gas cap." The end result of this process was the increased conservation of solution gas by natural processes. As a combined NIOC and consortium delegation presented at a 1965 symposium,

> This would be regarded as a major gas injection project if it had to be effected from the surface. If producing rates are not exceedingly high (as they are not) when compared with the rate of oil and gas segregation within the fissure system, the producing gas-oil ratio will continue to decrease, resulting in considerable conservation of the gas.

In other words, that fissured rock, when combined with the careful extraction policies that British Petroleum had long maintained, helped keep excess gas within the earth, a "fortuitous phenomena . . . perhaps unique to Iranian reservoirs" that enabled the "building up of vast reserves of natural gas which will be available for future utilization."[84]

But while these entangled factors reduced the amount of gas being brought to the surface, they did not eliminate the fact that significant quantities were still being discarded. With the Iranian government continuing to express discontent over the issue, in 1961, British Petroleum formed the Gas Study Committee to explore the prospects of gas utilization in the country. As had long been the case, the company's own interests were paramount, and one of the first options studied by the committee was the use of gas reinjection to boost oil production in southwest Iran's eight active oil fields. Of those sites, all save Ahvāz had their primary reservoirs within the Asmari limestone. They also all possessed significant gas caps, though those of Masjid-e Sulaymān and Haft Kel were secondary to decades of oil extraction rather than natural. More to the point, in the committee's estimation, such large volumes of dome gas had accumulated and persisted because the company's conservative production practices only lifted an average of 740 cubic feet of gas per barrel of crude, well below the 1,500 and 1,100 cubic feet per barrel averages of the United States and Venezuela. Though the ratio of gas to oil had increased as fields like Masjid-e Sulaymān began to reach the end of their productive lives, the committee was confident that the combination of restrained production and the Asmari's fractured limestone would "result in the retention in the reservoirs of a large fraction of the gas originally in place."[85] In addition, despite the growing success of gas repressurization practices around the world, the committee saw British Petroleum's experimental systems of the late 1940s as fundamentally unworkable due to that very same "extreme" fracture system. Indeed, while the disjointed rock matrix promoted the accumulation of free gas, it also hampered the ability of recycling to push more oil to the surface. The committee thus concluded that in fields with relatively poor rock permeability like Masjid-e Sulaymān and Naft Sefid, reinjection would only interfere with their natural gas drive and damage oil production. To have any meaningful benefit, any program of gas recycling would therefore need to be aimed at conservation rather than enhanced oil recovery. If, as had been proposed by NIOC, the decision was made to sacrifice future oil recovery at Masjid-e Sulaymān in favor of gas conservation, then it would be possible to use the reservoir's gas cap to store some forty million cubic feet of gas per day. Even then, however, the implications of a full gas recycling program were unclear, and the committee expressed reservations about the possibility that pressurizing the field above its natural limits could damage wells and stop oil production entirely. For that reason, and in light of the fact that gas from the field was already being used in company operations, they recommended against a program of recycling at Masjid-e Sulaymān.[86]

Most of southwest Iran's other fields were no more promising. At Haft Kel and Gachsārān, the committee found that oil production was driven primarily by water and not gas, a natural system with which repressurization would interfere. While too little was known about Gachsārān's geology to comment on the prospect of gas conservation there, at Haft Kel they thought it possible to reinject some 300 billion cubic feet of gas brought from Ahvāz. It would be an expensive choice, however, costing some $5.6 million, and with the recycling of Haft Kel's own gas thought to be even more expensive, the committee ultimately concluded that "surplus gas storage [at Haft Kel] cannot be justified on technical or economic grounds." Naft Sefid was no better, as an exceptionally large gas column undermined any benefit that gas recycling could have for enhanced oil recovery. Nor was the field a viable option for gas conservation, as its own output was already used in company operations while nearby Haft Kel's high-sulfur gas would "contaminate" Naft Sefid's valuable low-sulfur petroleum. Pāzanān, a field largely idle in the 1960s, was no better. With an "abnormally high gas reservoir pressure," recycling to the field was both prohibitively expensive and potentially dangerous. Only at Āghā Jāri, with its comparatively high reservoir permeability and weak natural drive, were the prospects of gas reinjection considered at all promising. While cost estimates for a recycling program were confidently predicted to be between $25 million and $30 million, the expected effect on oil recovery swung wildly between a 200-million-barrel loss and a 500-million-barrel gain.[87] With little to recommend at any of the fields, the members of the committee concluded that gas recycling was unlikely to have any beneficial effects on oil production in southwest Iran. They also found the potential benefits of reinjection for the sake of conservation to be even more suspect, depending heavily on speculation about future demand and market conditions. "Storage of gas in underground reservoirs is an expensive process," they wrote, "and a long-term basis requires the use of current income on projects for which no return will be realized for many years." To reinforce their point, committee members concocted a hypothetical ten-year program for the reinjection of 700 million cubic feet of gas into underground reservoirs. With an estimated initial outlay of $50 million, a yearly operating cost of $5 million, and a $0.02 production cost per thousand cubic feet, it was thought that over a twenty-year period a final sale price between $0.21 and $0.31 per thousand cubic feet would be needed to make a profit, a figure they believed to be uncompetitively high. It was thereby concluded that while

> surplus oil field gas can be stored without great difficulty from a technical viewpoint, there does not seem to be any economic justification for the operation, which is probably the reason it has never been done in the past.

For officials in British Petroleum, such findings, combined with the continued relevance of the Asmari's ability to keep gas underground, did little to counteract their belief that the entire idea of gas conservation was suspect. It was thus decided that "under these conditions, the Committee recommends against large capital investment designed primarily to save current associated gas production for future use by reinjection."[88]

***

Following the committee's recommendation, British Petroleum did not pursue any gas recycling projects in southwest Iran. For its part, despite the keen interest of Pahlavi officials in the idea, NIOC's first reinjection effort would not begin until 1976, years after natural gas would first come to significant consumption within Iran and well into the twilight years of the consortium arrangement.[89] It was an outcome entrenched in the tension between British Petroleum's prioritization of oil production and the Iranian government's desire to find and preserve a cheap source of energy for its industrializing ambitions. More than anything, however, it was the Asmari's fractured limestone that determined whether gas recycling could meet the goals that people set for it. In this way, while debate swirled around matters of commerce and development, at the heart of the issue was the question of what it meant to conserve gas within the warps and folds and fractures of southwest Iran's subterranean layers. While Pahlavi officials dreamed of using gas to power an industrialized society at some indeterminate point in the future, the "natural conservation process" of the Asmari's fissured limestone seemingly freed British Petroleum from needing to concern itself with preserving the gas it produced.[90] In this way, the company pursued a notion of 'good enough' conservation, one that saw the Asmari's natural preservation as sufficient, and the gas flared during oil operations as a normal part of doing business.

For British Petroleum, more important than finding uses for all that associated gas was resisting the Iranian government's pressures on the issue of gas utilization and recycling. Concerned about the costs of any such project, BP officials relied on their control of the geological and economic data of petroleum production to frame their operations as having made reasonable and sufficient efforts to preserve Iran's gas resources. The choices that were made by the company's leadership were thus imbued with the seeming concreteness of technical fact while being simultaneously reflective of the company's commercial priorities. Going further, by viewing natural gas through the lens of oil production, BP managers subordinated it to oil and obscured its distinct economic potential for Iran. In their view, gas had little intrinsic value, a marginally useful byproduct of the company's central mission of finding and exporting oil. It was an understanding that rendered natural gas functionally invisible, and far from incidental, this deliberate unseeing was crucial to maintaining the low overhead costs needed to furnish cheap oil to foreign markets. While company officials were theoretically amenable to the idea of supporting gas utilization in Iran, they were only willing as long as it asked little of them and their customers.

Pahlavi officials, in contrast, saw their country's natural gas resources as possessing tremendous economic and developmental worth. Reflecting that belief, over the years they engaged in a steady campaign to pressure British Petroleum to reduce the amount of gas it was discarding in the country's southwestern oil fields. More than a commercial and political dispute between two interested parties, the question of what to do with Iran's gas reflected the area's particular physical characteristics. Created by the heat, pressure, and anaerobic environments of the deep earth, trapped in rocky layers contorted by tectonic movement, southwest Iran's petroleum resources were distillations of its long natural history. Indeed, whether or not the Asmari's fractured limestone retained enough gas to make recycling a worthwhile investment was a decision rooted as equally in the geology of the region as it was in the different priorities of British Petroleum and the Iranian government. Sedimentary rock and reservoir pressures and the material properties of hydrocarbons rendered natural gas as either an uneconomical byproduct of oil or a substance with nearly unbounded potential to further Iran's industrialization. Such factors shaped the realm of possibility for gas exploitation, both enabling some uses and constraining others. With British Petroleum accepting such limits as the final word on gas utilization, it fell to Iranians themselves to create productive outlets for their country's gas resources. By the 1960s, that endeavor would be in full swing.

Two
Industrial Futures

In December 1964, speaking to the assembled delegates of the Seminar on the Development and Utilization of Natural Gas Resources, held in Tehran under the auspices of the United Nations Economic Commission for Asia and the Far East, Alinaghi Alikhāni, Iran's minister for the economy, presented natural gas as the developing world's great hope for the future. In his remarks, Alikhāni warned that booming populations were stretching the availability of "traditional raw material resources" to their limits. Overcoming such hurdles would require new sources of food and energy, and in Alikhāni's view, it was natural gas that was best poised to help the countries of the ECAFE region. In his estimation, gas could be a new fuel, burned in factories and in homes alike. It could also be a raw material, transformed through chemistry into new fertilizers that would feed the burgeoning populations of the world's developing regions. Monumental projects encompassing thousands of miles of distance, kilometers of rocky strata, and vast sums of money would be needed, but so long as members states embraced the "practical solutions" and "regional co-operation" that forums like the 1964 seminar were intended to promote, then their nascent and disconnected efforts to exploit the potential of natural gas could blossom into the foundations of prosperous societies.[1]

That Alikhāni was chosen to give the symposium's opening remarks was fitting. Not only did Iran sit atop some of the world's largest reserves of natural gas, but by the mid-1960s it also stood amidst a wave of projects aimed at making use of them. Driven by an explosion in the volume of the world's proven gas reserves in the four years prior to Alikhāni's speech—from some 660 trillion cubic feet to 1.6 quadrillion, approximately two-thirds of which was located in the Middle East—states around the world were beginning to look to gas as a potentially crucial new resource.[2] The 1964 symposium itself had grown out of the world's first major meeting on petroleum resources, hosted in Tehran in September 1962 by the National Iranian Oil Company.[3] At that meeting, gas had taken on unexpected importance, with U Nyun, executive secretary of the ECAFE organization, highlighting it as "increasingly important to several countries of the region." Over fifty papers on the production, storage, transportation, and consumption of natural gas were presented, and participants declared that the "new and substantial discoveries of natural gas in the ECAFE region, if utilized adequately, would play a vital role in accelerating the economic development" of their societies. Due to that "extreme importance," delegates to the symposium recommended that the U.N. Secretariat prepare a study of natural gas resources and their potential uses, with particular attention to be given to possible joint efforts among member nations. It was that study that would form the basis of the 1964 summit at which Alikhāni spoke.[4]

Iran's hosting of the summit reflected a growing enthusiasm for natural gas utilization among Iranian officials in the late Pahlavi period. By the 1960s, many had come to see the resource as key to building a modern society capable of drawing upon near limitless quantities of cheap energy. But making productive use of Iran's natural gas resources was far from straightforward, as standing in the way were the twin challenges of geographical distance and gas's physical volatility. Moving large volumes of gas between producing and consuming regions was a difficult and expensive proposition, one that undermined the economic viability of nearly all proposed outlets. Just as natural gas was entangled with the verticality of the rock column from which it was produced, so was its utilization intertwined with the topography and geography of the Earth's surface. Distance dictated where gas could be profitably sold, and in the first two decades after the Second World War, few regions were within reach, even within Iran itself. But natural gas could also be transformed via chemistry, and when it was, the land and soil of Iran itself became opportunities. Iran's first major gas project was for this reason a petrochemical facility, built to produce the synthetic fertilizers that were increasingly important to Iran's agricultural sector. Moved by train and truck and distributed as liquid and powder in rural communities across Iran, fertilizers condensed the utility of natural gas into forms that could be transported in economically efficient ways. At the same time, urban residents increasingly relied on liquid gas—mixtures of propane and butane—for heat and cooking fuel, a need met by private firms and the canisters they delivered by road. Over the course of the 1960s, these loose threads would begin to tighten into a robust web that bound ever more pieces of Iranian society to natural gas. Such strands were as much intellectual and cultural as they were material, for during that time gas became a key part of how many Iranians imagined their country's future. Theirs was a vision that lionized technological sophistication, economic prosperity, and national independence, bridging vast distances in order to root Iranian industrialization in the petroleum deposits of southwest Iran.

Distant Markets

In the early 1960s, natural gas utilization was a nascent phenomenon in much of the world, with little existing infrastructure to bring the resource to market. Unable to be stored in barrels or tanks like oil, natural gas required significant investment in pipelines to be transported, a material reality that had long restricted the movement of gas to small areas. The earliest uses of gas were thus often conceived in close connection to existing infrastructures for the production and transportation of coal. Starting in the early nineteenth century, "town gas," largely extracted from coal heated in oxygen-free environments, was used as a source of energy in many urban centers in Europe and North America. Distributed from centralized production facilities via wooden or metal pipes, such manufactured gas helped extend cities' productive hours by lighting streets, factories, public buildings, and, in some cases, people's homes. With hundreds of miles of pipes and thousands of companies and municipal systems supplying town gas, the resource became a major part of urban life, with many of the world's large cities relying on it by the end of the century. Town gas nonetheless remained a localized affair, with the limits of existing pipeline technologies keeping networks to a small size, and its utility was rooted in the relative ease of transporting coal. Unlike gas, which required transmission via closed systems to prevent leakage, coal was a solid substance that could be moved by piling it into wagons, trains, and ships. Sent to conveniently located gas works, coal could then be used to produce gas that was in turn distributed via pipe. In theory, natural gas could be substituted into those piped systems, but the same technical hurdles that kept town gas systems small also prevented the transmission of natural gas over long distances. For that reason, with few significant deposits of natural gas near urban areas, at the turn of the century the resource was largely unknown.

From its very inception, the possibility of gas utilization was thus bound up in the limits imposed by geographical distance. Prior to the 1950s, natural gas was almost exclusively used in the United States, with the country accounting for some 90 percent of all global production and consumption. Indeed, dedicated natural gas exploration was pioneered in the U.S. when the world's first gas well was dug in 1821 in Fredonia, New York. An open pit topped by a small wooden shed, the well provided fuel for lighting through crude pipes of hollow logs, sealed with tar and rags, that were connected to nearby shops and mills. Some fifty years later, and exactly as British Petroleum would do in Iran, early American oilmen working in western Pennsylvania treated the associated gas they lifted as an unwanted nuisance. By the 1890s, however, perceptions had changed, and many began selling gas for fuel to the nearby steel mills of Pittsburgh. Within a decade, natural gas use was widespread in Appalachia, and with the advent of new technologies for pipeline construction in the 1920s, the geographic range of natural gas utilization expanded much further again. At the time, the development of seamless steel pipes, the process of oxyacetylene welding used to join them, and the exploding energy needs of heavy industry came together to prompt the creation of new long-distance pipelines that connected the petroleum-rich Permian Basin of northwest Texas to Chicago and the industrialized cities of the American east. While the Great Depression and the steel demands of the Second World War suppressed pipeline construction between the early 1930s and the end of the war, the United States' roaring postwar economy gave the industry a fresh jolt. At times directly supplanting oil, American gas usage rose rapidly in the postwar era, and by the 1960s it accounted for one-third of all energy consumed in the country.[5]

Though American experiences had proved the viability of natural gas utilization early in the twentieth century, it was not until after the Second World War that the resource would be used in any meaningful way outside the United States. Beginning in the 1950s, however, a series of large gas finds and policy decisions quickly turned many European countries into major producers and consumers. Nowhere was this more true than in the Soviet Union, where the 1953 rise of Nikita Khrushchev to the Soviet premiership resulted in a new determination to make natural gas, as both a source of energy and petrochemical feedstock, a crucial driver of economic growth. Over the next decade, Soviet leadership sought to emulate the lucrative and fast-growing petrochemical industry of the United States, hoping to foster technological development, create new sources of revenue, and improve the living standards across the Eastern Bloc. But while the USSR possessed enormous oil and gas reserves in its Caucasian and Siberian territories, internal discord and ill-conceived plans hampered the effort to transform the country's chemical industry into a lucrative export sector. The initiative nonetheless resulted in the construction of numerous new pipelines and the widespread adoption of residential gas consumption, something that in later decades would be joined by sizeable and profitable exports to the countries of central and western Europe.[6] Indeed, by the 1970s, when Soviet gas shipments began to arrive, many European nations were no stranger to natural gas use. The continent's first major find had come in northern Italy during the Second World War, and by the early 1960s the country had Europe's most developed gas market. Driving the new attention to gas was a broad need for reliable supplies of energy that had emerged amidst postwar reconstruction. Accelerated by the continent's longest period of sustained economic growth in history, the two decades between 1950 and 1970 saw energy consumption rise at an average rate of 3.6 percent per year, much of it on the back of imported petroleum fuels like gasoline and diesel. Unlike in the industrial revolution of the nineteenth century, an era when most European states were able to tap large domestic deposits of coal, the postwar decades saw the geographical correlation between areas of significant energy demand and areas of significant energy production broken by the shift to oil. But oil's relative transportability and fungibility made it an ideal candidate for seaborne transport, and in this way the countries of western Europe became dependent on crude oil shipped from regions like the Middle East.[7] Nonetheless, despite the rapid rise in oil consumption, gas use, primarily in the form of coal and manufactured town gas, remained common in postwar Europe, though total usage was low in comparison to the United States, averaging a per capita rate of only five cubic feet per day versus seventy-seven.[8]

Even considering the finds of the 1950s, with Europe largely devoid of the major gas reserves that could be found in the Persian Gulf, distance, easily overcome in the case of oil, proved to be a very high hurdle for natural gas utilization. It was a geographical mismatch of continental scale, for moving natural gas between producer and consumer regions would require a network of pipelines thousands of miles in length. In 1951, the Bechtel Corporation proposed to create just that. Founded in San Francisco in 1898 and initially focused on the western United States, Bechtel began as a firm specializing in building railroads before diversifying into the design and construction of all manner of infrastructure around the world. By the early 1950s, the company had laid numerous oil pipelines in the Middle East, including what was at the time the world's longest, the recently completed Trans-Arabian pipeline that connected eastern Saudi Arabia to the Lebanese coast. Using that project as a steppingstone, Bechtel envisioned a gas pipeline stretching some 2,500 miles from the Persian Gulf—"the largest potential petroleum pool in the world"—to Paris, describing the potential project as a "modern, peaceful means for balancing some of the world's resources and promoting the general welfare." Estimated to cost $425 million and crossing nine countries in the Middle East, the Balkans, and central Europe, the pipeline would connect to existing town gas systems to deliver 500 million cubic feet of gas per day to 200 million people. An adjacent pipeline, built as part of $250 million second phase, could double the amount of natural gas delivered, while additional branch lines of up to 300 miles in length would be able to reach "all the industrial centers of Europe" and increase the population served to more than 331 million. Drawing at first from the fields of northern Iraq, the gas reserves of southwestern Iran, Kuwait, and eastern Saudi Arabia would be connected to the system as demand necessitated.[9]

For natural gas to be a viable source of energy, however, it would need to be offered at prices competitive with other forms of fuel. Bechtel believed that it could be done, and the company's analysts estimated that Middle Eastern natural gas could be delivered to end users at a cost of $0.50 per thousand cubic feet, well below the $1.00 per thousand cubic feet average it cost to produce town gas, let alone the $2.60 per thousand cubic feet that Parisian residential consumers were reported as paying.[10] But, as officials working for British Petroleum noted, the commercial viability of the pipeline rested on a number of questionable assumptions about its proposed route. According to BP's calculations, Bechtel's plan had an estimated cost of £60,800 per mile in the first stage and £55,200 per mile in the second. Those figures were deemed suspiciously low, as detailed survey work for the Middle East Pipeline project, a proposed line to transport Iranian crude oil to Syria's Mediterranean ports, had calculated construction costs to be £102,500 per mile. The MEPL route, moreover, crossed "easy country," while Bechtel's would need to traverse several heavily populated and mountainous regions.[11] Even more concerning were the geopolitical risks, and officials working for British Petroleum worried that the economic logic of the project might be undermined by "political issues," particularly the possibility that "embarrassing . . . [transit] Royalties . . . will be charged by the various governments over whose country the lines will pass." Indeed, as they argued,

> countries taking gas from the line will become more and more dependent upon regular supplies and great disorganization and hardship could be caused if a country, relatively near the source of supply, decided to hold the pipeline company to ransom for more money. The time is not foreseeable when racial and political relations will be so amicable and settled as to preclude such an event.[12]

Alternate routes, one that avoided the Balkans by crossing the Adriatic Sea and another that connected eastern Saudi Arabia with Paris via North Africa and Spain, were little better, adding considerable distance and difficult ocean crossings while not meaningfully reducing political risks.[13]

In the end, the Bechtel pipeline was never built. Though systems like it had been proven in the United States to be technologically viable, crossing thousands of miles and numerous countries exposed the project to insurmountable political and economic unknowns. Europe was simply not a commercially viable outlet for Middle Eastern gas in the first few decades after the Second World War. As one American delegate to the 1964 ECAFE summit wrote,

> the basic problem facing utilization of the large volumes of natural gas in the great oil producing areas of the Middle East is one of finding or developing a large market into which the gas can be moved at costs low enough to compete with manufactured gas and other fuels. Western Europe is already a large market for gas. But the Middle East is not well placed to supply this area in competition with the huge natural gas reserves in North Africa which are also pressing for market and with the newly-discovered big reserves in The Netherlands.[14]

Moreover, the "fallacy of the [Bechtel] scheme," as one BP official bluntly put it, was that "given a pipeline of this length and diameter, it could be used to much better advantage for oil than for gas."[15] By volume, crude oil had far higher energy density than natural gas, and a pipeline of any given size running from the Middle East to Europe could thus transport more energy more profitably if it carried oil. Depending on the precise technical and commercial choices made, British Petroleum estimated that moving gas to Europe would result in gross profits of up to £166 million that could be divided between "a reduction in rate to consumers, royalties, and profits." On the other hand, if thirty-two million tons of oil were sent through the pipeline per year instead, an amount the BP considered reasonable, the total could instead be as high as £512 million.[16] Distance was thus a very high hurdle when it came to utilizing the Persian Gulf's natural gas reserves, and it amplified the commercial impediments that had long kept the resource from profitable use. Technical feasibility and a demand for energy were simply not sufficient to overcome the yawning geographic discontinuities between the world's major producing and consuming regions in the early 1950s. In this way, distance defined the life of natural gas above the Earth's surface, just as depth and pressure did below, a result of the intermingling of commerce, engineering, and the physical and chemical properties of hydrocarbons.

Development in Iran

Animating the 1964 ECAFE summit at which Alikhāni spoke was a deep sense of urgency. Delegates saw natural gas as a resource possessing enormous economic potential for their societies, but one that was also under threat as associated reserves were progressively sacrificed to feed the world's demand for oil. No member nation was more pressed to solve this problem than Iran, for as the ECAFE region's largest petroleum producer it seemingly had the most to lose. Indeed, in the 1960s, the unrestrained flaring of Iran's natural gas continued apace, little having come of British Petroleum's measured efforts to find productive uses for it, the Iranian government's campaign to press the issue of recycling, or Bechtel's proposed export pipeline to Europe. Without a meaningful outlet, some 90 percent of southwestern Iran's associated gas was thus being burned for no economic gain, and hundreds of millions of cubic feet of Iran's hydrocarbon resources were disappearing into clouds of water vapor and carbon dioxide.[17] At the heart of this dilemma was the fact that natural gas, when measured by volume, had relatively little commercial value, and in order to be economically viable, enormous quantities would need to be exploited to cover the costs of either transportation or recycling. There was a third option, however, one that condensed the economic value of gas into something more easily marketed: petrochemicals. Officials working for both the Iranian government and British Petroleum had long noted the possibility of using gas as feedstock for the growing industry. In the late Pahlavi era, many new petrochemicals were coming to market that encompassed everything from industrial precursors to plastics to synthetic fibers to fertilizers. Such products were so diverse, voluminous, and cheap that by the 1960s petrochemicals had penetrated nearly all facets of life around the world. First developed in the United States in the 1930s, the industry had expanded rapidly to Europe and Japan in the decades after the Second World War. By the 1970s, private enterprise and state-owned firms all over the world were chasing the sizeable returns that could be made by transforming raw petroleum inputs into high-grade finished petrochemicals.[18] Intending to compete through the economies of scale that were made possible by the ready availability of inexpensive associated gas, the national oil companies of countries like Mexico and Saudi Arabia built large petrochemical facilities aimed at capturing markets for fertilizers, acids, and other commodity chemicals. In the 1980s, however, this veritable gold rush would result in significant overcapacity and a worldwide glut that largely scuttled the economic diversification schemes of many producer states. Such investments nonetheless proved crucial to meet the rising demands of their own societies.[19]

Iran too pursued such petrochemical dreams, and the sector became in the late 1940s an important piece of Iranian officials' modernizing ambitions. Recognizing the country's rapidly growing need for basic chemicals and, most especially, synthetic fertilizers, Morrison-Knudsen, hired to assess Iran's developmental prospects, had advised significant investment in the industry, seeing it as a crucial step in jumpstarting growth in the country's largely agrarian economy. Agricultural reform was central to the company's recommendations, and it urged a program of twenty-three distinct improvements ranging from new irrigation systems to altered soil management techniques to mechanization.[20] In practice, what Morrison-Knudsen described would prove to be an almost total transformation of Iran's economy and society, changes that embodied the "new consensus on development" that was coalescing in American policymaking in the decades after the Second

World War.[21] Strongly rooted in the progressive politics of the interwar era but drawing on older ideas of empire and Manifest Destiny, American policymakers sought to forestall the global spread of communism by proving liberalism's ability to raise standards of living and rebuild a world devastated by war. Taking an increasingly ideological form in an axiomatic belief in the superiority of mid-century American society and the benevolence of its government, this vision elevated large New Deal-era developmental projects like the Tennessee Valley Authority as exportable models for all nations.[22] In this way, and drawing upon the TVA's integrated example of infrastructural improvement and poverty alleviation, proponents of foreign development thought that the destabilizing forces of industrialization and market integration could be reconciled with liberal politics, thereby allowing impoverished agrarian societies around the world to be safely shepherded into industrialized modernity.

That the aims of such efforts so closely reflected American political norms and notions of industrialized mass consumption was no accident, for at root, development—or modernization, as the two words increasingly came to coincide during the period—was a homogenizing project aimed at remaking the world in America's image.[23] Announced as a foreign policy priority by President Harry Truman in his 1949 inaugural address, economic development was by the early 1950s given global reach. Truman's Point Four Program, as the initiative became known, was a wide-ranging policy of technical assistance and knowledge sharing, one that drew heavily upon private industry and academic expertise as well as state policy and financial resources. Increasingly formalized in the social sciences as modernization theory, developmentalist ideas became the bedrock of the world's new international order, not only animating institutions like the International Bank for Reconstruction and Development (later part of the World Bank), the Food and Agriculture Organization, and the World Health Organization, but also running as a powerful current through the global monetary system negotiated at Bretton Woods.[24] With personnel that hailed almost exclusively from the North Atlantic world, however, the grand developmental ambitions of many such organizations were frequently undermined by an inclination to discount the perspectives and experiences of those they sought to help.[25] Nonetheless, by the 1960s, modernization theory was driving American strategy in places ranging from Latin America to Southeast Asia to the Middle East. Such Cold War-era policies often hinged on programs that targeted agricultural development and expansion. Indeed, this was the era of the so-called Green Revolution, a term coined in 1968 by the U.S. Agency for International Development to denote the explosive growth in agricultural productivity that was enabled by new crop breeds, mechanization, and the increased availability of synthetic petrochemical fertilizers and pesticides.[26] In the eyes of many American modernizers, such techniques were crucial for improving the living standards of what were still largely rural populations around the world, thereby winning, it was hoped, support for governments friendly to American interests.

In 1950, Iran became the first nation to join the Point Four program, and for three years, more than a hundred American experts worked in the country to alter rural life through the introduction of new agricultural, educational, and medical techniques. But tangible results were limited, and their presence did not survive the 1953 coup d'état in Iran or the transition to the Eisenhower administration.[27] American involvement in the country nonetheless endured, a product of Iran's petroleum resources and the legacy of the Second World War. Prior to the 1940s, Iran had been marginal to American interests, and with the exception of two advisers hired earlier in the twentieth century to direct Iran's state finances, largely confined to private missionary and charity work.[28] The June 1941 German invasion of the Soviet Union changed the calculus of U.S. officials, however, and Iran became a major thoroughfare for Lend-Lease aid to the USSR. For that reason, during the war, Iranian sovereignty was severely circumscribed, for fearing Reza Shah's prewar flirtations with Germany, British and Soviet forces had invaded, deposed the shah in favor of his son, and commandeered the Trans-Iranian Railway. This wartime effort to ship considerable amounts of materiel to the Soviet Union brought large numbers of Americans to Iran for the first time, and with the onset of the Cold War in the immediate postwar period, U.S. attention only intensified as the country's petroleum resources and long border with the USSR grew in strategic importance. Iran's new significance was further underlined amidst the Iran Crisis of 1946, a nearly forgotten episode in which Soviet forces both failed to vacate Iran and overtly backed Azeri and Kurdish separatist movements in the northwest. It was an incident that had far-reaching consequences for the world, for when heavy international pressure finally compelled the Soviet Union to withdraw later that year, many American officials concluded that aggressive action was the best way to contain the global spread of communism. It was a lesson that would be applied in Korea in the 1950s and Vietnam in the 1960s.[29] Within Iran itself, that American Cold War strategy was translated into substantial American technical and financial assistance for the country, much of it directed toward the military and security services in an effort to harden the shah's rule against outside invasion and internal communist insurrection.[30]

With the exception of the brief Point Four mission, little of that aid was directed toward Iran's economic development, as American officials largely saw the country's oil revenues as sufficient for the task. For their part, in the late 1940s, Pahlavi officials sought to transition from the often ad hoc initiatives of the Reza Shah era to a system of formal developmental planning. Adopted in 1948 and based in large part on Morrison-Knudsen's recommendations, the First Seven-Year Development Plan (1949–55) promised significant investments in the country's agricultural, transportation, and industrial sectors. Expected to cost some $500 million, one-third of the money was to come from AIOC royalty payments while the remainder would be covered by a mix of funds from Iran's national bank and IBRD. The program's implementation would be overseen by the newly established Plan Organization, an office that would go on to manage Iran's developmental initiatives for the next half century. By the mid-1950s, however, the plan had proven to be a disappointment, at first hampered by the difficulty of creating a new government agency and then fully disintegrating amidst the turmoil of the 1951–54 oil nationalization crisis and Mohammad Mosaddeq's premiership.[31] Despite this failure, the 1950s saw the ascent of numerous modernizing technocrats into positions of influence within Iran's government. The movement was led by Abolhassan Ebtehāj, director of the International Monetary Fund's Middle East section and one-time head of Iran's national bank, who was appointed in 1954 to run a revamped Plan Organization. Drawing on the support of IBRD, the Ford Foundation, and the Harvard School of Government, Ebtehāj set out to staff the organization with young Iranian economists and engineers who had been educated in leading American universities. Trained in and adhering closely to the tenets of modernization theory and economic planning, these technocrats operated as an elite group, trusting in their expertise to guide Iran's industrialization through direct infrastructure investments and support for private enterprise.[32]

Such ambitious interventions were made possible by a rapid rise in the country's foreign receipts. Under the 1954 Oil Consortium Agreement, Iran's oil revenues grew rapidly, rising from some $34 million in 1954–55 to $181 million in 1956–57 to $437 million in 1962–63.[33] While much of that money was directed toward military expenditures, court patronage, and bureaucratic expansion, Iran's developmental programs also benefited. Slated to run between 1955 and 1962, the Second Seven-Year Development Plan initially earmarked $933 million to be spread across four sectors—agriculture, industry and mines, transport and communications, and social affairs. Drawn up by a group of experts handpicked by Ebtehāj and accepted unchanged by Iran's parliament, the plan found only mixed success in its effort to stimulate growth through private enterprise.[34] Stymied by a lack of clear objectives and little coordination among its investments, the Plan Organization quickly found that it had woefully underestimated the full cost of its program.[35] In 1960, moreover, despite seeing steady growth in the years after Mosaddeq's ouster, Iran had entered into a serious economic crisis due to a rapid and unsustainable rise in imports and a simultaneous (and sudden) drop in global oil prices. Driven in large part by the very developmental programs championed by the Plan Organization, this balance of payments crisis fatally undermined the modernizing ambitions of Iran's government. Bailed out by the International Monetary Fund, the country was forced to undertake austerity measures and consequently entered a two-year period of economic decline and social unrest. These events prompted the Plan Organization's modernizing technocrats to dispense with their free trade convictions and adopt protectionist policies, a shift that set the stage for Iran's rapid turn toward state-run industrialization in the 1960s and 1970s.[36]

Nitrogenous Transformations

Agriculture was ostensibly central to Iran's first and second seven-year development plans, with roughly a quarter of all expenditures dedicated to the sector. In practice, however, much of that money was committed to major dam-building projects that primarily benefited urban areas in the form of water and electricity.[37] Iran's rural regions nonetheless stood on the cusp of significant transformation in the 1950s, for as Morrison-Knudsen had argued in its 1947 report, modernizing Iran's agricultural sector was increasingly seen as a developmental necessity.[38] That could only happen, however, as part of a larger remaking of the country's agrarian economy. Beginning in the late nineteenth century, Iranian agriculture had become dominated by absentee landowners. Presiding over a vast sharecropping system organized around the ownership of entire villages, by the late 1950s some 80 percent of all arable land belonged to a mix of wealthy families (including the Pahlavis), *waqf* religious endowments, and the state.[39] In the years between Reza Shah's 1941 abdication and the reassertion of royal authority by Muhammad Reza Shah in 1953, a period marked by intense competition among Iranian notables, that economic clout was translated into political power as landowners used their authority to harvest votes in parliamentary elections. At the same time, such turbulence enabled the formation of mass political parties like the Soviet-aligned *Tudeh* party and Mohammad Mosaddeq's National Front. *Tudeh*'s membership was concentrated in Iran's growing urban classes, but the party's inveighing against the influence of wealthy elites was nonetheless instrumental in popularizing ideas of rural land reform in the country. For that reason, despite *Tudeh*'s violent suppression in the wake of Iran's 1953 coup, ideas like land reform lived on, coopted as part of the developmental policies of a shah intent on blunting threats to his restored rule.[40] Further, though they reflected very different understandings of how the problem should be addressed, both Morrison-Knudsen and Iranian advocates for land reform saw elite land accumulation as having brought Iran's rural society to the point of crisis. Prompted by the country's integration into global markets in the late nineteenth and early twentieth centuries, the cultivation of cash crops like cotton, opium, and tobacco had displaced much of Iran's staple grain production. That increasing orientation toward foreign markets accentuated the economic advantage of landowners, and Iranian elites largely pursued higher profits through ever greater rates of withholding, at times reaching more than 75 percent of total harvests.[41] Capital investment in infrastructure and new farming techniques was thus limited, and in 1960 less than 10 percent of holdings in the country were even partially mechanized. Because of this structural underinvestment, in the 1950s, Iranian agriculture remained labor intensive and left large areas fallow, conditions that led to low and declining per-capita yields between 1939 and 1961.[42]

While American development consultants like Morrison-Knudsen advocated for increased mechanization and commercialization, among Iranians land redistribution quickly became the central thrust of reforms. In 1951, Muhammad Reza Shah himself sought to win some popular legitimacy with the issue when, amidst the turmoil of the Mosaddeq era, he sold Crown estates and pressured wealthy landowners to follow suit. Their reluctance, however, meant that little changed until Iran's 1960 general election. It was then that the shah, buffeted by waves of unrest after overtly manipulating the results, appointed the reformist Ali Amini as prime minister and allowed the cause of widespread land reform to proceed.[43] For the shah, redistribution promised to build a base of support among the peasantry while also appealing to both Iran's urban intelligentsia and a Kennedy administration from which he hoped to extract more economic and military support.[44] But in this view he was out of step with influential members of his own government. Both Amini and Hassan Arsanjāni, the minister of agriculture, instead championed land reform as a means of improving the lives of rural inhabitants and breaking the power of wealthy landowners. The two men thus moved aggressively to further the cause of land reform, and in 1963 a formal process of redistribution was begun, though in the face of elite resistance and growing rural instability it was quickly watered down into a system of long-term tenancy agreements. While the change mollified landlords, peasants chafed under a scheme that offered them no ownership rights but demanded yearly cash rents. For their own part, modernizing Iranian officials saw the tenancy system as hampering the development of the country's agricultural sector, as now neither landlords nor peasants had any reason to invest in capital improvement. In the late 1960s, land redistribution was thus again altered to focus on sale rather than tenancy.[45] Declared complete in 1971, this third phase found mixed success, doubling the number of peasant proprietors in the country but ultimately leaving a million households, roughly one-third of Iran's rural population, without any real path toward ownership. Most of Iran's agricultural production thus continued to take place within the confines of semi-subsistence level farming, though mechanized estates that employed paid labor, typically owned by wealthy families or foreign investors, made considerable profit feeding Iran's growing cities. Large agribusinesses, on the other hand, including a high-profile venture in Khuzestān that involved significant foreign investment and the confiscation of the province's best lands, proved to be total failures. Thus, for all such sweeping intervention, Iran's total agricultural productivity remained low, and by 1968 the country had gone from being a net exporter of produce to a net importer, a difference covered by the country's petroleum revenues.[46]

All the attention that land reform received in Iran notwithstanding, beginning in the late 1940s, Iranian officials also engaged in a sustained effort to improve agricultural productivity through mechanization and the use of synthetic fertilizers. Viewing the sector as an important means of raising the living standards of a predominantly agrarian society, Iran's technocrats largely adopted the recommendations that Morrison-Knudsen had made for modernizing agricultural production in the country. But while mechanization and chemical fertilizers were often considered in tandem, their overall levels of adoption had diverged by the end of the 1960s. The use of new mechanical implements was slow to take off, impeded not only by the lack of incentive for landowners to invest in such machines, but also by the very low cost of labor. Rural populations had doubled in Iran between 1900 and 1960, not only weakening peasant bargaining power and cementing the dominance of wealthy landowners, but also undercutting much of the potential benefit of mechanization with the ubiquity of inexpensive familial labor. Through the end of the 1970s, there was thus relatively little uptake of mechanical farm tools, with the partial exception of tractors among peasant proprietors who used them to replace draft animals.[47]

Petrochemical fertilizers, on the other hand, proved to be far more successful, with usage rising from virtually nothing in the 1950s to more than 600,000 tons per year in the mid-1970s. Originally promoted by Morrison-Knudsen as a way to revitalize soils exhausted by centuries of farming, synthetic fertilizers quickly became integral to Iranian agricultural productivity. Smallholders were among the heaviest users, as the chemicals were one of the few reliable ways to increase the yields of their plots.[48] In the late 1940s, however, as Morrison-Knudsen was formulating its recommendations, domestic Iranian production was nearly nonexistent, with only one plant near Tehran supplying a negligible two tons of phosphate fertilizer per year. With no realistic prospect of increasing that figure in the short term, company experts advised an expensive import program, one designed to meet an overwhelming need to reinforce the depleted phosphorus and nitrogen (two essential plant nutrients) levels of Iran's agricultural soils. To offset the high costs that such a plan would place upon farmers, they advised that the "[Iranian] Government should give landlords and tenants one pound of fertilizer for each two they purchase."[49]

Though synthetic fertilizers were deemed essential to meeting Iran's developmental goals, the collapse of the country's first seven-year plan meant that their availability remained limited during the 1950s. The possibility of domestic production remained a tantalizing goal, however, and one that became increasingly intertwined with the country's efforts to find outlets for its natural gas resources. In 1957, the Montecatini company, an Italian chemical giant, was contracted to evaluate the prospects of natural gas utilization in southwest Iran. Recommending a program that combined irrigation, power generation, and new petrochemical facilities, the company focused on using gas in the region's agricultural sector, seeing it as the most straightforward way to aid a predominantly rural population. Some gas could be used directly, burned to pump water through an expansive new irrigation system and bring more land under cultivation. But deeming the nitrogen deficiency of Khuzestān's soils to be an acute crisis, Montecatini also envisioned boosting productivity through the heavy application of synthetic fertilizers like urea and, to a lesser extent, ammonium sulfate. But rather than import them as Morrison-Knudsen had suggested, the firm's analysts saw the need for fertilizers as an opportunity to also further the Iranian government's industrializing ambitions. Toward that end, Montecatini recommended building a petrochemical facility at Ahvāz and using gas from the oil field at Āghā Jāri to feed its production of nitrogen-based fertilizers and PVC plastics. The idea was deeply embedded in southwest Iran's geology, for not only did it seek to utilize the region's gas reserves, but the decision to focus on nitrogen also reflected the absence of the commercial grade deposits of phosphate rock needed to produce phosphorus-based fertilizers.[50] It was an ambitious plan, perhaps overly so, and despite advancing many of the developmental goals of Iranian modernizers, Montecatini's vision would never be pursued in an integrated way. Many of its pieces would, however, go on to find some version of independent life in the following years.

Foremost among them was the use of natural gas to produce synthetic nitrogen fertilizers, and in 1958 Iran's Ministry of Industry and Mines signed a contract with three European firms to design and build a facility for that purpose.[51] Sited thirty miles to the northeast of Shiraz in the town of Marvdasht and fed by a natural gas pipeline running 120 miles to the oil field at Gachsārān, the fertilizer plant was to be part of a large industrial complex that also included facilities for the processing of sugar and the production of cement.[52] Designed to make some 41,000 tons of diammonium nitrate, 40,000 tons of urea, and 1,000 tons of liquid ammonia each year, the Shiraz Chemical Fertilizer Plant was inaugurated in September 1963 at a ceremony attended by Muhammad Reza Shah and Charles De Gaulle, the president of France.[53] For development-minded Iranian officials, the facility's completion was a significant milestone, even if foreign companies had undertaken much of the actual design and construction work. The presence of the two national leaders thus reflected the plant's importance not only as a producer of badly needed nitrogen fertilizers—"one of the most important petrochemical products"—but also as the first stage of what was to be a broader effort to make Iran "self-sufficient in almost all major petrochemicals and plastic ingredients."[54] Indeed, as a supplier to Iran's domestic fertilizer market, the Marvdasht facility found considerable success, enough to prompt its expansion in the 1970s.[55] But having proven the feasibility of domestic fertilizer production, Pahlavi officials quickly turned their attention to developing the country's nascent petrochemical industry into something much larger. Toward that end, the National Petrochemical Company was established in 1964 as a subsidiary of NIOC, and it was charged with overseeing a sector that promised to not only support Iranian agriculture but also help diversify its economy away from raw commodity exports.[56] Thus while the fertilizer plant at Marvdasht had been built with domestic demand in mind, with petrochemicals also representing a potentially lucrative new source of revenues, many of Iran's later projects would be in large part oriented toward global markets. In its original recommendations, Montecatini itself had recognized this possibility when it suggested Ahvāz as the best site for a petrochemical facility, noting that the city had easy access to not only internal markets via existing road and rail infrastructure, but also the rest of the world through the Kārun River and the Persian Gulf. Even the decision to focus on urea over ammonium sulfate was driven in part by the possibility of export, as urea's higher nitrogen content meant potentially much lower shipping costs.[57]

This cognizance of petrochemicals' export opportunities was characteristic of Pahlavi development planning, for despite the emphasis that agriculture received in the work of consultants like Montecatini and Morrison-Knudsen, in practice the sector was often given short shrift by officials more focused on high-profile infrastructure projects and the creation of new industries.[58] For Iran, the great potential of petrochemicals lay in their condensation of natural gas's economic value into something more easily transported. Bechtel's unrealized pipeline had shown the enormous obstacles that direct natural gas exports would face, but the production of chemicals like urea transmuted the resource into new forms that were readily moved by road, rail, and ship. For that reason, in the 1960s, Pahlavi officials, building on the success of the Shiraz facility and reviving approaches considered by British Petroleum in previous decades, sought to strike foreign trade deals that would create profitable outlets for the country's gas reserves. At the 1964 summit on natural gas where Alinaghi Alikhāni spoke, the Iranian delegation, by far the largest at the conference, proposed a scheme whereby the country would supply nitrogenous fertilizers to the entirety of the ECAFE region. At the time, worldwide demand for such products had grown 50 percent over the previous seven years, but Asia, Africa, and Latin America, despite accounting for some 60 percent of the world's arable land, only consumed a fifth of global supplies. Emphasizing potential economies of scale, the Iranian delegation argued that centralizing production in Khuzestān could reduce the cost of fertilizer production and thereby help feed the region's rapidly growing populations. Envisioned was a regional network, at the center of which would sit a large petrochemical facility that took cheap natural gas from southwestern Iran's oil fields to produce substantial volumes of ammonia that would then be shipped abroad for direct use or further processing. For the Iranian delegates, this arrangement promised to be a mutually beneficial one. Buyers could both enjoy the advantages of centralized production and maintain some control over the final form of their fertilizer supplies while Iran would find a guaranteed outlet for its otherwise unused natural gas reserves.[59]

In the end, however, this proposed plan for a regional fertilizer network also proved to be too ambitious. But the long record of failed petrochemical schemes belied just how attractive Iran's inexpensive and underutilized gas resources were. Combined with the country's relative geographic proximity to the growing societies of South and East Asia, southwest Iran proved to be a very attractive investment prospect for American chemical concerns like BF Goodrich, DuPont, and Allied Chemical, all of which signed major development and marketing deals with the Iranian government in the 1960s and 1970s.[60] Though Iranian planners had dreamed of using such projects to supply as much as 10 percent of the world's petrochemical products by 1980, many failed under the weight of delays, spiraling costs, global overinvestment in the sector, and the rapid rise of petroleum prices after the 1973 oil crisis.[61] All the while, however, Iranian consumption of petrochemicals grew rapidly, and soon the country's internal market was outstripping the ability of facilities like the Shiraz Chemical Fertilizer Plant to meet. In response, joint ventures that had been originally intended for export markets were reoriented to meet surging domestic demand. The most notable of these was the Shāhpur Petrochemical Complex, a large joint venture between Iran's National Petrochemical Company and Allied Chemical Company. Built in the late 1960s in the port city of Bandar-e

Shāhpur, by 1970 the facility was using gas from the oil field at Masjid-e Sulaymān to produce a wide variety of fertilizers and commodity chemicals. By 1974, despite Allied's withdrawal from the project, Shāhpur had become a major domestic supplier, each year producing some 800,000 tons of chemicals, 60 percent of which were fertilizers, from a daily feedstock of 177 million cubic feet of high-sulfur gas.[62]

But Shāhpur's success reflected more than the rapid rise of fertilizer use, as it was also a product of the unstable underground geographies of southwest Iran's petroleum fields. From its very conception, the Shāhpur plant had been intertwined with an unexpected availability of high-sulfur gas in the Masjid-e Sulaymān field. For decades, British Petroleum had lifted oil from the field's Asmari layer, a limestone stratum dating to the Oligo-Miocene period some thirty-five to fifteen million years ago. With the Asmari nearly depleted by the early 1960s, BP had opted to try and extend the life of the field by exploring for oil in deeper strata. First finding a small Eocene-era pool, the far more significant discovery was a large deposit of sulfur-rich natural gas in rock dating back some 174 to 163 million years to the Middle Jurassic. Connected for the first time by British Petroleum's well, when the corrosive high-sulfur gas of the Jurassic layers caused the shaft's casing to rupture, the three petroleum deposits, each with their own characteristics, were for the first time combined into one. Moreover, once the reservoirs were joined, gas began to migrate upward from the high-pressure Jurassic layers to the Eocene and Oligo-Miocene deposits, a process that resulted in the unforeseen presence of very large volumes of natural gas within Masjid-e Sulaymān's existing production horizons. With both methane and sulfur being valuable raw materials for the production of petrochemicals, Iranian officials quickly seized the opportunity to advance their ambitions and press British Petroleum to support the construction of a new petrochemical facility.[63]

The creation of the Shāhpur facility was thus the product of a vast assemblage encompassing everything from Pahlavi development policies to the commercial concerns of British Petroleum to the long natural history of southwest Iran's sedimentary layers. Changes in any aspect of that assemblage would thus reverberated far and wide, and as petroleum reservoirs changed under the influence of oil operations, so too did the course of Iran's industrialization. Put differently, the Shāhpur plant and its success as a producer of petrochemicals were social and political consequences of unexpected and contingent encounters deep beneath the earth. But while the history of the facility was exceptional in the clarity with which such connections were demonstrated, the entire project of harnessing Iran's natural gas reserves was cut from the same cloth. The choice to rest Iranian agricultural productivity on the expanded use of synthetic fertilizers was one that reflected the ambitions of Pahlavi officials and the expertise of their hired consultants, but it was also one that was made possible by nature's fundamental uncontrollability. As would become increasingly visible in the following decades, Iranian industrialization was not made through a straightforward domination of nature; rather, it was created through simultaneously altering and deepening the connections between people and the natural world. Petrochemical fertilizers weakened Iranian agriculture's connection to surface life cycles by opening marginal areas to planting, lessening the need to rotate crops and let land lie fallow, and boosting yields beyond what could be obtained unaided. At the same time, natural gas, as the raw material of nitrogenous fertilizers, created a new set of bonds, ones that extended cultivation across thousands of feet of vertical distance and millions of years of time.

Condensable Fractions

For Montecatini, petrochemical fertilizers were the central pillar upon which Khuzestān's agricultural sector could be improved. But what synthetic fertilizers could not do was address Iran's growing demand for large volumes of cheap energy. It was a challenge that existed not just in cities and industrial zones, but also in the rural stretches where the company focused its attention. Indeed, as Morrison-Knudsen had noted a decade earlier, Montecatini saw the energy needs of southwest Iran's rural residents as contributing significantly to the land's poor fertility, reporting that "brush-woods and cattle manure . . . [were] used as fuel for cooking and heating in villages," a practice that led to an "impoverishment of soils which has been continuing for centuries."[64] By the late 1950s, however, that dependence on brush and manure was unusual, as by that point charcoal, largely produced in thousands of small and inefficient kilns, had become the country's form of household fuel. That shift had substantial environmental implications, for despite Pahlavi-era efforts to restrict the use of mature trees, Iran's growing population and the difficulty of enforcement made charcoal production an important impetus for forest loss in the twentieth century.[65] Over the course of the 1960s, however, even as overall energy consumption rose, charcoal use declined, dropping from some 450,000 tons to 300,000 tons per year in the mid-1970s. Though the 1963 nationalization of Iran's forests had been in part responsible for that drop, far more important was the increasing use of petroleum fuels like gasoline and kerosene.[66] Indeed, in the first half of the twentieth century, the consumption of oil products rose steadily in Iran, with kerosene eventually coming to serve as urban households' primary source of energy. From just 70,000 cubic meters in 1927 to more than 800,000 in 1957, the year Montecatini delivered its report, demand for kerosene had grown rapidly. By 1969, that figure would more than quadruple again, reaching 2.5 million cubic meters.[67]

But while oil consumption was a boon for Iran's forests, it also increasingly conflicted with the export commitments upon which state coffers had come to depend. Domestic oil use more than tripled during the 1950s, from some 1.0 million to 3.5 million tons, and in the early 1960s, NIOC officials, basing their estimates on the upcoming Third Development Plan, predicted that total yearly consumption would reach as high as fourteen million tons by the end of the decade. Among Pahlavi officials, however, that forecast sparked as much alarm as it did celebration, for by that point, with more than 90 percent of domestically consumed oil products refined at Ābādān and then shipped northward by rail, they were already grappling with overwhelmed distribution systems and endemic shortages that threatened to choke off the country's economic growth. Even more worrisome was a notable and accelerating imbalance in the consumption of different fuel oils, one that threatened to either force domestic rationing or impede exports. As was reported in early 1963,

> Analysis of the prior situation of the consumption of the four major products [diesel, kerosene, gasoline, and fuel oil] shows that the consumption of diesel primarily, and kerosene secondarily, have increased more than the other major products. This rapid increase has now made the proportion of products derived from crude oil unbalanced. In this context, if measures are not adopted . . . Iranian exports from Ābādān refinery will be fundamentally damaged.

At odds here were the two basic demands placed upon NIOC: provide inexpensive fuel in support of the country's industrialization—"amongst the most important of duties that have been placed on the National Iranian Oil Company"—and also facilitate the oil exports that increasingly underpinned the country's economy. In an effort to alleviate the strain, NIOC first sought to develop new areas outside the concession area and diversify its sources away from southwest Iran's consortium-operated fields. While D'Arcy's original concession had granted him the exploration and production rights for all of Iran save for the country's five northernmost provinces, the renegotiations of the early 1930s had reduced that area to some 100,000 square miles along the Iraqi border and Persian Gulf coast. That region, encompassing the Zagros Mountains and their foothills, was home to the vast majority of Iran's petroleum reserves, but scattered across the remainder of the country were a number of commercially viable deposits. These pools of oil and gas were available to NIOC for exploitation, potentially serving as a way to increase domestic fuel supplies while also leaving exports untouched. Toward that end, in the early 1960s, the company chose the "economical" and "technically correct" option of building a new oil refinery near Tehran and connecting it to the Alborz field, a deposit discovered in 1952 on the outskirts of Qum some seventy-five miles to the south.[68]

Situated approximately 300 miles to the north of the epicenter of Iran's petroleum industry, the Alborz field's utility to NIOC was rooted both in the presence of oil and its location in central Iran. Not only was the field outside the legal reach of foreign oil firms, but with no realistic export potential, its exploitation could also be fully dedicated to the capital region's rapid industrialization. Discovered alongside the equally sizeable Sarājeh gas field, Alborz lay within the Qum Formation, a deposit of sedimentary rock dating to the same Oligo-Miocene period as southwest Iran's prolific Asmari Formation. The two had nonetheless been deposited in different environments and had been subject to different geological forces in the ensuing eons, a divergence that had resulted in both distinct petroleum compositions and the Alborz's much higher reservoir pressures. Combined with a complex system of thrust faults—areas where older rocks have been pushed above younger—that high pressure made exploration of the field a slow and uncertain process. It would thus be more than a decade until the field was ready for production, and it would not be until the late 1960s that a new refinery and pipeline system, designed and built by a consortium of European and American firms, was complete.[69] It was nonetheless an important step in the Pahlavi state's effort to provide cheap energy in the country, as the Alborz field's location near Tehran sharply reduced the transportation costs of the rapidly growing volumes of oil being consumed there.[70] Indeed, for the first time, NIOC was independently using petroleum deposits to supply a major market, and as such, the exploitation of Alborz marked a turning point in the country's developmental history. The decision to build a system of oil production, transport, and refining around the field expressed the intense and competing demands that had been placed upon Iran's national oil company. Finding and developing new petroleum resources enabled the continued growth of energy consumption in Iran; it also furthered efforts to build an industrialized and independent society. But for all the project's success, it simply could not supply enough fuel to meet Iran's present and future demand. For that, Pahlavi officials would increasingly turn toward the utilization of natural gas.

As late as the 1970s, kerosene was much more readily available in urban areas than rural, leaving residents of places like Khuzestān to strip the land in search of fuel.[71] In the late 1950s, Montecatini had recommended that finding substitutes for plant and animal fuels be made a priority, settling on "liquid gas" (*gāz-e māyeh*) as a way to both increase the region's agricultural fertility and make use of Iran's gas reserves.[72] Liquid gas, more broadly known as liquefied petroleum gas, is a condensed blend of propane and butane, two hydrocarbons that occur naturally in petroleum deposits and are constituent parts of both crude oil and natural gas. Either produced during petroleum refining or separated from oil and gas as they are extracted from the earth, the propane and butane that make up liquid gas are both higher in energy and less volatile than methane (the predominant hydrocarbon of natural gas), making the product far easier to handle as a form of fuel. As such, liquid gas is not dependent on massive structures like pipelines to move in useful quantities, a characteristic that in 1950s Iran made it an attractive choice for supplying energy to urban areas and scattered rural populations alike. Indeed, once it had been compressed into canisters, very often sized for a single household's weekly needs, liquid gas could be easily distributed via the country's existing road and rail networks. Montecatini thus proposed supplying Khuzestān's population with five- and ten-kilogram canisters of liquid gas, expecting that the province's existing road network, though in many places unpaved, would be sufficient for distribution. That gas would come from nearby petroleum operations. With as many as 80,000 households believed to be potential customers, and a predicted average consumption rate of fifty kilograms per year, the company argued that the availability of liquid gas from the Āghā Jāri and Lāli oil fields was more than enough to meet the region's energy requirements for decades to come.[73] Despite the comparative ease with which liquid gas could be distributed, however, making and filling canisters required specialized equipment that did not yet exist in Iran. It was for that reason that, prior to the mid-1950s, the bulk of the propane and butane produced in Iran had been flared, with only small amounts being used by industry employees in Ābādān and Ahvāz.[74]

By the time Montecatini was preparing its report, however, liquid gas was already beginning to be used in Iran's major urban areas. Fostering that growth in liquid gas consumption was Sherkat-e Butān, a private firm that had been founded in 1953. Beginning operations the following year, the company used tanker trucks to transport liquid gas from the refinery at Ābādān to Tehran, where it was dispensed into imported canisters for distribution. Though Butān's emblazoned vehicles would eventually become a recognizable part of daily life in Tehran, the new fuel was at first not widely embraced. Few residents of the city possessed the necessary appliances to make use of it, and many more worried about the possibility of deadly accidents, viewing the use of a pressurized canister of flammable gas to be tantamount to placing a bomb inside their homes. Even among early adopters, learning to adapt existing lifeways, most especially cooking techniques, was a lengthy process marked by significant trial and error. In response to these hurdles, Butān undertook extensive advertising and educational campaigns, seeking to convince Iranians of the safety of liquid gas while also teaching them best practices for its use. In doing so, the company built upon the fact that, for many of the country's urban residents, liquid gas use was attractive because of its comparative cleanliness. Both charcoal and kerosene were notorious for producing large quantities of smoke when burned, something that had made detached kitchens a widespread norm and forced people to prepare meals in what was often the dirtiest space in their home. Foreshadowing environmentalist arguments in favor of natural gas utilization in the following decades, liquid gas was touted for its clean combustion and sold as a means of liberating Iranian households from the irritating byproducts of other fuels. Growth in the new fuel's use was nonetheless slow in the 1950s, largely confined to the homes of a small number of wealthy urban residents.[75] For such families, however, the new fuel's cleanliness both brought kitchens inside and helped foster new lifeways that more closely mirrored the mid-century American ideals that were being promoted in the popular press, textbooks, and home economics programs.[76] By the late 1950s and early 1960s, however, with Iran's burgeoning oil wealth and modernization efforts increasing the ranks of the country's urban middle classes, the use of liquid gas had grown significantly, with many households embracing the use of new gas appliances and the prosperous convenience they represented.[77] Butān, by that point having expanded into the fabrication of gas canisters and appliances like gas stoves and water heaters, quickly became a household name across much of Iran. Nor was it the only firm to do so, for as Iranians embraced the use of liquid gas, over time dozens of new competitors entered the market. With so many entrants, the industry grew rapidly, with total consumption rising from 2,590 tons in 1960 to 572,000 in 1979.[78]

Though widely adopted in the late Pahlavi era, liquid gas would eventually be in large part supplanted by piped natural gas. The use of liquid gas as a common household fuel was thus in many ways a transitional step in Iran, enabling the utilization of gas energy without the decades-long process and significant public investments in new infrastructures that natural gas use would require. It nonetheless provided many of the same benefits, including relative cleanliness, domestic availability, and a lack of competition with oil exports. As such, propane and butane were the "condensable fractions" of natural gas, as Montecatini's experts had put it, and just as petrochemical fertilizers were understood as concentrating the value of Iran's gas reserves into more easily exploitable forms, so too did liquid gas.[79] At heart, and in comparison to natural gas, the utility of liquid gas lay in its ease of transport, and dozens of private firms were able to use Iran's existing road and rail networks to profitably supply the fuel to communities all across the country. It was a disparity rooted in the materiality of hydrocarbons, for while liquid and natural gas overlapped, the difference of a few carbon and hydrogen atoms drove large variations in the physical practicality and commercial viability of their use as sources of energy. But nature also played a role in limiting the long-term utility of liquid gas, for propane and butane, dwarfed by the volume of methane within the very same deposits, comprised only a small percentage of the total hydrocarbons available in Iran's petroleum reserves. For that reason, liquid gas was always a comparatively limited resource within Iran, and its utilization did not fully realize official ambitions for exploiting the country's gas reserves. The industrialized future that Pahlavi officials imagined, dependent as it was on the availability of vast quantities of cheap energy, could therefore not be built upon liquid gas. Indeed, even as liquid gas consumption grew rapidly in the 1960s and 1970s, significant effort continued to be dedicated to finding outlets for the natural gas that was still being flared in the fields of southwest Iran.

Imagined Futures

For all the material differences between liquid and natural gas, in Pahlavi-era Iran there was often a great deal of rhetorical slippage between the two. Grouping them under the allied banners of "gas," "gas fuel," and "gas energy," this conceptual mixing represented not only the significant overlap of their physical natures and perceived benefits, but also the social and political meanings that were beginning to adhere to them. Beginning in the 1960s and continuing thereafter, hydrocarbon "gas," in all its forms, came to embody the developmental aspirations that many Iranians held for their country. It was a view that was actively promoted by state officials. In their speeches and policies, in their interviews with the national press, and in the pages of ministerial publications, an imaginative tale of Iran's future was narrated with gas. It was a dramatic story of national progress, one told through a developmental lexicon that found consistent expression in a triumvirate of image, word, and statistic. Writing that "every policy issue . . . has a culture," William A. Gamson and Andre Modigliani have argued that political positions are expressed through "media packages"—collections of "metaphors, catchphrases, visual images, moral appeals, and other symbolic devices"—organized through a central "frame" that offers a "number of different condensing symbols that suggest the core frame and positions in shorthand."[80] In this way, "gas" became a grammar through which claims about Iran's future were articulated, expressing not just the particulars of gas exploitation but also the broader policy orientations that underlay it.

Political performance framed nearly every aspect of gas energy's representation in official media, and a symbolic vocabulary linking the benefits of gas to Iranian modernization quickly coalesced as projects were frequently presented in similar, sometimes even identical, ways. Whether it was the soaring rhetoric of rapid economic growth and increased living standards; the striking photographs of sophisticated technologies and towering constructions; or the presentation of statistical data on the enormous distances traversed and volumes of gas moved, the spectacle of gas utilization was routinely used as a tool of political legitimation for the shah and his government. Structuring these representations was a sociotechnical imaginary of gas that began to emerge in the 1960s, one that elevated the resource as a manifestation of a promised future marked by prosperity, national independence, and modernity. Sheila Jasanoff has observed that "performances of statehood in modernity are increasingly tied to demonstrations and to public proofs employing scientific and technological instruments," and nowhere was that more true than in late-Pahlavi Iran.[81] With development and economic growth having become explicit policy goals in the late 1940s, the following decades saw the Pahlavi state increasingly seek to bolster its position through what Cyrus Schayegh has termed a "politics of material promise." First emerging out of a crisis of insufficient electricity supply in Tehran in the 1950s, the restored Pahlavi monarchy aimed to solidify its shaky post-coup position by pledging to use industrialization and new infrastructures to meet Iranians' growing consumerist demands. Less directed policy than a series of uncoordinated but overlapping programs and promises, massive projects like hydroelectric dams came to be publicly celebrated as physical embodiments of the state's ability to foster industrialized prosperity.[82] It was an effort that intensified in 1963 when Iran embarked upon a series of top-down social reforms termed the White Revolution. Formally named the Shah and People's Revolution, the sprawling developmentalist initiative was intended to forestall a communist uprising and reconcile the continued existence of the monarchy with a 'modernity' that seemed to have little room for such political systems.[83] Encompassing everything from rural land reform to women's enfranchisement to the provision of healthcare and education, the White Revolution was at heart an effort to bind lower and middle class Iranians to the shah's rule through improved living standards and promises of a prosperous future. Less aggressively secularizing than the social policies of Reza Shah, the White Revolution's constituent programs were allied to ongoing efforts to support the adoption of consumer lifestyles among Iran's rapidly urbanizing population. Toward that end, Pahlavi officials also accelerated their efforts to promote industrialization and stimulate the country's economic growth, a push that was formalized in the country's Third Development Plan (1962–68).[84]

Unlike Iran's previous developmental plans, the Third sought to shift focus away from state-run industry and massive infrastructure in favor of private enterprise and small projects like water wells and secondary roads. In practice, however, monumental works of engineering and construction continued to dominate Iran's developmental investments. By 1966, three enormous new dams had been built and three-quarters of the program's entire budget for road construction had been spent on major highways. Rather than fostering small- and medium-scale private enterprise through the extension of credit, as had been planned, the majority of funds earmarked for industrial expansion had been dedicated to state-owned projects for the production of steel, machinery, and petrochemicals.[85] That such ventures received the lion's share of investment reflected the overwhelming symbolic importance of scale in the eyes of the shah and his officials. In their view, massive infrastructures like dams and pipelines embodied Iran's accelerating modernity, proof of the country's ability to shed its past weakness and enter a new era of power and prosperity. "Westernization" was the country's "welcome ordeal," Muhammad Reza Shah wrote in 1960, but his vision was not primarily rooted in secularization or particular political systems. It instead rested on the mastery of science and technology, of taking the world's latest inventions and making them Iran's own, of "adjusting the technology to [Iran's] culture and [Iran's] culture to the technology."[86] Animating Iran's developmental drive was thus a valorization of technology, particularly enormous structures like dams, refineries, and pipelines. Such monumental edifices were closely associated with the Pahlavi monarch, and official articulations of Iran's utopian future created a sociotechnical imaginary that not only prioritized the immense but also revealed a vision of what the country's social order should be in the modern age. In this tale, prosperity flowed downward from the top of Iran's social hierarchy, with the shah and his state officials depicted as patriarchal figures providing for Iranians of all social classes. Iranians themselves were largely reduced to passive consumers and their labors erased, the void filled by the materiality of infrastructure itself and the titanic and seemingly autonomous agglomerations of metal and concrete that were birthed and directed by Iran's political leadership and expert classes.

Beginning in the 1960s, gas energy became an ever more central means by which Iranian sociotechnical imaginaries were articulated and the politics of development communicated. Such official conceptualizations found extensive expression in the publications of institutions like the Ministry of Petroleum and the National Iranian Oil Company. In newsletters, magazines, and books aimed at both industry employees and the general public, Iranian gas projects were prominently imagined, described, and celebrated within a broader context of industrializing modernization. Chief among those publications was *Nāmeh-ye Sanʿat-e Naft-e Iran*, "the monthly journal of the personnel of the Iranian oil industry," which reported extensively on the plans and projects of the Iranian petroleum industry between 1962 and 1979.[87] Always celebratory in nature, the mostly unsigned articles emphasized the achievements of NIOC and its subsidiaries, often framing them in global terms by assessing Iran's petroleum industry in relation to those of other nations. Using articles that described enviable qualities of industrialized societies or explained some remarkable scientific advancement, the magazine created an implicit developmental hierarchy that cast the wealthy countries of the North Atlantic as models to emulate and those in the Global South as potential beneficiaries of Iran's growing capabilities. Even more than its brisk modernization, it was natural gas and NIOC's efforts to harness it that set Iran apart from much of the world, providing the country an opportunity to catch up with developed nations and become a leader for the rest. In this way, according to *Nāmeh*, natural gas was the means by which Iran could both attain and prove its modernity, in the process building a society that exhibited technological sophistication and benefited from the widespread availability of cheap energy.

In early 1963, in an article on Tehran's new oil refinery, the magazine articulated the potential benefits of gas for the first time, describing it as an important supplemental fuel and spur for economic growth. Though focused on oil, the article signaled that NIOC's intensifying efforts to meet Iran's growing energy needs would also utilize gas, and the discovery of the Sarājeh gas field four years earlier was thus celebrated as "one of the greatest achievements of NIOC." Natural gas held great promise for Iran, *Nāmeh* stated, for in utilizing it

> 1) the air pollution of Tehran will be reduced; 2) a cheaper fuel will be available to consumers; 3) the life and durability of fuel-burning devices will be increased; 4) the cost of repairing and maintaining fuel-burning devices will be lessened.[88]

In this way, by describing the potentially significant economic and environmental advantages of gas over other fossil fuels, *Nāmeh* underlined the budding importance of the resource in NIOC's energy imaginary. A year later, in an article titled "The Breadth of the Domain of Natural Gas," (*Vas'at-e Dāmāneh-ye Masraf-e Gāz-e Tabi'i*) the point was made in even sharper terms. Highlighting a 356 percent increase in gas consumption in Iran between 1960 and 1964, the article used "large industrial" countries like the United States as referents to describe a vision of gas utilization woven through the entirety of modern society. It described the use of gas for applications like the heating of chicken coops, the firing of steel and glass furnaces, and the baking of bread in industrial ovens. Gas was a raw material for the chemical industry, a feedstock for plastics, synthetic rubber, synthetic fabrics, fertilizers, explosives, pharmaceuticals, and dyes. It purified the interiors of people's homes, for "in modern (*jadid*) kitchens the use of gas-burning stoves has eliminated the smoke and filth that accompanied older stoves and turned kitchens into clean and beautiful rooms." Far from being a simple turn of phrase, the aesthetics of gas were fundamental to how the resource was understood in 1960s Iran. There was beauty in the cleanliness of a modern kitchen, as there was in the spectacle of conspicuous energy consumption. *Nāmeh*'s article marveled at the use of gas flames for the "ornamentation" of restaurants and shops in the United States, as well as the eruptions of fire and water in Hollywood productions. For the article's anonymous authors, gas was not a source of energy to be hidden behind a mask of electricity or a raw resource to be tucked away in distant factories; it was instead to be celebrated as an embodiment of Iran's national aspirations, its use both a goal and means of Iranian modernization.[89]

Nāmeh's lauding of gas was not confined to words. Illustrations and photographs, not always explicitly tied to an article's textual discussion, were also crucial means by which NIOC's hopes and expectations for gas were communicated. Going beyond mere visual aid, such depictions required that a substance imperceptible to the naked eye be made somehow visible. Gas, intangible outside extreme conditions, thus came to be represented by both its end uses and the technologically advanced refineries and pipelines with which it was produced and transported. In the late Pahlavi era, this aesthetic condensation of gas became politically meaningful, for the coalescing of a technological sublime marked by massive assemblages of metal, money, and expertise signaled the significance of gas as both a material substance and a modernizing impulse.[90] In this way, gas infrastructures, when shown in close relationship with the landscapes that surrounded them, became a vehicle whereby aesthetic qualities were used to make claims about how society could and should be structured.[91] While deeply embedded in the particular social and political circumstances of Iran as an overtly industrializing society, such representations drew upon a globalized idiom of petroleum and energy. All over the world, whether in developed societies or otherwise, official conceptualizations of petroleum have often exalted technology through "fetishizing futuristic, machine-cool images of efficiency and enablement" that helped make oil and gas "synonymous with development, modernity, and 'the good life.'"[92] Images of Pahlavi Iran's budding gas infrastructures strongly channeled such themes, employing a visual grammar that rested upon the monumental scale of gas energy's physical infrastructure, its techno-scientific advancement, and the intimacy of domestic kitchens. Flanking the article's opening two-page spread, the photographs in *Nāmeh*'s 1964 article "The Breadth of the Domain of Natural Gas" operated in just such a manner, connecting the interiors of Iranian homes to petroleum operations in the country's southwest. Working to portray gas as a clean and technologically sophisticated source of energy, the first photograph depicted the Ābādān refinery's loading point for liquid gas, a place where tankers from companies like Sherkat-e Butān could fill their holds before beginning the long trek to Iran's northern cities. (figure 1).[93] Like many representations of petroleum around the world, the unadorned image emphasizes the infrastructure of gas utilization, centering gleaming storage tanks and ordered delivery lines while the refinery's cracking towers and smoky flares rise in the background. Devoid of workers, drivers, or any other people, the photograph leaves the industrial apparatus to stand alone and claim responsibility for the availability of gas, seemingly clear proof of the impending arrival of a future of abundant gas energy.

The force of that ostensible visual evidence was given weight by the nature of photography itself, a medium that exists simultaneously as art and record. Roland Barthes has advanced the notion of a photograph's considered *studium*—that which is perceived "as a consequence of my knowledge, my culture"—and its intuitive *punctum*, "what I add to the photograph and what is nonetheless already there." For Barthes, while meaning is created in the space between a photograph and its viewer, such images also possess an irreducible quality of depicting something

وسعت دامنه مصرف گاز طبیعی

FIGURE 1. **Opening Two-Page Spread of the Article "The Breadth of the Domain of Natural Gas" from *Nāmeh-ye San'at-e Naft-e Iran*.**
Source: "Vas'at-e Dāmāneh-ye Masraf-e Gāz-e Tabi'i," 8.

that is, or once was, real. Photography thus had "evidential power" as a medium, for it was in the "*arrest* of interpretation that the Photograph's certainty resides: I exhaust myself realizing that *this-has-been*."[94] But while uncritical acceptance of photography's ability to capture reality is unwarranted, it is necessary to recognize that, as Susan Sontag writes, "photography does not simply reproduce the real; it recycles it," and that through "photographic images, things and events are put to new uses, assigned new meanings."[95] Barthes and Sontag direct us to see photography as inhabiting an uncomfortable space between naïve assumptions about unmediated access to other times and places and what T. Jack Thompson terms a fully "viewer-centered hermeneutics" that equally assumes all meaning is created in the minds of viewers.[96] Photographs thus gain their power in the interplay between image, subject, viewer, framing devices, and even "the context in which [the image] is seen."[97] Through photography, the machinery of the Ābādān refinery served not only its technical function of dispersing liquid gas to waiting tankers, but also as a reminder of the sophisticated technology that underlay the new fuel's use. If gas energy was the emerging lifeblood of modern societies, as the article argued, then Iran had already mastered the foundations of it.

Opposite the image of the refinery, *Nāmeh*'s article went further by presenting a second photograph of a smartly dressed young woman cooking over a white kitchen stove (figure 1).[98] Her coifed hair, pressed knee-length dress, bare arms, and colorful apron decorated with the months of the Gregorian calendar evoked contemporary notions of prosperous domestic life in suburban America. The graceful lines and groomed appearance of the model were further combined with the hygienic space she inhabited to emphasize the cleanliness that gas could bring. The cropping out of much of the stove itself, in contrast to the imposing structure of Ābādān refinery, projected a message of invisible convenience within the home. Though gas use was made highly visible in Pahlavi Iran, petroleum's general social and cultural invisibility, an unseeing that enables consumers to ignore the political and environmental violence that oil and gas use require, nonetheless remained part of its appeal in the country.[99] Indeed, the tension between the visibility of gas as infrastructure and its invisibility as a domestic fuel was central to its place within the sociotechnical imaginary of gas that was coalescing in late Pahlavi Iran. For Iranians, gas energy was both a highly sophisticated technical undertaking and an important enabler of new lifeways of comfort and consumption—a towering accomplishment of the Pahlavi state and a simple domestic tool.

Nothing represented that productive tension more than 'modern' women preparing meals in clean domestic spaces, and the trope was common in *Nāmeh* during the 1960s. Such images and their import were rooted in the Iranian government's postwar promotion of consumerism and domestic femininity. Beginning in the late nineteenth century and increasing thereafter, women's place in Iranian society had become a symbolically charged site where conflicts over modernity, tradition, religion, and secularism played out. Though often reduced to the question of veiling by outside observers, such debates ranged far wider, touching upon everything from family law to suffrage to education and employment.[100] As wives and mothers, women sat at the center of nationalist debates in Iran, largely excluded from civic participation but nonetheless charged with ensuring the nation's future through their childrearing duties. In the Pahlavi era, that notion of patriotic motherhood was intensified through an official cult of domesticity, one that sought to teach women to raise patriotic male citizens and deferential female companions.[101] With the advent of Iran's participation in the United States's Point Four program in the 1950s, those efforts were explicitly tied to the domestic norms of mid-century America. Women were thus given instruction on new modes of hygiene, food preparation, aesthetic taste, and the norms of mass-market consumerism, a reorientation that shaped everything from women's dress to the use of space in people's homes.[102] In that context, images of women were a powerful tool for communicating the promised benefits of gas energy. Such was the case in one 1966 article that presented a drawing of a fashionable woman cooking at her gas stove as well as a photograph of a young girl manipulating the controls of another.[103] In other instances, such images were tied directly to the sophisticated technologies of the modern age, appearing alongside jet aircraft, automobiles, and factories.[104] No less than the photographs of gas energy's massive infrastructures, these images telegraphed ideas about modernity and the ambitions of Pahlavi officials for Iran within it. It was a vision that was as much social as it was technical, and they articulated a top-down, state-directed notion of gas energy where NIOC was tasked with building the foundation of a new Iran of imposing machinery and clean kitchens. The juxtaposition of fashionable unveiled women with refineries and pipelines thus bridged the distance between domestic life and the monumental edifices of industry being erected around the country, a visual representation of the "Breadth of the Domain of Natural Gas" that included the entirety of a modern society.

***

In his 1961 book *Mission for My Country*, Muhammad Reza Shah placed petroleum at the very center of his aspirations for rapid industrialization and social change, writing that the "growing role of oil and gas presents us Persians with a great opportunity" to both "help raise the standards of living both here and all over the globe" and reassert the country's "national sovereignty."[105] Indeed, by the early 1960s, Iran was entering an era marked by an intensifying focus on modernizing national development. Responding to the shah's ambitions and aiming to quickly industrialize the country and foster the adoption of consumer lifestyles, Pahlavi officials increasingly sought ways to both maintain oil exports and provide the substantial quantities of cheap energy that their plans required. Toward that end, officials working for the National Iranian Oil Company turned to new petroleum fields and oil refineries that lay outside the concession area where foreign firms held sway. At the same time, Pahlavi officials continued to search for ways to harness Iran's vast reserves of natural gas, by then having come to the conclusion that any future of gas exploitation would need to be created by Iranians themselves. For that reason, leaving behind decades of fruitless effort to press firms like British Petroleum to utilize or preserve the country's gas resources, in the 1950s, Pahlavi officials worked to create new industries in which to use the associated gas that was being discarded in the country's southwest. Distance was a formidable barrier to natural gas utilization, however, and Iranian officials turned to petrochemicals in order to both condense the value of gas into more economical forms and sate Iran's growing hunger for synthetic agricultural fertilizers. Petrochemicals represented an opportunity to utilize gas, build a lucrative export industry, and further the country's agricultural modernization. What they did not do was offer new sources of energy. For that, and at the same time, numerous private companies began transporting and marketing liquid gas in Iran's cities. Though liquid gas use would prove to be an intermediate step toward the widespread adoption of piped natural gas in later decades, what nonetheless emerged was a new pattern of energy consumption that both accustomed Iranians to using the new fuel and increasingly bound their lifeways to the country's gas resources.

But for Pahlavi officials, gas also held meaning beyond its ability to supply energy, serving political ends as much as it did technical and economic ones. For that reason, the edifices of steel and concrete used to capture, purify, and transport gas were framed as proof of the Pahlavi state's ability to make Iran prosperous. The shah's 1961 words, emphasizing as they did the intertwined goals of development and national independence, reflected the imagined futures that many Iranians were beginning to articulate through the medium of gas. Even more than oil, gas came to represent profound imaginings of Iran's future and was therefore fertile ground for political meaning-making. Expressed in the pages of publications like *Nāmeh-ye San'at-e Naft-e Iran* was a sociotechnical imaginary that positioned gas as the structure upon which Iran's future would be built. It was a vision that saw Iranians drinking deeply from the natural wealth that lay beneath their feet, burning it as fuel for an ever-intensifying modernity. The assemblages of technology, carbon, and ambition that structured Iran's sociotechnical imaginary of gas were the first pieces of what would in later decades become a gas system that encompassed nearly all of Iran. But the coalescing threads that connected Iranians to underground petroleum pools also linked them to their surrounding environments in new ways. Their nascent use of gas was a product of political imaginings and new technologies, but it also manifested the hybridity of fossil fuel infrastructures. In urban areas, hygiene motivated the adoption of liquid gas among urban consumers, as the comparative cleanliness of gas combustion rendered it a boon for indoor air quality. In rural areas, distance and poor soils drove the use of natural gas as a feedstock for synthetic fertilizer production, products that changed the literal nature of the ground on which they were spread. In this way, in the 1960s, Iranians began to restructure their society around a system of fossil hydrocarbon use that both altered and reflected the land of their country. Rather than weaken Iranians' ties to the natural world, gas utilization instead strengthened them. In the coming years, as the country's consumption of energy intensified and enormous new natural gas refineries and pipelines were built to sustain it, those bonds would grow only closer.

Three

National Arteries

Recounting in the summer of 1965 his recent experiences traveling through Europe and North America, Manuchehr Eqbāl, chief executive of the National Iranian Oil Company, spoke enviously of the natural gas industries he had seen up close. Accompanying the shah on a royal tour organized by a group of petroleum companies, Eqbāl had visited France's "immense" Lacq gas field and toured the productive petrochemical facilities of Norway, experiences that prompted him to publicly lament the "32 million cubic meters of gas" that Iran "wasted daily."[1] Indeed, in the mid-1960s, a moment when the Shiraz fertilizer plant was operating at full capacity and Iranians were making ever more use of liquid gas in their daily lives, the vast majority of natural gas produced in Iran continued to be discarded for little productive gain. But it was also a moment when the possibility of utilizing natural gas on a wider scale began to be considered more concretely by modernizing Pahlavi officials. Their developmental plans, increasingly formalized and ambitious, were built around industrialization and, in practice, the creation of large state-owned industrial enterprises. Their extensive need for cheap fuel and feedstock would spur concomitant outlays for infrastructure across the country, with dams, ports, highways, and pipelines all receiving significant investment. Spurred by this industrializing developmentalism, the 1960s would prove to be a turning point for Iran's natural gas resources, the period when the first concrete steps were taken to transform them from waste into the country's primary source of energy.

For many Iranian officials, piped natural gas was seen as potentially far more beneficial than liquid gas for the country's developmental goals. But it was also much more difficult to transport in useful volumes, requiring extensive networks of pipelines to bring it to consumers. Undeterred by the enormity of such a project, in the 1960s figures like Eqbāl advocated for the construction of a pipeline system that would move gas from southwest Iran's petroleum fields to the cities of Isfahan, Kāshān, Qum, and Tehran. In their telling, such a pipeline would be the foundation upon which a new era of increased energy consumption would be built, powering factories and homes across an immense swath of the country's most populated regions. Arguing that it would be "one of the greatest successes of the Iranian petroleum industry," Eqbāl envisioned natural gas both meeting the energy requirements of "all the factories of Isfahan" and being used for generating cheap electricity in towns and villages all along the pipeline's route.[2] Far from idle dreaming on the part of NIOC's chief executive, by the end of the decade a pipeline project fulfilling such aspirations would be on the verge of operation. That system, the First Iran Gas Trunkline, or IGAT-1, would in the following decades prompt the transformation of natural gas from something used in a small corner of Iran to a resource consumed across the country. IGAT-1 was an immense and complicated endeavor, spawning a primary pipeline more than a thousand kilometers in length, numerous branch lines, gas separators, gas compressors, a gas gathering network, gas distribution grids, a radio communication system, a pipe mill, and a natural gas refinery. Its design and construction involved state ministries, NIOC, the newly established National Iranian Gas Company, foreign governments, paid engineering consultants, and international contractors. Originally conceived to supply fuel for Iran's growing urban industrial sector, IGAT-1 was expanded and made possible when the Soviet Union agreed to take natural gas as payment for help in building Iran's first steel mill. From its inception, and in multiple ways, the pipeline project thus explicitly tied together natural gas and industrialization in Iran, coming to serve as both a developmentalist program around which other industries would be built and the central means by which energy would be conveyed to them.

For all the foreign involvement in the project, IGAT-1 remained in many ways an Iranian undertaking. Iranian officials proposed the idea, arranged its financing, in large part determined its contours, and oversaw the program's execution. That visibly Iranian disposition, whatever the full truth of it might have been, was hailed by Pahlavi officials as a nationalist triumph. Indeed, as a sprawling amalgamation of technology, expertise, and commerce, IGAT-1 was lionized in official and public media as a demonstration of the state's ability to foster industrialization and stimulate economic growth. The project was moreover used on the part of the Pahlavi government to lay claim to the mantle of a conservative but nonetheless very real politics of anticolonialism, one rooted not in formal independence, something Iran had never lost, but in the ability to control and benefit from the country's natural resources. But IGAT-1 had important ties to the natural world too, as its design and operation were formed within the constraints that were imposed by gas composition, mountainous topography, and pressure differentials. That hybridity, a meeting of human ambition and natural possibility that was mediated by technology, strongly shaped both the functioning of the pipeline system and its political significance. For all the claims to a modernizing conquest of nature that Pahlavi officials asserted, and IGAT-1 was often presented as enacting exactly that, the pipeline system was in reality an assemblage built from complex interactions of human and nonhuman forces. Seeking to make a new source of cheap energy available for their industrializing society, many Iranian officials would spend the 1960s navigating that intricate knotting, working to translate their developmentalist ambitions into the steel and concrete of natural gas infrastructure.

Vital Arteries

By the mid-1960s, petroleum pipelines had come to sit at the very heart of Iran's developmental ambitions. Speaking in January 1966 at the inauguration of a pipeline for Tehran's new but as yet unfinished oil refinery, Manuchehr Eqbāl emphasized NIOC's commitment to building Iran into the ranks of major world nations. Born in Mashhad in 1909 and educated as a physician in Iran and France, Eqbāl was a consummate political insider, having risen through the ministerial ranks before serving as prime minister in the late 1950s. Stepping into NIOC's top post in November 1963, a position in which he would remain until his death in 1977, Eqbāl was a trusted and loyal courtier of Muhammad Reza Shah.[3] It was during his tenure at NIOC that much of Iran's gas infrastructure was conceptualized and constructed, and his frequent speeches and wide-ranging press interviews were often used to express official ambitions for Iran's gas resources. Hailing the new oil pipeline as one of Iran's "vital arteries" and the "ultimate cause (*mojeb-e nahāyat*) of honor and glory," Eqbāl declared that its completion meant "another step in the direction of the nation's economic development" had been taken. He went on, stating that with

> the implementation of this plan, the length of pipeline in the country assigned only for . . . the supply of domestic consumption is over 3,600 kilometers. Iran, in terms of its oil and gas pipeline network, [now] stands in the same rank as several important countries of the world.[4]

Evident in Eqbāl's statement was an implicit developmental, perhaps even civilizational, hierarchy, one rooted in the existence of large-scale infrastructures dedicated to the exploitation of petroleum. In his estimation, Iran could climb that hierarchy by building new pipelines, a vision that directly linked national greatness to intensified fossil fuel consumption. But for all their prominence as physical embodiments of Iranian officials' industrializing ambitions, the significance of pipelines in Iran lay most immediately in the simple need to move fuel in a cost-effective manner.

With some 130 miles separating the Shatt al-Arab from D'Arcy's explorers at Maydān-e Naftān, pipelines had been integral to petroleum operations from nearly the moment oil had been discovered in 1908. Though it took two arduous years to complete, by 1911 Iran's first pipeline had been built to ferry Masjid-e Sulaymān's dark crude oil to a new refinery on the island of Ābādān. By 1945, Anglo-Iranian's network had grown dramatically, stretching a total of 1,320 miles and carrying nearly twenty million tons of oil per year.[5] But AIOC's system was dedicated entirely to its export operations, leaving Iranians to depend on oil products moved via comparatively expensive road, rail, and pack animal routes. For the consultants at Morrison-Knudsen, hired by the Iranian government to draw up a developmental roadmap for the country, the high fuel prices that resulted were a significant obstacle to economic growth. "Iran needs more and cheaper fuel to stimulate its industries and promote the comfort and health of its people," the company stated in its 1947 report, laying much of the blame on oil products that were "more expensive than fuel should be." To solve this problem, Morrison-Knudsen suggested the construction of several new oil and gas pipelines in order to transport fuel as cheaply as possible from Khuzestān to Iran's major cities.[6] Demonstrating a preoccupation that would appear time and again in the history of Iran's energy infrastructure, the company's proposals were oriented around feeding Tehran, the country's political and economic center of gravity.[7] Morrison-Knudsen considered three potential pipelines for supplying oil to the capital region. All of them would run from Khuzestān through the cities of Khorramābād, Arāk, and Qum but differed in their total throughput and whether they would carry crude oil or refined products.[8] Considered to be the best option for delivering the lowest prices to end consumers, a pipeline carrying crude oil to a new refinery in Tehran's southern outskirts was ultimately selected for construction.[9] It was at the inauguration of that pipeline that Manuchehr Eqbāl spoke in early 1966, its operational commencement alongside Tehran's new refinery promising an imminent increase in the availability of fuel oil, kerosene, and motor oil in the region.[10]

But oil products had never been Morrison-Knudsen's primary focus, and the consulting company had not suggested the prioritization of new oil pipelines or refineries in its study. It had instead advocated for the construction of a natural gas system along the lines of those beginning to appear in the United States.[11] Aimed at meeting the energy needs of Iran's fast-growing cities, the main pipeline for this proposed system would extend more than 500 miles from AIOC-operated fields in the southwest through Lāli, Dezful, Andimeshk, Khorramābād, Burujerd, Arāk, and Qum before finally terminating in Tehran. Branch lines would serve Isfahan, Yazd, Hamedān, Kermanshah, Qazvin, Tabriz, and Rasht while a secondary line would connect the same fields to Shiraz. In Morrison-Knudsen's estimation, natural gas could largely replace fuel oil and kerosene in those cities, thereby meeting the bulk of their demand for energy. Existing road and rail transport routes would continue to be used for gasoline and diesel shipments. All told, the project was predicted to cost some $70 million, and by the 1960s, Morrison-Knudsen suggested, up to 118 million cubic feet of gas could be moved per day, enough to serve two million people, or just under 9 percent of Iran's total population.[12] Though there would prove to be a long lag between the report's publication and the start of the IGAT-1 project that it inspired, the firm's plan was a seminal moment in the history of Iranian natural gas utilization. Going beyond ideas for petrochemical production and other schemes focused on Iran's southwest, Morrison-Knudsen's experts had for the first time articulated a vision for natural gas use that was national in scope. It was moreover one that focused on the widespread provision of fuel to both industrial and residential consumers. As the report explained, for industrialization to take root in Iran, large supplies of cheap energy were needed, and there was no source more readily available than the country's natural gas reserves.

Morrison-Knudsen's proposed natural gas pipeline was, however, built on what the Anglo-Iranian Oil Company argued were a series of faulty presumptions and oversimplifications. Concerned by the implications of the consulting firm's recommendations for its ongoing dispute with the Iranian government over the fate of Khuzestān's natural gas, AIOC had sought and received advance notice of the study's conclusions.[13] Reflecting its longstanding resistance to finding productive outlets for the associated gas it produced, Anglo-Iranian aimed to cast doubt on the feasibility of building domestic gas markets within Iran. Indeed, one company official wrote that a

> comparison between conditions in the United States and in Persia cannot be correctly drawn, as the extent to which natural gas could be used for domestic purposes in Persia is very doubtful and it is unlikely that the small industrial user, who at present uses fuel oil in heavy oil engines, bath-houses, bakeries, etc., would convert apparatus for the use of gas.

Moreover, he went on to write, 118 million cubic feet of gas per day was roughly equivalent to 1.0 million tons of fuel oil per annum, or more than four times Iran's entire yearly consumption of the heavy oil product. In the late 1940s, it was not clear when, if ever, Iranian society would demand so much energy, and while reducing the scope of the gas pipeline was possible, doing so would likely undermine the project's economic viability.[14] Even more concerning for Anglo-Iranian, however, were Morrison-Knudsen's estimates on the volume and quality of natural gas available to feed the line. Consulting experts had based their recommendation on an assumed daily production of at least 120 million cubic feet of associated natural gas from AIOC's fields at Haft Kel, White Oil Springs, Masjid-e Sulaymān, and Lāli. AIOC doubted such a figure was realistic, arguing that an average daily production of more than 100 million cubic feet was unlikely, and that in any case it was impossible to extend availability estimates past 1955. Oil extraction was moreover expected to soon peak in these fields, reducing the long-term potential for associated gas production. That shortfall could be offset by drawing from Āghā Jāri and Gachsārān as well, but only at the cost of an additional $25 million. Anglo-Iranian officials further argued that Morrison-Knudsen's assumption of an average hydrogen sulfide content of 5.5 percent was likewise faulty, proving to be nothing more than the simple mean of the highest (11 percent at Lāli) and lowest (trace amounts at White Oil Springs) levels.[15] Even the energy content of the available gas could not be assumed to be uniform or stable, as each field had different reservoir compositions and pressures.[16]

Made clear by Anglo-Iranian's critiques was the fact that Morrison-Knudsen had largely ignored the materiality of petroleum and the technical demands of its production. For the hired consultants, a natural gas pipeline was a technological and economic object, a thing of steel and financing that had been proved possible in the United States and was thus seemingly transferable to Iran. None of AIOC's objections made the pipeline a technical impossibility, but they did underscore the challenges of translating developmental plans, no matter how expertly conceived, into functioning systems. Conceptually divorced from the oil and gas pools that it would draw upon, Morrison-Knudsen's proposed pipeline system existed in an analytical space that treated the natural world as something external to the operations of energy infrastructures. But in reality the opposite was true, and as the events of the following years would show, any natural gas pipeline would be strongly shaped by nature's contours. Nearly two decades would pass between Morrison-Knudsen's proposal and the beginning of work on the IGAT-1 pipeline, and throughout that period the challenges identified by Anglo-Iranian remained in force. Eventually, however, just such sprawling pipeline system, drawing upon southwest Iran's petroleum fields, crossing rivers and arid plains and snowy mountain passes, would come to provide cheap energy for Iran's increasingly industrialized cities. It did so not as a superimposition on the landscape, but as something built within and responding to the specific constraints imposed by nature and technology. For Iran's budding gas infrastructure, the natural world could not be treated as a mathematical average or a cordoned-off externality; it was instead an inseparable piece of the whole, and one that would prove to have tremendous influence. Before any of that could come to pass, however, questions about how to finance the project would need to be answered.

Soviet Steel

Between the publication of Morrison-Knudsen's 1947 report and the mid-1960s, the possibility of building a large natural gas pipeline was left unfulfilled. Despite considerable official interest in the idea, the proposal had fallen victim to both the reticence of British Petroleum (as Anglo-Iranian was renamed in 1954) to support gas projects as well as the financial and engineering challenges of building such a system in Iran's rugged terrain. The spread of gas systems around the world nonetheless remained a topic of discussion within the country. In 1964, for example, *Nāmeh-ye San'at-e Naft-e Iran* reported that "the consumption of natural gas in highly-industrialized countries grows daily," that "in America, the importance of natural gas is now greater than that of coal and second only to oil in supplying fuel and power," and that the United States "had constructed 687,000 miles of pipelines for the transportation of natural gas."[17] In Iran, the pipeline to feed the Shiraz Chemical Fertilizer Company and other industrial units in Marvdasht, built in the early 1960s, had proven the feasibility and utility of natural gas to the state's developmental programs. Such successes fueled the ambitions of officials like Manuchehr Eqbāl, who argued that a "cross-country (*sarāsar*)" natural gas pipeline would be "another big and impressive step" in Iranian industrialization, one intended to deliver "inexpensive fuel and power for cities, their surrounding areas (*ghasabāt*), and . . . along the [pipeline's] path."[18]

Notable in Eqbāl's remarks was his explicit emphasis on energy rather than revenues, a position that contrasted with a contemporaneous global debate that linked petroleum rents to decolonization. In recent years, scholars have explored the political significance of oil in the early postcolonial period, emphasizing the close connection between disputes over revenue sharing and the movement for sovereign rights and economic decolonization in the Global South. Iranian officials like Fuād Rouhāni, adviser to the shah and OPEC's first Secretary General, and Jamshid Āmouzegār, economist and eventual prime minister, were influential participants in that transnational dialogue. Animating them and their counterparts in Latin American and the Middle East was a desire to wrest control of the oil trade away from the multinational firms that had long dominated the sector. Focusing on oil as the bedrock of global trade and industrialized life, such figures were united by a deep sense of righting historical wrongs, and they aimed to seize economic as well as political emancipation for their societies. Often hailing from elite backgrounds, advocates like Rouhāni and Āmouzegār worked the halls of the United Nations and the panels of international summits to build consensus with representatives from commodity-exporting states all over the world. Between the early 1950s and the mid-1970s, they would find real if somewhat attenuated success, giving rise to a proliferation of oil deals more favorable to producer states, the creation of numerous state-owned oil companies, the 1960 founding of OPEC, and the flexing of newfound economic muscle during the oil crises of the 1970s.[19] While the refiguring of global economic power would never be as complete as these advocates hoped, the period did see a reduction in the influence of the major oil firms and an undermining of the colonial patterns of extraction they were seen as upholding. Indeed, though it would not meaningfully alter the reality of foreign control over the country's oil resources for some time, the 1951 creation of the National Iranian Oil Company was part of that long process, acclaimed as a step toward the promise of putting Iran's petroleum resources to the benefit of all Iranians.

But while fierce intellectual and political battles were waged over oil, natural gas was never in similar contention, a result of its comparative novelty as a useful resource and the difficulty of its transport. Manuchehr Eqbāl's ambitions thus embodied a desire among Iranian officials to, in part, sidestep the fierce debates over oil and instead harness Iran's natural gas reserves as "inexpensive fuel" for their developmental initiatives.[20] In this vision, pipelines were drivers of "rapid economic progress" as well as the connective threads of a state system dedicated to providing for the welfare of Iranians. In Eqbāl's telling, it was NIOC that had worked to heed the "managerial leadership" of the shah and make real his "eloquent idea" of directly harnessing Iran's petroleum resources to provide "happiness and a better life for the general public." As he said in a 1967 interview, the

> efforts and labor I and all the personnel of the Iranian oil industry are ever [focused] on implementing the commendable orders [of Muhammad Reza Shah] to make the best use of the largest source of Iran's natural and God-given wealth. With the extension of the distribution network of oil and gas pipelines, the vital material [*mādeh*] of petroleum and its essential derivatives will, with ease and abundance, reach the hands of our compatriots in the farthest points of the country.[21]

Eqbāl rooted his vision in what he understood to be the evolving needs of an industrializing society. Development indeed required financing and expertise, much of which Iran obtained abroad, but it also needed lasting access to substantial volumes of cheap energy. That could be found at home, but largely on Iran's geographic periphery, far to the south of rapidly growing cities like Tehran and Isfahan. For Eqbāl, connecting the two regions was thus a priority, and as Morrison-Knudsen had proposed two decades earlier, the best way to do so was with a pipeline, one that would eventually run from southwest Iran and its rich petroleum deposits to the country's central and northern cities and, from there, to the Soviet Union.

Built in the late 1960s, the first Iran Gas Trunkline and its branching distribution system would go on to form the backbone of Iranian natural gas utilization. IGAT-1 was designed by the Iranian Management and Engineering Group, a subsidiary of the British engineering services company Sir Frederick Snow and Partners, and was originally envisioned as a twenty-eight-inch pipeline carrying 315 million cubic feet of gas per day between Ābādān and Tehran.[22] That sole focus on domestic deliveries, however, quickly gave way to one that combined service to Iranian cities and the Soviet Caucasus, a change rooted in broader policy shifts aimed at diversifying the country's foreign affairs away from the United States. Alarmed by the fall of the Hashemite monarchy in neighboring Iraq and skeptical of the American commitment to the survival of his own leadership, in the late 1950s Muhammad Reza Shah began to seek better relations with the Soviet Union. Though negotiations in 1959 for a non-aggression pact between the two states failed under pressure from Iran's Western allies, leading to an extended period of intense recriminations, within a few years relations were again stabilizing. In 1962, their rapprochement was formalized with an agreement whereby Iran would never host American missiles in exchange for Soviet recognition of the monarchy's durability. Though the shah had never seriously considered leaving the Western camp, the accord represented his commitment to an independent foreign policy, and it prompted new pledges of military aid from the United States while firmly establishing Iran as a rising regional power. The détente moreover buried acrimonious memories of Russian, and later Soviet, invasions and meddling, setting the stage for new economic agreements between the two countries in the mid-1960s.[23] Of those deals, the most important for Iranian industrialization and gas exploitation was the Soviet-Iranian Agreement of Cooperation, signed on January 13, 1966. As part of the arrangement, Soviet negotiators agreed to provide assistance for the construction of a steel mill near Isfahan, a machine tools plant in Arāk, and the primary IGAT-1 pipeline. Rather than hard currency, payment would come in the form of gas exports for a fifteen-year period, a compromise that substantially boosted the viability of a national natural gas system in Iran. Indeed, the new export requirement greatly expanded the size and scope of the IGAT-1 project while also committing an external power to its completion. On the other hand, the change did not shift the general understanding that the pipeline system's primary beneficiaries would be Iranian cities and their budding industrial sectors.[24] For the Soviet Union, the benefits of the deal never rested on an absolute need for Iranian gas, though the agreement did negate an impending requirement to build a pipeline from the country's primary fields in western Siberia to the Caucasus. Far more important in the minds of Soviet officials, negotiating only a few short years after the Cuban Missile Crisis and the Sino-Soviet split, was that they had gained closer ties to an important regional ally of the United States. For their part, Iranian officials had finally found a partner willing to help them realize their ambitions to further their country's development through new steel and natural gas industries.

The promise of a steel mill was itself the culmination of a decades-long effort to bring the industry to Iran. As early as 1935, Iranian officials had sought backing for such a project, though time and again they had been frustrated by the unwillingness of Western governments, doubting Iranian steel's competitive viability, to provide either technical or financial aid. By the mid-1960s, that frustration had boiled over, causing Pahlavi officials to turn to the USSR to realize the "long-standing national goal."[25] An October 1964 Soviet offer of aid in exchange for oil concessions in northern Iran was flatly rejected, but later proposals that took Iranian commodities as payment were considered to be more "tempting."[26] This was no small thing, for the long persistence with which the Iranian representatives had pursued the possibility of a steel mill reflected its political import within the country. Indeed, though the shah reassured American diplomats wary of Soviet intentions that any deal would be struck on a "pure" economic basis, the issue had become a highly charged political issue.[27] A month after the initial Soviet offer, Ahmad Mirfendereski, Under Secretary of the Iranian Foreign Ministry, stressed the Iranian public's deep well of feeling on the subject. At that time, Prime Minister Amir-Abbas Hoveydā had given remarks to the press that had led the American Charge d'Affaires in Tehran, Stuart W. Rockwell, to believe that an agreement between Iran and the Soviet Union was imminent. In commenting to Mirfendereski that he "hoped [Hoveydā] had a long spoon" if he wished to "sup with the devil," Rockwell queried whether the prime minister's remarks had been a "trial balloon" to assess public opinion. Rockwell reported that

> Mr. Mirfendereski replied that if this had been Mr. Hoveydā's purpose, he would find public reaction overwhelmingly favorable to a Soviet-built steel mill. In Iran, as in other underdeveloped countries, a steel mill is a symbol of prestige and national independence, and "the people" want one very badly and don't care who builds it. If Egypt and Turkey can have a steel mill, why should not Iran? The people, said the Under Secretary, are getting tired of the fact that nothing is done about a steel mill except talk about it. The government has to take some account of this widespread feeling. The people don't care whether a steel mill is economic or not—they just want one.

In Mirfendereski's view, for many Iranians the lack of a steel mill carried meanings that went well beyond any economic or industrial benefits it might bring. It told a story about their country and themselves, one that left them backward and small. As Rockwell further noted,

> Throughout the conversation Mr. Mirfendereski stressed the emotional aspect of the Iranian people's desire for a steel mill, saying that it was not a reasoned or reasonable thing. The lack of a steel mill is one of the things that makes Iranians feel inferior and under the thumb of foreign nations. The acquisition of a steel mill would be a sign of independence and coming of age. Many people in the government have

> the same attitude and that is why, although it has been delayed and delayed, the project for a steel mill does not die but keeps coming closer to possibility.

Whether or not Soviet aid came bundled with strings and threats—and Mirfendereski noted that "the majority" felt it to be "insulting" that American officials worried that a few hundred Soviet technicians could destabilize the country—the potential danger was of secondary importance to seizing the chance to build a functioning steel industry.[28]

While Mirfendereski's words elided disagreement within the Iranian government as to the wisdom of the plan—some feared a fate similar to Egypt and its dependence on continuing Soviet technical aid for the Aswan Dam—the agreement, formally signed in late 1965, proved to be a crucial moment for Iranian industrialization.[29] With Soviet design help and the lending of technical advisers and skilled workers, construction on the steel plant began the following year in Isfahan. By 1972, the mill was operating at full capacity with an annual output of some 550,000 tons of steel, though plans for expansion were delayed first by insufficient supplies of coal and then again by the 1979 revolution.[30] With the success of the steel plant, Iranian officials were poised to seemingly fulfill their twin developmental goals of building a steel industry and putting Iran's natural gas to productive use. They had moreover found a solution to a supposed national deficiency that had taken on a very high public profile. Indeed, the close links that Mirfendereski drew between mill, the use of natural gas exports to pay for it, and the supposed wishes of the "Iranian people" were telling. Whether or not Iranians really desired for their country to possess a steel mill, and whether or not Mirfendereski truly had the knowledge that they did, the Under Secretary implicitly rooted Pahlavi legitimacy in the completion of such a project. With most Iranians expecting to see little direct benefit from the steel mill, and its commercial viability still unclear, his was a perspective that positioned industrialization as a goal worthy in and of itself, one that advanced the nation whether or not any given project was optimally efficient. In this vision, national progress was rooted in the heft of heavy industry—in the command of steel—but it was also one that was ultimately underpinned by natural gas.

Pipeline Industrialization

In the mid-1960s, despite nearly two decades having passed, Iranian officials continued to draw upon Morrison-Knudsen's recommendations when discussing the contours of any new natural gas system in Iran. That changed, however, with the gas-for-steel agreement with the Soviet Union and its imposition of a sizeable new export requirement. Tasked by NIOC to accommodate the additional obligation, IMEG expanded the original design of the IGAT-1 system, enlarging the diameter of the main pipeline from twenty-eight inches to forty-two and extending it to the Soviet border at Āstārā. Natural gas itself would be sourced from the oil fields at Gachsārān and Āghā Jāri, with purification, dehydration, and refining to take place at a new facility to be built near Behbahān, a small city in southeast Khuzestān. From Behbahān, gas would be transported northward, providing fuel for Isfahan, Kāshān, and Qum before continuing on to the Soviet Caucasus via Qazvin and Rasht. Large branch lines would serve Tehran and Shiraz, while small spurs could theoretically feed the towns and villages of the pipeline corridor.[31] With IMEG handling design responsibilities, in 1965 the National Iranian Gas Company was created as a subsidiary of NIOC in order to centralize all affairs related to the country's gas reserves and manage the rapidly evolving IGAT-1 project. It would be staffed by employees transferred from the national oil company as well as foreign consultants.[32] Overseen by a board headed by the prime minister, NIGC and its parent were largely arms of the Iranian state, often explicitly charged with supporting the broader developmentalist goals around which government policy was increasingly organized. The company's formal portfolio encompassed mid- and downstream operations, including the refining, transportation, sale, and distribution of natural gas once it had been separated from crude oil.[33] In practice, in the late 1960s, that meant undertaking market research and feasibility studies while also supervising the firms hired to build the IGAT-1 pipeline and its supporting infrastructure.[34] But the pipeline project was so large and so important to the country's developmental plans that many ministries, even the Pahlavi state as a whole, had a sizeable stake in its ultimate outcome. In light of that significance, a "high commission" comprised of NIOC's managing director, the head of the Central Bank of Iran, and the director of the Plan Organization was created to ensure the project's success.[35] This group exercised considerable control over the program, and most of the major decisions for IGAT-1 would be made by them.

The close involvement of high-level government officials made clear the significance of natural gas utilization to many Pahlavi officials' developmentalist ambitions. IGAT-1 was to be a critical piece of Iran's infrastructure, facilitating the consumption of cheap energy in towns and cities across a broad swath of the country. But that close connection to the state's programs of industrialization also caused significant uncertainty as different interest groups vied to influence the size and scope of the project. With IMEG actively redesigning the pipeline system in early 1966, the Pahlavi Foundation, representing the shah's personal business interests, proposed a joint venture with Bechtel Corporation to build it as a forty-inch line. Alinaghi Alikhāni, Minister of the Economy, proposed instead a larger forty-eight-inch line able to move much larger volumes of gas. Still others advocated that the pipeline route be altered to include the city of Tabriz as well. Driving this debate were not just differing developmentalist visions, but also a lack of reliable information on realistic future demand within Iran. While the Soviet take was set by contractual terms, because many expected consumers did not yet exist or had no experience relying upon natural gas, accurately predicting domestic consumption and its rate of growth was much more difficult. Nor was the future availability of associated gas known with full certainty. Indeed, IGAT-1, and the national program of industrialization that it was intended to support, in large part continued to rely upon a system of oil production that lay outside the Iranian government's ability to influence. In the 1960s, Iran's oil industry remained under the control of international oil firms, and there were worries that their decisions, made with their own interests in mind, could undermine the IGAT-1 program. Any changes in oil production put at risk the expected availability of associated natural gas, something that would in turn imperil both NIGC's ability to provide service to domestic consumers and fulfill its export requirements to the Soviet Union. Such an outcome would reverberate far and wide in Iran, for by forcing payment for the USSR's work on the Isfahan steel mill to come from Iran's limited hard currency reserves, it would in turn threaten the financial viability of all of the country's developmental plans.

Far from an idle concern, and with no clear way to bridge the gap between the promised $286 million in Soviet work credits and what was projected to be the project's final cost of $1.0 billion, Pahlavi officials paid considerable attention to IGAT-1's financing and its implications for the government's budgets. An initial proposal for national bonds was scrapped for fear that a lackluster reception would be interpreted as a public repudiation of the project and, potentially, the Pahlavi state as a whole.[36] With foreign funding deemed the only real alternative, the project's large credit requirement thus limited the potential origin of construction materials and contractors to those countries also willing to lend money.[37] More challenging yet was the limited pool of qualified firms. IGAT-1's main line, being larger than most existing systems and crossing very difficult terrain, was in many ways at the forefront of pipeline engineering in the 1960s.[38] There was no domestic capacity for such a project, for Iranians, long subject to British Petroleum's racialized system of labor discipline, had been largely excluded from the skilled technical and managerial aspects of oil production.[39] Despite a history of labor activism, pressure from the Iranian state, and the de jure nationalization of the industry in the early 1950s, by the mid-1960s there had been little chance for NIOC, and by extension NIGC, to build the expertise needed for an ambitious project like IGAT-1.[40] Their role would be supervisory and political, often mediating between the developmentalist desires of Pahlavi officials and the recommendations of foreign experts like IMEG and hired pipeline contractors.

Complicating matters was the very ambitious schedule that the Iranian government demanded.[41] So as to meet an expected operational start in 1970, pipe for crucial southern sections of IGAT-1's primary line would need to be ordered no less than six months before the December 1966 start of construction. More northern portions of the line required similarly long lead times, and in all cases extended construction periods were expected, in places reaching up to two years or more. For that reason, rather than unfolding sequentially, the pipeline and its associated systems for gas gathering, treatment, and distribution would have to be built concurrently. Construction would also have to start without a full consensus having been reached on either the project's aims or design.[42] Nonetheless, in the summer of 1966, IMEG's redesign called for the pipeline to be comprised of four major sections. Two forty-two-inch diameter segments would connect Behbahān to Isfahan and Isfahan to Sāveh, with spurs to Kāshān and Qum. A thirty-eight-inch section would link Sāveh to Āstārā with a major twenty-eight-inch branch line serving Tehran.[43] Rather than give the project in its entirety to a single contractor, NIOC and IMEG opted to subdivide the project and its construction responsibilities.[44] Exactly mirroring the original plan to transport gas to Iranian cities and its later revision to include the Soviet Caucasus, the most salient division occurred at Sāveh. Marked by a physical change from forty-two-inch to thirty-eight-inch diameter pipe, the shift also represented an important political transition between a southern section built by a constellation of European firms—primarily Entrepose of France and Williams Brothers of both Germany and Great Britain—and a northern one turned over to the Soviet Union.

But other aspects of the project remained in flux. Despite the primacy of foreign contractors in the construction of the main pipeline, some Pahlavi officials sought to use the project to foster domestic industrial capacity. In the 1970s, after the operational start of the pipeline, that drive would come primarily in the form of gas distribution networks.[45] Before then, it was expected that the project would generate sizeable new orders for domestic manufacturers of products like metal containers, plumbing components, and heat exchangers.[46] Much more exciting, however, was the possibility of establishing Iran's first pipe mill and making within Iran the vast quantity of steel pipe that the IGAT-1 system would require. It was a proposal with a substantial long-term upside, for not only could the mill be used to support IGAT-1, but after the program's completion it could also be turned toward the growing global market for pipe as a profitable new export industry. It was an idea that carried considerable risk, however, for the mill would need to be built quickly if it was to support the pipeline's construction. Far more than just a question of technical feasibility or scheduling, the choice of whether to pursue domestic pipe manufacturing went to the heart of the sometimes conflicting impulses that animated the IGAT-1 program as a whole. At stake were its fundamental parameters, particularly the pipeline's potential carrying capacity, and the extent to which the needs of the program should be subordinated to the country's larger developmentalist goals. Importing finished steel pipes was considered to be a safe choice that posed little overall risk to the project's timely completion. On the other hand, it was also more expensive and did little to further Pahlavi officials' broader ambitions, factors that soon prompted a "firm policy decision" in early 1966 to pursue domestic pipe production.[47] But building a new pipe mill and utilizing its output posed considerable danger to keeping the IGAT-1 project on time and within budget. Indeed, IMEG, asked to review competing proposals, deemed the winning bid from Los Angeles-based Torrance Machine and Engineering Inc. to have an unrealistically aggressive nine-month construction schedule.[48] Even more suspect, in the consulting experts' view, was the entire idea of building a pipe mill at all. As IMEG argued, from "the perspective of the economy of the IGAT-1 plan . . . ordering the factory [was] not justified." On the other hand, if the "encouragement and advancement of industries" more broadly was considered to be an important goal of IGAT-1 and should "the government be willing to tolerate damage and danger to the project," then a pipe mill could be "justified."[49]

The choice of whether or not to build the mill reflected IGAT-1's embedding within a larger matrix of industrializing ambitions. The provision of cheap energy was a critical plank of that endeavor, but it was not the only one. Iran's Third Development Plan, running from 1962 to 1967, had earmarked considerable investment for the country's industrial sector, much of it initially intended to support private enterprise but ultimately absorbed into large state-run projects like the steel mill at Isfahan, a machine plant in Tabriz, and joint petrochemical concerns with the likes of Allied Chemical and B.F. Goodrich. The upcoming Fourth Development Plan, to encompass the period from 1968 to 1972, went further and aimed to reduce foreign dependence through import substitution industrialization.[50] Whatever risks the plan might have had, the pipe mill proposal fit neatly within that paradigm, and at a high-ranking meeting held on June 16, 1966, attended by, among others, Prime Minister Amir-ʿAbbās Hoveydā and representatives of

IMEG and NIOC, it was decided to accept Torrance's proposal. But even this choice was subject to the whims of developmentalist ambition, and when it was found that the plant could be modified at little cost to also produce forty-eight-inch diameter pipe, it was suggested that IGAT-1 be again redesigned to make use of the larger size. The idea, supported by the prime minister, was intended to bolster the possibility of the mill becoming a lucrative source of exports, for Iran would then be in possession of one of the very few facilities in the world able to produce pipe of such size. With no immediate external market for such a product, however, and despite a Soviet pledge to take the extra gas that an enlarged line would carry, the change was ultimately rejected on the grounds that it overly endangered the IGAT-1 program. Indeed, in the end, even the unmodified mill was over budget and behind schedule, and very little of the pipe it produced was used to build IGAT-1. All told, the Ahvāz pipe mill only manufactured some 70,000 of the required 500,000 tons of pipe, failing in its primary aim of supporting the pipeline project and forcing the Iranian government to turn to imports from Europe and Japan at significant extra expense.[51] The mill was nonetheless indicative of the sometimes conflicting hopes that Iranian officials placed on Iran's budding natural gas infrastructure.[52] Whether aimed at providing cheap energy or a market for new domestic industries (or both), IGAT-1 was to be a potent tool with which to create an industrialized future for the country. The specificities of how best to achieve that outcome, or even what it should look like, were nonetheless at times hotly contested. In the 1960s and 1970s, the Pahlavi state's top-down developmentalist orientation was deeply entrenched, but that did not translate into uniformity of thought or action among those charged with implementing that vision. What was created was often the product of significant uncertainty and intense negotiation and compromise, a reality that would encompass not just officials' interactions with each other and outside institutions, but also the natural world.

The Nature of Supply and Demand

At the Seventh World Petroleum Congress, held in Mexico City in April 1967, two NIGC representatives spoke on IGAT-1, describing it as the "first major step towards utilising [Iran's] wasted natural resources" and something that would "make Iran the first country in the Middle-East to utilise indigenous natural gas resources on a large scale." With some 1,300 million standard cubic feet of associated natural gas being produced each day (MMSCFD) in southwest Iran—the vast majority of which was "flared to waste"—they proclaimed that the project would "do much to help the economic growth and progress of Iran" through the provisioning of cheap energy. As they explained, construction was to happen in two phases. The first, with a targeted completion date of January 1970, was to fulfill export requirements to the Soviet Union as well as projected industrial demand in 1974. A second phase, to be started in the mid-1970s, was to focus on expanding gas gathering and distribution networks, aiming to meet what Pahlavi officials expected would be rapidly growing domestic energy needs.[53] Left unsaid during the presentation was that in the spring of 1967, despite the program's aggressive construction schedule, IGAT-I's route, size, and source of gas had not yet been finalized. Soviet negotiators continued to press for gas deliveries well in excess of the roughly 1,000 MMSCFD that the initial agreement stipulated, asking for as much as double that figure, while also pushing for the pipeline to run through easier terrain near Zanjān and Tabriz rather than along the western coast of the Caspian Sea. Iranian negotiators resisted these modifications, particularly the change in route, arguing that they would strain the project's supply of pipe and threaten its speedy completion. Even more salient for them, however, was the prioritization of domestic consumption, and they insisted that any alterations be aimed first and foremost at serving "cities and villages along the route . . . in the west of Iran" rather than Soviet requests.[54]

Designed for a total carrying capacity of 1,650 million standard cubic feet of natural gas per day, of which 100 MMSCFD was dedicated to powering the line itself, IGAT-I's export deliveries were to begin in 1970 at 600 MMSCFD and rise to 1,000 MMSCFD by 1973. A contractually obligated amount, exports represented the pipeline's year-round base load. The remainder, 550 MMSCFD, equivalent to roughly 220 quadrillion joules of energy per year, would be available for use within Iran. In the late 1960s, Iranian energy consumption was rising rapidly, growing nearly 46 percent (350 to 510 quadrillion joules per year) between 1965 and 1969.[55] The majority of that energy was derived from oil, some 84 percent in 1967–68, and the resource accounted for more than 77 percent of residential, commercial, and industrial energy needs; 100 percent of transportation and agricultural requirements; and 72.5 percent of electricity generation. Residential and commercial customers were responsible for some 44.3 percent of all energy demand in Iran, industrial applications approximately 22.5 percent, and transportation 21.9 percent. At the time, natural gas met very little of the country's energy requirements, supplying only 1.3 percent of total energy usage. With a yearly production of 21.3 billion cubic meters in 1967—equivalent to 788 quadrillion joules, or nearly double Iran's total energy demand of approximately 430 quadrillion joules—natural gas therefore had the potential to both remake and expand the country's energy landscape.[56] Owing to its size and concentration of industry, nowhere would that be more true than Tehran, and it was expected that the city would be the largest domestic taker of gas from IGAT-1. As such, it was subject to the most study by NIGC planners, and they projected that the city's total demand would start at 95 MMSCFD in 1970 and steadily rise to 276 MMSCFD in 1979, the final year for which predictions were made. Much of the gas consumed would be used to generate electricity, 81 percent in 1970 and 46 percent at the end of the decade, while much of the remainder would be dedicated to industrial facilities. Residential service, projected to begin in 1973, was not expected to ever account for more than a quarter of Tehran's total natural gas usage. In all cases, rates of gas consumption would not be static, fluctuating both hourly and from a seasonal low of 160 MMSCFD in the summer to 416 MMSCFD in the winter.[57]

Despite such intricate analysis, predicting domestic gas consumption remained a difficult problem throughout the lifespan of the IGAT-1 program, with NIGC officials contending not only with uncertainties about demand, but also with politically charged questions about how to best put natural gas to use. Planning was entrusted to a team led by Mohsen Shirāzi, a man of long experience in the oil industry and eventual chief executive of NIGC. Shirāzi and his team envisioned a future of widespread natural gas use within Iranian society, from factories and power plants to homes and small businesses. Their imagined future proved to be controversial, however, with others in the Pahlavi government pushing for more limited utilization. Arguing that electricity was a more economical form of supplying energy, Mansour Rouhāni, the head of the Ministry of Water and Electricity, strenuously objected to the possibility of residential service, arguing that natural gas should be used only for industrial applications and electricity generation. For their part, Shirāzi and NIGC disputed the minister's contention that electricity was more economical, arguing that the country's power stations were inefficient, and that 70 percent of their energy input went "up the smokestack." NIGC thus calculated that it would require the equivalent of all of Iran's current electricity generating capacity just to meet the energy needs of the 238,000 homes and businesses that the company planned to serve in Tehran. Even if that were possible, Shirāzi contended, some 99 percent of all homes in the city were wired for only five amperes of electricity, well below the sixty that the average family was expected to require. Any plan to rely entirely on electricity for Tehran's residential energy needs would thus necessitate an expensive and time-consuming program to refurbish and expand the city's power grid. Such a requirement, Shirāzi and his team argued, nullified any notions of natural gas as being less economical. Despite Shirāzi's strenuous objections, however, Rouhāni and the Ministry of Water and Electricity were able to block NIGC's plans for six months, finally backing down on the shah's orders only after a group of consultants from Stanford University, hired to break the deadlock, found in support of NIGC's position.[58]

On the other hand, while some like Rouhāni rejected the use of natural gas in residential settings, others within the Iranian government opposed its use by industry. Fathollāh Nafisi, a senior NIOC director charged by Eqbāl to evaluate NIGC's planning, argued that while gas could readily replace kerosene and diesel for small residential and commercial consumers, a scheme that would limit Iran's need for additional oil refining capacity, it should not be used in place of fuel oil by factories and power plants. His fear was that a rapid expansion of natural gas service to such consumers, a task that was much faster and easier than slowly connecting tens of thousands of individual homes and businesses, would upset the product balance in the country's oil refineries. For each barrel of oil processed, Iranian refineries could only transform half into middle distillates like kerosene and diesel, extensively used by residential consumers, while about a third became the heavy fuel oil that factories and power plants depended upon. If natural gas quickly replaced the need for fuel oil but was much slower in replacing middle distillates, then that balance could be thrown off, leaving large and accumulating quantities of fuel oil on NIOC's hands. There were, however, ways to address the issue. The extra fuel oil could be exported, albeit at what would likely prove to be depressed prices, or Iran's oil refineries could be reconfigured to yield different product ratios. Neither option was straightforward, and Nafisi argued that their prohibitive costs would quickly overwhelm any potential economic benefit from utilizing natural gas. As with NIGC's dispute with the Ministry of Water and Electricity, Nafisi's objections stalled work for months, prompting Manuchehr Eqbāl to order two studies to evaluate the relative economic benefit of widespread natural gas use in comparison to oil. When they predicted a net $100 million gain during the first decade of IGAT-1's operations, Nafisi was overruled, and NIGC's original plan to provide gas service to all of Iranian society was approved.[59]

While the contours of Iran's nascent natural gas infrastructure were contested in the late 1960s, that uncertainty did not extend to the Pahlavi state's commitment to providing cheap energy. Indeed, the question at hand was not whether industrialization was desirable, or even whether natural gas should be used to support it, but how it could best be used to do so. There was little initial consensus among officials, as inter-ministerial competition and "administrative feudalism" were powerful and complicating forces on all levels of planning and developmentalist policymaking in the 1960s.[60] Gas energy, for all its possible advantages, was entering a functional energy sector, and its use was potentially destabilizing to established interests both within and outside Iran's government. Mansour Rouhāni's arguments, for example, had reflected his ministry's official stance, but it was also a position that had been quietly backed by Mohsen Khalili, owner of the Sherkat-e Butān liquid gas company, who had slipped Rouhāni a report upon which he based his arguments.[61] The provision of energy via piped natural gas threatened both their organizations, and it was only the direct intervention of the shah, on the advice of hired consultants, that their intransigence was overcome. Disputes about whether gas should be consumed directly or transformed into electricity thus reflected not just technical concerns, but also political and commercial ones. Whatever ideas some, like Shirāzi, may have had about making natural gas Iran's primary source of energy, in the late 1960s its utility as a resource was not yet fully established within Iran. Doing so would be a lengthy process, one that turned as much on broader perceptions of expert analysis and national interest, as well as the raw exercise of political authority, as the specificities of how energy was best consumed.

But questions regarding IGAT-1's supply of gas were no more settled in the spring of 1967 than questions about demand had been. While the pipeline system's original design, focused only on internal Iranian consumption, had envisioned using otherwise discarded associated gas from Iran's southwestern oil fields, it was not clear whether such volumes were sufficient to meet both anticipated domestic demand and the new export commitments to the Soviet Union.[62] Indeed, the 1,300 million standard cubic feet per day that was being produced was less than 80 percent of the amount needed to reach IGAT-1's total capacity of 1,650 MMSCFD. With exports expected to ramp up quickly between 1970 and 1973, planning for the system's long-term requirements was thus a high priority.[63] The expected exploitation of southwest Iran's oil fields posed political and legal problems, however, for while NIOC was entitled to use any natural gas produced there for domestic purposes, that right, as stipulated in the 1954 concession agreement governing Iran's oil sector, did not automatically extend to exporting it. IGAT-1's shipments to the Soviet Union thus needed consortium approval, and while the firms were willing to grant it, in compensation they insisted on being able to collect and sell the heavier hydrocarbons—mostly propane and butane—that would be stripped out of the pipeline gas. The demand did not sit well with Iranian officials, who complained that firms like British Petroleum, which had ignored for decades their desire to harness gas, were now claiming a market that Iran had developed on its own.[64] But with two important national projects on the line, IGAT-1 and the steel mill at Isfahan, there was little they could do, and after months of acrimonious negotiations NIOC largely acceded to the consortium's terms.[65]

Initial plans for IGAT-1 had anticipated drawing upon a cluster of seven fields surrounding the town of Behbahān in southeast Khuzestān (figure 2).[66] A change to the main pipeline's route shifted that center of gravity to the northwest and the village of Bid Boland, the place where IGAT-1's gas refinery and processing facilities would ultimately be built.[67] Selecting source fields was a difficult task, however, as planners needed to balance construction costs, refining costs, gas volumes, gas compositions, gas pressures, and the long-term reliability of production. Many options were considered, and though it was among the most expensive, in the end it was decided to use associated gas from the fields at Āghā Jāri, Mārun, and Ahvāz, a configuration that could consistently yield up to 1,705 MMSCFD of gas.[68] But raw natural gas could not be extracted from the earth and introduced directly into the pipeline. Despite their long treatment by scholars as functionally homogenous, oil and gas are not singular substances, and they become standardized products only after significant human intervention. Nor are petroleum deposits identical in their compositions and reservoir conditions, generally holding not only different mixtures of hydrocarbons but also impurities like water, carbon dioxide, and hydrogen sulfide (H_2S). Such impurities were a major complication, for not only were hydrogen sulfide and carbon dioxide corrosive, but under the right conditions water vapor could crystallize with heavier hydrocarbons and block the pipeline. IGAT-1's design thus included a set of specifications for "pipeline quality gas" that tightly controlled composition, energy content, pressure, and temperature (figure 3).[69] None of southwest Iran's raw natural gas met such standards. Gas from Āghā Jāri, for example, contained more than ten times the allowed levels of H_2S and was only 79 percent methane. Mārun's gas, on the other hand, had very little hydrogen sulfide, but its methane content fluctuated between 79 and 87 percent.[70] Transforming extracted gas, a naturally variable substance, into a standardized product thus required a complicated multistage process. After being separated from oil near the wellhead, gas was largely, but not entirely, stripped of propane and heavier hydrocarbons. With the separated liquids shipped to

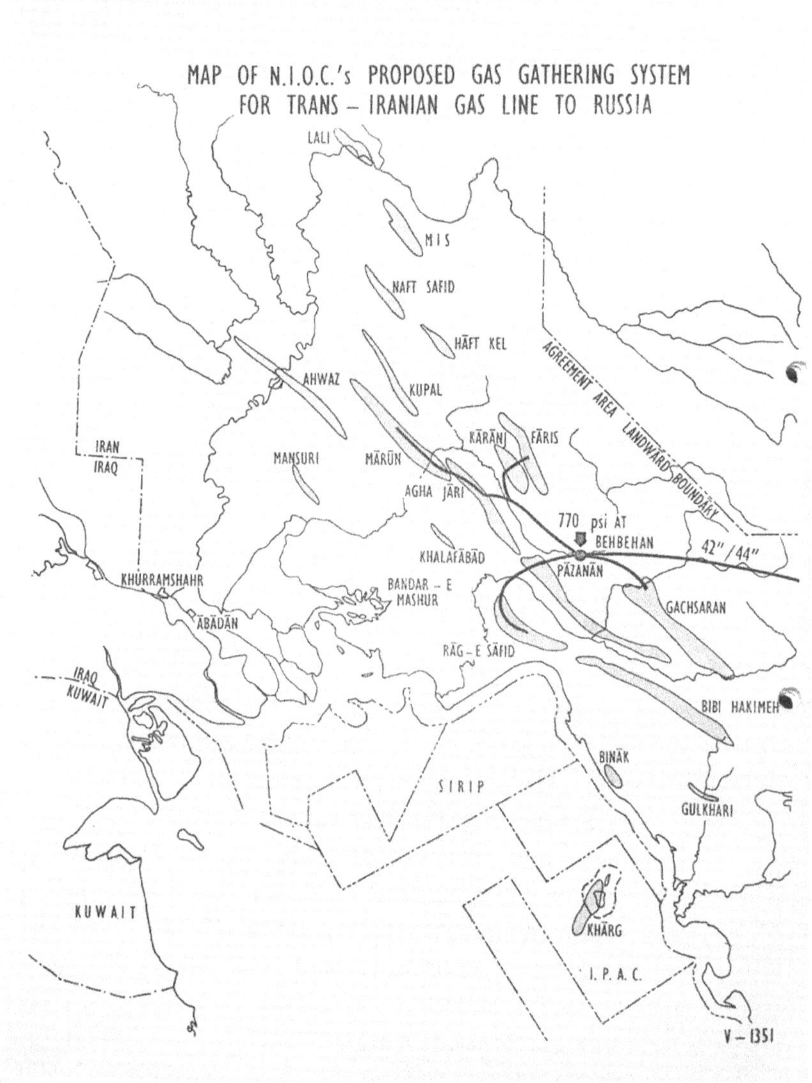

FIGURE 2. **Map of NIOC's Proposed Gas Gathering System for Trans-Iranian Gas Line to Russia.**

Source: "Review of Gas Matters," p. 4, figure 2; *Natural Gas and LPG, Iran (1965–1966)* (23312), BP Archive, University of Warwick. © BP plc.

Methane	83 mol %
Ethane	12"
Propane	3.5"
Butane	1"
Pentane	0.01"
Gas Gravity	0.66"
Dew Point	–10°C
Calorific Value (net)	1060 Btu/SCF
Carbon Dioxide	1% max.
H_2S	0.44 grain/100 SCF

FIGURE 3. **IGAT-1 Gas Specifications.**
Source: Naghavi and Manoochehri, "Iranian Gas Trunkline Project," 262.

Māhshahr for export, as the consortium had demanded, the remaining lean gas was sent to Bid Boland for purification and dehydration. From there the treated gas was injected into the main pipeline for transport northward.[71]

As a refined product, IGAT-1's natural gas was thus never fully "natural" in any real sense of the word. The specifications that governed its treatment were moreover in some ways arbitrary, having been set as part of Iran's contractual negotiations with the Soviet Union. But there were also material limits that those specifications challenged. As part of the original agreement, Iranian negotiators had promised an energy content of 1,060 BTUs per standard cubic foot, a notably high figure in comparison to similar contracts elsewhere in the world. While Pahlavi officials had thus given a de facto discount to the USSR, one they repeatedly tried and failed to rescind, their more serious challenge was adhering to that specified energy content while also maintaining the low hydrocarbon dew point needed for pipeline operations. A product of both temperature and pressure, dew point marked the transition between a hydrocarbon's gaseous and liquid state, and the heavier the hydrocarbon, the higher its dew point. For IGAT-1's gas to meet the specified energy content, more propane, butane, and pentane would need to be retained in the mixture than was the norm, thereby inevitably raising the combined dew point of the pipeline gas.

This, in turn, raised the possibility that hydrocarbons could condense in the line and form damaging blockages. The issue was compounded by Iranian negotiators' initial failure to include any flexibility in the gas specifications, an important oversight as the variability in extracted gas could make meeting pipeline requirements difficult. Under certain conditions, it could prove to be impossible. Fearing that such difficulties could disrupt the system in its entirety, as late as 1969 NIGC officials expressed anxiety over the project's gas specifications, eventually succeeding in having some small amount of flexibility included in the contract for gas delivery.[72]

Construction on IGAT-1 began in December 1967, marked by a ceremony in Āghā Jāri where the shah himself spoke.[73] In Khuzestān, Williams Brothers Germany, hired to build the gas gathering network, was busy clearing land. Farther north, Entrepose, the primary pipeline contractor south of Sāveh, was doing the same. Amidst the high peaks of the Zagros, Williams Brothers Great Britain was beginning work on what was expected to be the most difficult section of the line. At Bid Boland, Costain, a British company, was preparing to build a gas refinery designed by the American firm Pritchard Corporation.[74] The status of the pipeline's northern segment, however, turned over to the Soviet Union, was much less certain. At issue was the amount of work granted to the Soviets and what they would be paid for it.[75] Despite the IGAT-1 program's aggressive schedule, negotiations between the two countries had dragged on for nearly two years as Soviet negotiators sought to use the pressure of time to extract higher payments.[76] With talks nearly deadlocked as construction began, it was only their elevation to the ambassadorial level that enabled any further progress to be made. A compromise was eventually reached in the spring of 1968, with Iranian negotiators increasing the amount they were willing to pay and the Soviet ambassador declaring that "friends of the Soviet Union" should not "pay the penalty" for what had come to be seen as an unrealistically high initial ask.[77] It was an accord driven more by geopolitical concerns than economic ones, as Soviet leadership ultimately chose to prioritize good relations over economic gain. Nonetheless, while Iran and the Soviet Union were able to find a resolution, the long negotiations proved costly to IGAT-1, delaying the start of construction on its northern section from January to July 1968 and putting the project behind schedule.

But other issues remained in play. In the original 1966 agreement, Iran had agreed to buy from the USSR the thirty-four enormous compressors needed to move gas through IGAT-1's main pipeline. That provision had quickly come into question, however, as both IMEG and NIOC engineers understood Soviet capabilities in the field to be inferior to what could be found in Europe and North America.[78] Soviet negotiators knew this as well, and to keep Iran from looking elsewhere, they opted in March 1968 to supply the compressors on credit with repayment to come as part of IGAT-1's gas exports.[79] Far from ancillary, compressors and the repressurization they enabled sat at the heart of the pipeline's functioning. Compression allowed more gas, and thus more energy, to be packed into the line and transported by it. Even more important, in everything from extraction to transmission to delivery to consumers and the Soviet border, pressure was the motive force that pushed gas through the network. As it moved, periodic recompression was necessary to contend with the effects of distance, friction, and elevation gain. For that reason, between its sprawling size and requirement to cross mountainous terrain, IGAT-1's functioning depended upon high gas pressures. Maintaining sufficient pressure was thus a critical aspect of the pipeline network's design and operations, and in many ways IGAT-1 was at root a mechanism for managing gas pressures across vast distances. It was a task that began while petroleum still lay beneath the surface, for within the earth, whether associated with oil or not, natural gas is innately pressurized, and Iranian petroleum reservoirs are often characterized by notably high gas pressures.[80] Much of southwest Iran's early oil production had thus relied upon a natural "gas drive" that forced crude oil toward the surface when a deposit was tapped. Over the intervening decades, British Petroleum had put considerable effort into managing southwest Iran's reservoir pressures, aiming to prevent damaging overexploitation and the premature depletion of that gas drive (see chapter 1). Further, seeking greater control over the composition of extracted petroleum, the company had developed a multistage process for separating associated gas and oil, a development that smoothed out reductions in pressure and made the process more predictable.[81] Natural gas pressures were, however, no more uniform than compositions, varying widely between, or even within, petroleum fields. Āghā Jāri, for example, yielded gas at pressures ranging from fifty to 400 PSI, whereas Mārun was expected to run as high as 1,000 PSI. Each field and well supplying IGAT-1 thus used tailored collection systems, sometimes boosting and sometimes reducing gas pressures to a standardized 850 PSI before sending it on to the refinery at Bid Boland.[82]

Once purified and dehydrated, gas entered IGAT-1's main pipeline at an operating pressure of nearly 1,100 PSI.[83] Pressures were not uniform within the line, rising and falling between compressor stations as the route traversed what was often very difficult terrain. The most formidable challenge was the Zagros range, a mountainous region stretching nearly a thousand miles from southeastern Anatolia to southern Iran that was capped by peaks reaching more than 14,000 feet high. With scouting done on foot and horseback over the course of 1966, it was found that possible routes for the pipeline were few and challenging. In the end, it was decided that from Bid Boland the pipeline would run east through the Zagros foothills, crossing two rivers and passing through the very narrow Nāli and Pirzāl gorges. Pirzāl marked the start of the route's most arduous section, for after recompression the pipeline turned north and began a very steep thirty-five-mile climb to its highest point at more than 10,000 feet above sea level. This demanding sector, and the following 250-mile stretch through rugged alpine valleys, strongly shaped IGAT-1's design and gas specifications. Lofty mountain passes and frequent changes in elevation meant that high operating pressures and frequent recompression were needed, which in turn influenced hydrocarbon dew points and thus the composition of pipeline gas. Large seasonal variations in temperature compounded the issue, as did the relative inefficiency of compressing gas at high elevations. Higher pressures demanded stronger steel, and so did the need to use lighter pipe in harsh mountain terrain. Compressor stations could not be optimally spaced, as few suitable sites existed in the rocky Zagros. For that reason, south of Tehran, IGAT-1's pipeline gas was repressurized six times and while also feeding branches to Isfahan, Kāshān, and Qum. Near Sāveh, the pipeline split into a large

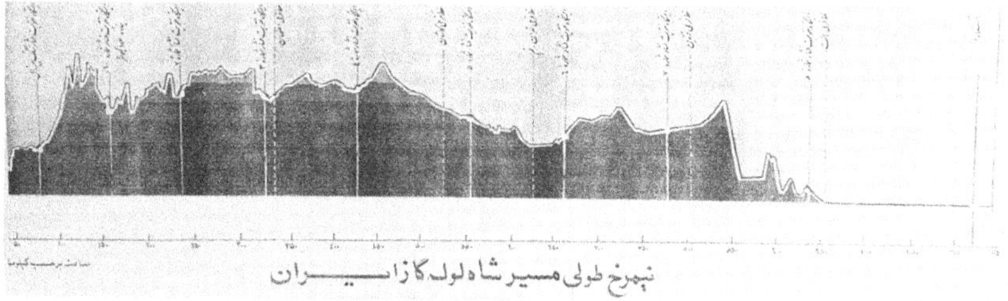

FIGURE 4. **Elevation Profile of IGAT-1.**
Source: Sherkat-e Melli-ye Naft-e Iran, *Shāhluleh-ye Gāz Iran* (Tehran[?]: Enteshārāt-e Ravābat-e ʿOmumi-ye Sanʿat-e Naft-e Iran, n.d.), 10.

thirty-inch branch to Tehran while the main line, reduced to a diameter of forty inches, continued toward the Soviet border via Qazvin, the Sefid Rud river valley, and the city of Rasht.[84] North of the capital region, in order to cross the Alborz mountains, gas was recompressed an additional three times before reaching Āstārā.[85]

All told, IGAT-1's main pipeline was nearly 700 miles in length, crossed two mountain ranges, seven major rivers, 150 minor waterways, thirty-four highways, and, in four places, the Trans-Iranian Railway.[86] Considered at the time to be some of the most difficult pipelaying in the world, the line crossed arid terrain and wet, regions that were hot and others that were cold, traveled under some rivers and over others, and in several places ran alongside existing infrastructure while in others it passed among high peaks rarely visited by anyone.[87] It was a route that had been chosen as the "shortest possible distance between the Southern oil fields and Astara" that also enabled economical transport to major industrializing cities like Isfahan and Tehran.[88] It was also a path selected for its potential ability to send gas to western cities like Hamadān and Kermanshāh should the deal with the Soviet Union ultimately fall through.[89] In that sense, IGAT-1's traversal of the Iranian landscape was dictated by the concerns of industrialization and international relations. But IGAT-1's route was also shaped by the demands of nature, of mountain chains and elevation gain, and the pressures needed to overcome them. Fifteen percent of IGAT-1's total length was considered to be difficult terrain for the pipeline, and those stretches did more to shape the pipeline's material reality than all the rest. Compression enabled more fuel and energy to be delivered per unit volume, but the actual pressures chosen were mandated by the need to push gas up the steep slopes of Iran's mountain ranges. Moreover, IGAT-1's pipeline gas, neither wholly natural nor synthetic, was itself produced at the intersection of petroleum's material properties, the technical requirements of gas transport, and the commercial agreements that structured the project. For all the sophisticated technology and expert knowledge that was deployed to control it, natural gas was never a mere extension of the project and the human ambitions it embodied. Natural gas was a volatile substance, and no accumulation of knowledge could transform it into something altogether docile. The project's gas specifications could thus never be an unbending imposition on the natural world, and the fact that Iranian negotiators had initially agreed to such terms, likely out of ignorance, made no difference to the fact that some flexibility was necessary to maintain the operational integrity of the pipeline system.[90] This lack of control was echoed in NIGC's demand planning, as competing political and commercial interests sought to orient the pipeline program towards their own ends. Connected across hundreds of miles by a thread of steel pipe, these forces—conflicting imaginaries of energy usage and unruly natural realities—shaped both the design and function of Iran's gas infrastructure. IGAT-1 was not only a creation of developmentalist ambition, political authority, natural materiality, or even technical feasibility; it was instead a product of all these things, an assemblage of human and nonhuman factors that promised to link ancient hydrocarbon deposits to consumers hungry for cheap energy.

National Arteries

In October 1970, nearly three years after construction began, IGAT-1 commenced operations. With large volumes of natural gas now flowing northward to Iran's major cities and the Soviet border, the future of cheap energy that Iranian officials had for decades worked towards was on the cusp of becoming a reality. But the IGAT-1 project was also over budget and ten months behind schedule, bedeviled by unrealistic expectations, discord among its participants, and unforeseen events. Prolonged negotiations between Iran and the Soviet Union had not only reduced the scope of the latter's involvement but also delayed construction in the north by six months. The movements of pastoralist herds had slowed work and at times damaged prepared ground, as had the unstable earth of the rice fields and sodden forests along the Caspian coast. Flooding in the Zagros in 1968 had likewise impeded pipelaying while river crossings proved at times to be more demanding than expected.[91] Frequent design changes and complicated financial arrangements had injected significant uncertainty into the project while a "very compressed" schedule had allowed little margin for error. Even more serious was the failure of the pipe mill, an outcome that had put the project even further behind schedule and notably raised its costs. Indeed, outlays (minus Soviet work credits) had doubled from IMEG's initial estimate of $350 million to nearly $700 million, a final tally that had reportedly angered the shah.[92] For all its escalating costs and delays, the IGAT-1 project was nonetheless considered to be a success. As Sa'id Naqavi, a senior NIOC engineer, wrote, "unlike what is common for plans similar to IGAT-1, from the beginning . . . [Iranian officials] point[ed] less to aspects of economics and profits and . . . more in the direction of correct implementation and timeliness."[93] What mattered most for officials like Naqavi was the fact that the pipeline project had made natural gas available in cities across Iran, a view that rooted the pipeline's significance primarily in its ability to further the state's aims for industrializing development. Nonetheless, by the end of the 1970s, the project had proved itself to be far from a financial catastrophe despite its inflated price tag. While domestic gas rates had been expected to be somewhat higher than those offered to the Soviet Union, and thus more important for recouping investments, the quadrupling of oil prices after 1973 shifted the balance. With gas prices rising from $0.308 to $0.57 per 1,000 cubic feet and again to $0.76 in 1978, Iran ultimately earned some $1.2 billion for the seventy billion cubic meters of gas it sent to the USSR, enough to pay for the roughly $1.0 billion final cost of the pipeline project and put to rest concerns about its financial wisdom.[94]

But the official significance of IGAT-1 went much further, and it came serve, both before and after its completion, as a key pillar of the Pahlavi state's politically legitimating developmentalist discourse. Much as had been true with liquid gas in the early 1960s, such viewpoints were expressed by Iranian officials in their speeches, in their interviews with the press, and by state publications. Foremost among those official outlets was again *Nāmeh-ye San'at-e Naft-e Iran*, and between 1967 and 1978, the magazine frequently celebrated the IGAT-1 project as making an industrialized future of cheap energy possible for Iran. As part of that process, the pipeline itself came to be layered with meanings rooted in, but nonetheless transcending, the materiality of its existence. Indeed, the physical scale of IGAT-1 was a core rhetorical tool for Pahlavi officials, and it was frequently trumpeted through photographs of pipeline systems and formulaic recitations of numerical data on the project's size and scope. Using the evidentiary force of quantification as "a way of making decisions without seeming to decide," such articulations sublimated what were fundamentally political choices about the state's developmentalist programs into detailed discussions of technical requirements and achievements.[95] Numerical data was thus a key part of official discourse on IGAT-1, working alongside photographs to concretize the developmentalist claims being made through gas infrastructure. Published a month after construction began, one article in *Nāmeh* introduced the project using an avalanche of figures, casting the project as a sophisticated technical undertaking aimed at bringing the benefits of natural gas to Iranians. Describing the system in exacting numerical detail, the article reported section length, pipeline diameter, the gain and loss of elevation, the thickness of pipe walls, the size and power of compressors, and operating pressures. It described the various branch lines that served Iranian cities: a sixteen-inch pipe to Shiraz transporting seventy million cubic feet of gas per day (5 percent of the system's total); another sixteen-inch line transporting 100 million cubic feet of gas daily to Isfahan and the steel mill (6 percent); two six-inch diameter pipes to Kāshān and Qum for the daily transmission of 6.5 million cubic feet (0.35 percent) and seven million cubic feet (0.4 percent) of gas, respectively; a thirty-inch diameter pipe to Tehran moved 276 million cubic feet per day (20 percent); Qazvin would receive 22.9 million cubic feet, Rasht 23.7 million, Bandar-e Pahlavi 5.9 million, and Tabriz 47.5 million (a total of 4 percent for the "northern cities"). Smaller towns and villages along the route would get a combined 2.5 million cubic feet per day. Natural gas itself was subject to similar treatment, and the article went to great lengths to describe the volumes of associated gas that were produced from each field and the specifications to which they would be purified. More than an empty statement of dry statistics, the prominent recitation of these figures simultaneously announced the ambitious scale of the IGAT-1 project, underlined the techno-scientific expertise needed to make it a reality, and promised an impending energy-rich future for Iranian society.[96]

Large figures like those describing the dimensions and carrying capacity of the network—thousands of kilometers of pipeline, hundreds of millions of cubic feet of gas—reinforced the monumentality of gas infrastructure that publications like *Nāmeh* created with their photographs. It was an evocation of technological grandeur that embraced the physicality of gas infrastructure while effacing many of its human and social aspects. In what would become a characteristic aspect of the magazine's coverage of IGAT-1, *Nāmeh's* rhetorical performances largely erased those who designed and built the system. The magazine portrayed IGAT-1 as a nearly autonomous system, one that sprang forth from the ambitions of Iranian officials and operated with seemingly self-managed efficiency. At stake in the article was not the directing of people and organizations but the command of the pipeline as a material and technological system. It was the rolled steel of the pipes, the gleaming towers of the refinery at Bid Boland, the turbines to generate compression force, and the land to be blasted, gouged, and smoothed into submission that was the stuff of concern. The article promised a near future of technical mastery, one defined not by fashion or democracy or religion, but by an ability to tame rock and metal in order to erect the "vital arteries" that delivered energy to Iranians. Allied to such claims was a forceful assertion of IGAT-1's Iranian character. Indeed, the 1968 article in *Nāmeh* that introduced IGAT-1 scarcely mentioned the deep involvement of foreign experts and contractors, even going so far as to downplay the fact that 61 percent of the gas moved by the pipeline was destined for the Soviet Union. While the map that accompanied the article depicted the pipeline's route and carefully indicated IGAT-1's source fields, compressor stations, and cities to be served, left conspicuously blank and unlabeled was the Soviet Union.[97] For influential figures like Manuchehr Eqbāl, pipelines were a concrete expression of civilizational modernity, and it was thus crucial for Iranians to be able to lay claim to them. Reflecting that perspective, in *Nāmeh*, IGAT-1—measured, mapped, photographed, and described in all its physical imposition—was presented as material proof of not just Iran's technological advancement, but also the Pahlavi state's role in driving it.

The rendering of IGAT-1 as a symbol of a particularly Iranian techno-modernity continued throughout the pipeline's construction. In mid-1968, six months after the start of construction, *Nāmeh* published an overview of the project's pipelaying operations that explicitly tied the monarchy to the system and its developmentalist potential. Depicted was a visual grammar that highlighted the figure of the shah and influential personages like Manuchehr Eqbāl and Sa'id Naqavi, presenting IGAT-1 as a product of their vision and expertise. In contrast, as was common in depictions of petroleum infrastructure around the world, those working on the line were anonymous, their faces covered by welding masks and their experiences suppressed by a broader discursive emphasis on technology and energy consumption.[98] Reiterating the shah's "emphatic" order that Iran's natural gas—"this great source of fuel and energy"—be put to productive use, the article highlighted not only the pipeline's materiality but also its intimate connection to the terrain that it traversed. From initial surveys to the leveling of the ground to the digging of trenches and the creation of a sand-based substrate for the pipe, the land of Iran was made part of IGAT-1's story. The influence of topography was shown in both extreme and mundane circumstances, from the hazardous passages of the Zagros and Alborz to the straightforward crossing of the country's central plains. Photography was again an essential part of that depiction, with one image, focused on a section of pipe connecting the Mārun field to the refinery at Bid Boland, showing the line dipping and curving through southwest Iran's rough terrain (figure 5).[99] In dwarfing the small human figures standing near it, the pipeline's primacy is asserted while its ostensible domination over the landscape—scraped, gouged, and marked by people and their

FIGURE 5. **A Section of Pipeline from IGAT-1's Gas Gathering Network.**
Source: "Pishraft-e Sari-ye Sākhtemān-e Tarh-e Shāhluleh-ye Gāz" in *Nāmeh-ye San'at-e Naft-e Iran* 7, no. 2 (Tir 1347), 5

heavy machinery—is likewise claimed even as the dry hills quietly influence the line's contours. According to *Nāmeh*, this was IGAT-1: a thing of steel and dirt and rock, a sprawling system joining northern and southern Iran, a marriage of technology and land that proved the Pahlavi state's mastery of the former and ability to conquer the latter. It was a stark portrayal of developmental prowess, standing in contrast to global norms that more often depicted pipelines as "efficient, remote from human habitation, and ostensibly noncontaminating."[100] *Nāmeh*'s article clearly depicts the violence that construction wrought on the landscape, joining an aesthetic tradition in which infrastructures are used to assert a technological capacity to tame the land in the service of industrialization and a future arrived.[101] Whether shown poised above a trench or spanning a river, such was the weight of IGAT-1 in Iran's official discourse, not only a promise of future energy use, as important as that was, but a manifestation of nature's harnessing and the national progress it enabled.[102]

While IGAT-1's main pipeline was the project's centerpiece, the official grammar of technological monumentalism radiated outward to other aspects of the program. Whether in photographic and textual descriptions of the gas refinery at Bid Boland or the pipe mill at Ahvāz, the same preoccupation with size and technological sophistication was evident, inviting Iranians to witness the effectiveness of Pahlavi developmentalist policies. One 1969 article framed the gas refinery at Bid Boland as a point of national pride, comparable in significance to the oil refinery at Ābādān, a facility that had once been celebrated as the world's largest. Writing that it would "encompass the largest and most modern purification units in the world," the article presented the refinery as evidence of Iran's rapid advancement, an argument it supported through quantification: five purification units, to later rise to nine; 240 million cubic feet of gas refined per day; "giant" thirty-two-meter purification towers, each weighing 200 tons; and 150,000 kilowatts of electricity consumed. The familiar photographic tropes appeared as well, with one highlighting the presence of dignitaries at the refinery's groundbreaking and two more depicting massive towers and refining equipment that dwarfed minuscule workers.[103] Nor was this discursive construction of IGAT-1 confined to *Nāmeh*. Using similar textual and photographic idioms, numerous other official publications evinced a preoccupation with monumentality, quantification, and political legitimation. Published in the mid-1970s, NIGC's *San'at-e Gāz-e Iran* ("The Iranian Gas Industry") epitomized in its sixty colorful pages the rhetorical currents at work. Aimed at making "visible" the "great and brilliant phenomenon" of the White Revolution and a new era of "happiness and prosperity for the Iranian people," the book fielded an impressive array of statistics, spiking trend lines, maps, descriptive passages, and full-color photography to demonstrate the integral role of gas in the country's industrializing development and rapid economic growth.[104] Children's books employed a similar developmentalist vernacular. Published in 1973 and aimed at elementary students, *Gāz* ("Gas") focused on IGAT-1's enormous pipeline and refinery, explaining that prior to the system's creation, "all of Iran's natural gas disappeared" and that "the design of IGAT-1 prevents the waste of gas." It described the Bid Boland refinery as "one of the world's largest gas refineries" and the "center of Iran's gas." It told a tale of the shah traveling to the Soviet Union to make a deal so that "the great wealth that had for long years gone to waste . . . [would be] placed in the service of the country's economy and development." Illustrating the book were five photographs of IGAT-1's machinery.[105] Nor was unreality a barrier to the use of such rhetorical techniques, as books like *Naft va Zendegi* ("Oil and Life") often described projects that only existed on paper. Published in 1972 as part of a U.N. initiative to promote publishing in developing states, the work compiled four expert presentations from Iran's petroleum industry. Aimed at the "lovers of study and research," "academics," "educators," and "all the people of Iran," *Naft va Zendegi* was lavishly illustrated with maps, charts, and photographs, combining them with extensive numerical data to give descriptions of unbuilt pipelines, refineries, and gas networks a sense of solidity that they otherwise did yet not possess.[106] In doing so, the publication did more than report the unrealized plans of Iranian officials. As was true in most official depictions of IGAT-1, it instead momentarily condensed an imagined future of cheap energy into something more tangible, providing readers with a sense that not only was such a prospect possible, but inevitable.

***

Designed and built in the late 1960s, IGAT-1 was to serve as the backbone of a new national system for supplying cheap energy. Reflecting the rising energy demands of Iranian industrialization and stretching more than a thousand kilometers from the petroleum fields of southwest Iran to the Soviet border at Āstārā, the project brought together, and condensed into material form, the swirling political, economic, technological, and natural forces that shaped the Pahlavi state's developmentalist policies. But natural gas was not singular, and neither were its uses, and Iranian officials entered into fierce debates about whether it should be consumed directly or used to generate electricity. The decision to promote direct consumption by citizens and industry alike, a choice supported by expensive foreign consultants, made gas into a symbolic representation of what was promised to be a near future of technological sophistication and high energy consumption. Indeed, within their imaginaries of gas utilization, Pahlavi officials tied their country to a vision of progress measured in refinery capacity built, thousands of kilometers of pipeline assembled, and millions of cubic feet of gas produced and consumed. Infrastructural systems like IGAT-1 were thus made into highly visible and aestheticized expressions of the state's developmentalist ambitions, symbols of an industrialized modernity that Iranians were actively building. Further animating Pahlavi officials was an understated but nonetheless very real anticolonial agenda, one allied to international efforts to capture higher oil rents but nonetheless also oriented toward the domestic exploitation of the country's natural gas resources. To utilize southwest Iran's associated gas meant overcoming the opposition of foreign oil majors, a state of affairs that sharpened the liberatory edge of Iran's developmentalist ambitions. That Iranian officials worked to fulfill those hopes by striking a deal with the Soviet Union was thus understood as an assertion of sovereignty and a hedging of geopolitical bets by a country that had been long stymied in its quests to build a national steel industry and utilize its natural gas. But for Iranian modernizers, the construction of gas infrastructure was beneficial not only as a way to provision fuel but also as a catalyst for establishing new industries in the country. Whether it was through the construction of a steel mill in Isfahan, a pipe mill in Ahvāz, or the manufacture of smaller items for the pipeline project, IGAT-1 was understood as an opportunity for furthering Iran's economic growth. While Iranian officials' industrializing dreams existed independently of gas and predated its exploitation, over the course of the 1960s and 1970s the two would become increasingly intertwined as the IGAT-1 system was designed, built, and operated. It was a development that would ultimately make gas the energetic foundation of modern Iranian society.

As both a tool of industrialization and an explicit measure of civilizational worth, the Pahlavi state's embrace of natural gas and its infrastructures echoed the efforts of others around the world to organize and augment their power through the logics of resource extraction.[107] Whether focused on politics, technology, economic development, or some combination of them, official discourses of the IGAT-1 project were built on declarations of politically legitimating control. Such articulations of mastery extended to the pipeline's relationship with the natural world, and publications like *Nāmeh-ye San'at-e Naft-e Iran* celebrated the conquest of land and topography with an extensive textual and photographic record of engineering success. But that purported domination of nature was always fraught, and it was never as fully realized as official publications portrayed. IGAT-1's design and operation bent to the interlinked demands of technological and natural materiality, and every aspect of Iran's new pipeline system was shaped by the properties of gas, earth, steel, and geology. IGAT-1's viability was built upon the availability of large volumes of associated gas and the utility of gas drive in oil production; its gathering and refining operations were sculpted by the compositions and pressures of that gas; its pipelines bent and curved and climbed and descended as they struggled to traverse Iran's rugged mountain landscapes; its operating pressures and gas specifications were molded by elevation and climate; its physical form and heft were tied to the strength of steel; and its utility was determined by the energy content of different hydrocarbons. A sprawling assemblage of human, technological, and natural forces, in the 1960s and after, IGAT-1's hybridity sat at the heart of Iranian development, rooting the country's industrializing society deeper into the land than it ever had been before. For the first time, a network of pipes connected Iranians to the petroleum pools of southwest Iran, interweaving their daily lives with a technical system that would continuously respond to the mandates of topography and seasonal change. In the following years, as NIGC extended IGAT-1's reach and gas consumption rose in the country, those connections would grow ever tighter, a product of Iranians' hunger for cheap energy and the natural and material phenomena that shaped the design and operation of the pipeline system.

Four
A Ghoul at the Gates

In the summer of 1972, Muhammad Reza Shah, seeing the effects of industrial emissions on the skies of Shiraz, ordered that all the city's "smoky machinery, particularly [its] brickmaking kilns," be converted to use natural gas. The shah was not the first to note the city's growing clouds of smoke, for residents of Shiraz had long seen their area's brickmakers as unrepentant polluters who needlessly used cheap and dirty fuels like old engine oil, raw crude oil, mazut, sawdust, and trash.[1] Despite their history of complaints, it was the shah's order that finally spurred official action, and in response, Manuchehr Piruz, the Fārs provincial governor, wrote to Shiraz's mayor to note that the National Iranian Gas Company stood ready to aid the city in the "prevention of air pollution."[2] Nor was Shiraz the only Iranian city to receive such directives or to look toward natural gas a potential solution. More than a year later and some 500 miles to the north, Lt. Gen. Mohsen Hāshemi-Nezhād, commanding officer of Iran's Imperial Guard, reported to Gholāmrezā Nikpey, the mayor of Tehran, that the shah had observed voluminous clouds of smoke rising from the numerous asphalt plants west of Mehrābād Airport. Concerned that these emissions could affect air quality across "all of Tehran," the shah had demanded that all "these workshops and furnaces . . . change location."[3] Gently resisting the edict the next day via urgent letter, Nikpey instead promised to oversee a series of modifications to the factories. Rather than relocate, the asphalt plant owners pledged to install exhaust scrubbers and swap their primary source of energy, first from fuel oil to diesel, and then, as they themselves requested from NIGC less than two weeks later, to natural gas.[4]

These two episodes, separated by time and distance, were linked by a growing concern among Iranians for the markedly worsening air quality of their cities. Fearing that the worst was yet to come, in the late Pahlavi era, an increasing number came to see air pollution as a central challenge for their industrializing society. That sense of dread was well founded, for to step foot in Tehran today is to be submerged in a world of hazy, choking air. Thick layers of smoke and smog bear down on the city, obscuring vision, irritating eyes and throats, shuttering schools, and emptying public spaces. A city long marked by steep wealth and climatic inequalities, recent decades have seen the respiratory lives of Tehran's residents diverge as those with means undertake an increasingly expensive scramble up the slopes of the Alborz Mountains in search of clean and cool air. But such acutely poor air quality was not an inevitable outcome of Iran's twentieth-century industrialization, and for a time it appeared that such a fate might be avoided. Repulsed by the polluted cityscapes of the industrialized world and seeing them as a dire warning for their own future, in the 1960s and 1970s, many Iranians came to understand environmental degradation as threatening the fundamental validity of their country's modernizing project. In response, commentators wrote frequently of "the environment" (*mohit-e zist*) and a suite of threats to it. Most often, however, they focused on the increasing visibility of soot, smog, and haze in the country's skies, making such visible pollution emblematic of the dangers that lurked within the Pahlavi state's developmentalist policies and prompting government officials to seek out ways of avoiding the poisoned future that many foresaw. In doing so, they would come to vest their greatest environmental hopes in a new source of energy: natural gas.

The decline in air quality in Iranian cities was a direct result of the country's accelerating industrialization and its accompanying increase in energy consumption.[5] As was true for a broad swath of the developing world, Iran's total energy usage rose rapidly in the late Pahlavi era, growing some 350 percent between 1965 and 1979, from 350 to quadrillion to 1.58 quintillion joules, a near tripling of the per capita rate of 14.2 gigajoules to 42.4.[6] In comparison, world energy consumption rose only 81 percent over the same period (156.1 quintillion to 283.3 quintillion joules). There were great discrepancies hidden within that total, however, for while some developing states like Brazil (311 percent, 990 quadrillion to 4.06 quintillion joules) and China (211 percent, 5.53 quintillion to 17.18 quintillion joules) also saw a rapid intensification of energy usage, others like India (106 percent, 2.23 quintillion to 4.26 quintillion joules) and Pakistan (67 percent, 300 quadrillion to 500 quadrillion joules) posted lower, if still notable, figures. Nor was such divergence confined to the Global South, for while energy usage in some industrialized societies increased briskly, as in the United States (49.5 percent, 51.98 quintillion to 77.7 quintillion joules), in others like the United Kingdom (12.8 percent, 8.34 quintillion to 9.41 quintillion joules), it did not.[7] In Iran, skyrocketing energy consumption was driven by a rapidly expanding economy, one that grew some 11.4 percent per year in the 1960s and 6.5 percent in the 1970s. Though such figures included the country's sizeable oil exports, they nonetheless also reflected the increasing importance of its non-oil sectors, particularly industry.[8] But there was no industrialization without copious amounts of fuel, and in Iran, prior to the construction of the First Iran Gas Trunkline, that largely meant oil products. Climbing at an average yearly rate of 11.3 percent, oil usage expanded rapidly between the late 1950s and early 1970s, eventually totaling some 11.1 million tons (510 quadrillion joules) in 1971 and 30.2 million tons (1.29 quintillion joules) in 1979.[9] It was the effects of that intensifying fossil fuel consumption, the future of cheap energy that Pahlavi officials had worked to build, that Iranians began to notice, not only in their country's growing wealth and influence, but also in its increasing troubles with urban air pollution.

While there is little surviving air quality data from the late Pahlavi period, made clear in contemporaneous accounts was the increasing intrusion of atmospheric pollution into the lives of Iran's urban residents. Their escalating worries, periodically backed by the shah's personal interest, intersected with the new availability of natural gas that IGAT-1 had enabled. Seeing in gas the possibility of a cleaner future that did not necessitate a retreat from their developmental ambitions—a squaring of modernization's environmental circle—a broad set of Iranian officials folded notions of ecological preservation into their visions of industrialized prosperity. Indeed, by the 1970s, dominant beliefs about the necessity of conquering nature were beginning to be challenged within government circles. As a countercurrent to the state's overwhelmingly developmentalist orientation, desires for land and wildlife conservation had long percolated within Iran's elite social and political circles, giving rise to a series of laws that protected threatened game species and nationalized the country's forests and rangelands. By the last decade of Pahlavi rule, that undercurrent had blossomed into greater recognition of the utility of environmentalist projects for both asserting state power in peripheral spaces and contributing to a nationalizing identity rooted in restorationist beliefs about Iran's past imperial glory.[10] Iranian officials, seeing both environmental problems and their potential solutions as outgrowths of society's existing structures, began to pursue "ecological modernization," aiming to leverage the state's growing administrative and infrastructural capacities to prevent pollution and environmental degradation.[11] In this way, long before Iran's first organized environmentalist movements of the late 1990s and early 2000s, air quality concerns animated a sustained engagement with ecological protection among Iranian state officials.[12] In their minds, if air pollution was caused by rising fossil fuel consumption, then switching fuels could potentially reduce its severity.

But Iranian officials were also cognizant of the role that uncontrollable natural forces played in the issue, particularly in places like Tehran. Underpinning Iran's encounter with air pollution were topography, climate, and chemical composition, significant nonhuman factors that helped make the country's cities into hybrid environments.[13] For scholars, accounting for such hybridity means tracing the complex and contingent histories, both anthropogenic and natural, that combine to produce environmental health and decay. Viewed through that lens, Iran's natural environments were more than inert fields upon which modernizing officials exercised their will; they were instead influential forces in their own right, contributing greatly to the formation of polluted urban spaces. Iranian experts, working in institutions like the National Iranian Gas Company and the Plan Organization, implicitly understood their cities as hybrid, even if they did not use that kind of language, and they saw natural phenomena like mountain ranges and atmospheric inversions as significant factors in both creating the problem of air pollution and in constraining the range of potential solutions. Committed to a policy of industrializing development but incapable of altering the pollution-intensifying effects of terrain and climate, Iranian officials sought instead to reduce emissions through the differing combustion properties of fossil fuels. For these figures, salvation thus seemingly lay in natural gas, possessing as it did a double identity as Iranian industrialization's best source of cheap energy and as a fuel that promised to protect the country's urban skies.

Like Two Unfiltered Cigarettes

In the spring of 1972, two Italian engineers arrived in Shiraz at the invitation of Mohsen Qahremāni, the owner of a regional brickmaking firm. Distressed by the growing clouds of smoke and soot in the city's skies, and moved to action months before the shah ordered his compatriots to follow suit, Qahremāni asked the pair to adapt his facilities to use natural gas. Employing pipes that had been imported at great cost from abroad, the engineers completed their work in the early summer. Within days of their departure, however, the facility's new natural gas system lay in ruins, the expensive pipe unable to withstand the 1,000-degree heat of the kilns. Between damaged equipment and production stoppages, the firm's financial losses were considerable. But even more keenly felt was the project's failure to reduce emissions, and in a letter to the provincial governor, Qahremāni lamented his company's forced return to the "old materials of oil and smoke."[14] His enterprising spirit on the subject notwithstanding, Qahremāni was not alone in his worries about air pollution or in his conviction that transitioning to natural gas would help mitigate it. Indeed, by the time he sought out the services of the Italian engineers, the potential environmental benefits of natural gas had already been a topic of public discussion for more than a decade in Iran.

In the 1960s and 1970s, Iranians' growing environmental concerns coincided with a global wave of anxiety surrounding the health effects of industrial pollution, irradiated nuclear fallout, and the widespread use of persistent toxins like DDT. Epitomized by Rachel Carson's 1962 book *Silent Spring* and the public outcry it helped galvanize, the era saw the emergence of grassroots movements and new regulatory mechanisms for environmental protection. Pahlavi-era Iranians, or at least elite ones, were active contributors to that broad transnational effort, particularly as it was instantiated under the auspices of the United Nations. Driven by perennial fears of resource exhaustion and Malthusian limits, the postwar period saw the rise of numerous models for state-led environmental preservation and conservation, two approaches that were in practice often conflated.[15] Animated by a desire to better manage the country's natural resources, Iranian officials began in the late 1940s to take part in United Nations symposia on environmental issues. Such programs were largely aimed at the generation and sharing of technical and scientific knowledge on the environment, and for many years they were dominated by the presence of experts hailing from the countries of the North Atlantic. Iranian officials nonetheless enthusiastically joined initiatives like the Arid Zone Programme (intended to make desert regions more productive) and the 1971 conference on the Conservation of Wetlands and Waterfowl that Iran hosted in the city of Ramsar.[16] Within Iran, such notions of responsible stewardship gave rise in 1956 to the Game Council of Iran and its mandate to regulate hunting in protected regions. Reorganized and expanded in 1967 as the Game and Fish Department, in 1971 the section and its work were folded into broader efforts of environmental protection under the newly created Department of Environmental Conservation. Building upon earlier efforts, the new department sought to protect wildlife, revive damaged ecosystems, and promote outdoor recreation, opting in many cases to aggressively limit traditional activities like grazing and woodcutting in the country's six national parks and thirty-five wildlife preserves.[17] But for the first time, the department was also charged with the broader mission of preventing environmental degradation, a change that reflected shifting international notions of environmentalism and humanity's relationship with nature.[18]

Often seen as an inflection point in the history of our understandings of the global environment, the 1972 United Nations Conference on the Human Environment for the first time elevated the voices of a large number of participants from the developing world. Better known as the "Only One Earth" conference after the "officially unofficial" book that guided the symposium's agenda, the meeting was organized to foster international cooperation on postwar-era concerns like the overexploitation of natural resources and the ecological effects of rapid industrialization.[19] Though marked by a cacophonous mixture of differing opinions and perspectives, for many of the experts in attendance, particularly those from the Global North, cooperation meant technocracy and the assertion of an ostensibly apolitical ability to balance human needs and ecosystem health. But having so recently fought to secure political independence and sovereign control over their natural resources, many leaders from the Global South resisted what they saw as neo-colonialist notions that they should remain poor and refrain from pursuing the kind of resource-intensive economic development that had made industrialized countries wealthy. Figures like India's Indira Gandhi thus asserted a right to modernize as they saw fit, expressing a willingness to consider limits only if direct transfer payments were made in compensation. In its debates about governance and sovereignty, the Stockholm conference consequently reinforced existing state-centered models of development and environmental protection even as new coordinating institutions like the United Nations Environment Programme were created.[20] But "Only One Earth" was more than a meeting of state delegates, as it also attracted numerous nongovernmental organizations and a raucous contingent of radical thinkers and activists. Among them was a man by the name of Mohammad Taghi Farvar, an American-trained Iranian biologist who endeavored to highlight the environmental challenges of societies in the Global South. In Farvar's mind, far too little attention had been given to the steep environmental costs of large-scale models of development, and it caused him to question the fundamental premises of top-down modernization. Though some fiercely criticized him as needlessly politicizing what they believed should be technocratic questions of environmental protection, Farvar was far from alone at the Stockholm conference.[21] The resulting clashes have come to be seen as one of the symposium's enduring legacies, marking a point where older conservationist approaches began to give way to broader critiques of industrializing development. But Farvar's beliefs resonated closer to home as well, and in the following years, as he returned to Iran and took up a position within the Iranian bureaucracy, it became clear that many of the country's modernizing officials had likewise begun to question the environmental violence that development had wrought.

In 1974, Iran's Department of Environmental Conservation was again renamed, this time simply to the Department of Environment. Its writ was moreover broadened again, now including everything from managing national parks to mitigating air and water pollution to preventing the overuse of pesticides to maintaining the "beauty of the natural environment." As with many Iranians writing on the subject after the 1972 conference in Stockholm, those working within the department, including Farvar, saw pollution and environmental degradation as intimately linked to "progress." It was a strong connection, but also one that Iranians could break with sufficient care, for pollution in the country had "not yet reached the horrifying level that fully industrialized countries face." Most significant for many such writers was air pollution, the "irritating poisonous clouds and polluted vapors" that had developed in Iran's urban areas.[22] Inspired by the U.S.'s 1963 Clean Air Act, and facing intensifying public calls for laws combating the issue, a series of air quality regulations were proposed, including new rules on where polluting facilities could be located, the height of their smokestacks, the kinds of fuels they could use, and limits on their total allowable emissions.[23] In response to public agitation, in 1974 Iran's parliament passed the Environmental Protection and Enhancement Act, part of which gave the Department of Environment a specific goal for the "preservation, restoration, or enhancement of the quality of the ambient air in Iran for optimum social and economic benefit."[24] The law was strengthened in 1975 with the passage of the National Clean Air Regulations, a decree which empowered the department to set new national standards for air quality, fuel compositions, and total emissions.[25] Such laws were not as effective as hoped, however, for the work of regulating air quality and emissions was hampered by the lack of "local data" on air quality across much of the country. Without access to detailed information, the department was forced to establish broad regulations "based on . . . general experiences" rather than specific observations and targeted interventions. It was thus decided that the best course of action to protect urban air quality was to use the "best practicable technology to prevent emissions at the source."[26] In Iran, that would come to mean, in theory and in practice, gas energy.

The positioning of natural gas as a way to mitigate air pollution was not a sudden development in the early 1970s. For years, the Ministry of Petroleum, the National Iranian Oil Company, and the National Iranian Gas Company had all highlighted the resource as having both economic and environmental benefits.[27] One article published in *Nāmeh-ye San'at-e Naft-e Iran* in early 1963, part of a series of five printed between 1961 and 1966, argued that the widespread use of natural gas would result in "the air pollution of Tehran . . . [being] reduced" while "a cheaper fuel [was also made] . . . available to consumers."[28] By the mid-1960s, urban air quality had moreover become a significant public issue. First to sound the alarm was Jahānshāh Sāleh, a trained physician and the Minister of Health, whose public warnings a decade earlier had been mocked in the country's press and ignored by a government that saw environmental concerns only through the lens of wildlife protection.[29] By 1963, however, opinions had begun to shift, and in that year the Supreme Council of City Safety declared that Tehran's air was "not fit for respiration."[30] Three years later, Sāleh, by then head of the University of Tehran, cast the problem as having reached crisis proportions. Speaking to a group of Iranian meteorological and medical experts, he argued that if immediate action were not taken, then "within the next ten years smoke will carry us into the sky, and we will be destroyed!" Already, he went on to say, "if you watch Tehran's dawn or dusk from a high point, instead of the clear and beautiful sky Tehran was famous for all over the world, you will see only a black layer of smoke that makes breathing difficult."[31] Sāleh was far from alone in his views, and between 1967 and 1971 more than a hundred meetings and seminars were held on the issue of air pollution.[32] Articles in major national dailies like *Ettelā'āt* reinforced such charged depictions with dramatic photographs of smoggy skies and headlines that screamed "The Air of Tehran is Intolerable" and "The Air of Tehran Threatens!"[33]

Industry publications like *Nāmeh* steadily reported on the issue, frequently comparing the need to mitigate air pollution to the necessity of ensuring potable water.[34] At the same 1966 seminar where Sāleh spoke, the director of the Health Society of Tehran advised that

> The [once] blue and beautiful sky of Tehran pollutes this city's land. If no action is taken to prevent an increase in the level of air pollution, it will not be long before life in this city will be impossible.[35]

Senator Sadeq Rezāzādeh Shafaq, speaking seven weeks later to the assembled Committee to Combat Air Pollution, voiced the group's consensus that Tehran's quick industrialization had "polluted and poisoned the city's air . . . [and] made breathing intolerable."[36] One 1967 NIOC report forecast that without significant measures Tehran's air quality would continue to deteriorate, and while residents had "slowly become habituated to inhaling dirty and polluted air," it was causing increased rates of pulmonary disease and cancer.[37] Other high-ranking officials expressed similar sentiments, with some like 'Abdul Rezā Ansāri, the Minister of the Interior, stating that "the prevention of severe air pollution has the attention of the government" due to its effects on public health.[38]

When Mohsen Qahremāni and Muhammad Reza Shah bemoaned the poor air quality of Shiraz in the summer of 1972, their preoccupation with what several other factory owners described as the "smoke of crude oil" was thus commonplace.[39] All the complexity of air pollution, its causes, and its effects notwithstanding, for many Iranians in the 1960s and 1970s "air pollution" held a particular and tangible meaning rooted in the smoke of fossil fuel combustion and the visible uncleanliness it produced. "In the not-so-distant past," one writer lamented in *Ettelā'āt* in the summer of 1971, "old Tehran . . . had pleasant and attractive weather," but smoke now threatened to "suffocate" the city. The article went on to describe the "strange" and "alarming" state of the air, noting that in

> San Francisco, an hour spent walking in the streets is equivalent to smoking one unfiltered cigarette. But if you wander for an hour in Tehran, it is just as if you smoked two unfiltered cigarettes! This is the actual face of Tehran, a city that for most of the provinces is a visionary one, but visionary for smoke!

FIGURE 6. **Cover of a Collection of Air Quality Data Depicting Industrial Pollution Obscuring Mt. Damavand from the View of Those Below.**

Source: Sāzmān-e Hefāzat-e Mohit-e Zist, *Āmār-e Āludegi-ye Havā-ye Tehran, Seh Māheh-ye Sevvom-e Sāl-e 1357* (Tehran[?]: Enteshārāt-e Sāzmān-e Hefāzat-e Mohit-e Zist, 1357[?]), cover.

As something that could be seen, smelled, and felt, time and again the shifting, heterogeneous, "gloomy" clouds that lay over Iran's cities were described by observers as "smoke." Magazines and newspapers reinforced that understanding with photographs of smoggy haze, smokestacks discharging heavy fumes, and black exhaust pouring from motor vehicles. Accompanying *Ettelā'āt*'s article was a black-and-white photograph of thick haze smothering the city; captioned "Tehran is drowned in smoke," the image demonstrated that for city residents air pollution was not an abstract mixture of invisible gases but a conspicuous presence in their lives.[40] Indeed, smoke's ubiquity was frequently expressed through a sustained visual grammar of smokestacks, plumes of smoke and fire, and murky clouds that blocked eyesight. One striking image, published in 1971 alongside an article on NIOC's efforts to fight air pollution, foregrounded notions of impending peril through its stylized composition. Shot directly into a low, haze-distorted sun and employing shades of crimson interposed between areas of inky darkness and white brilliance, the photograph depicted a factory's "fumes" and "flares" coiling upward into a pitch-black sky, the ominous mood transmuting the image into a warning of dire threat.[41] A year later, as part of its reporting on a symposium dedicated to air pollution, the magazine again employed this graphical idiom, highlighting notions of uncleanliness with an image of smokestacks disgorging murky clouds of smoke and soot into a sickly yellow sky.[42] Other depictions, both photographic and illustrated, were more explicit, contrasting smoky pollution with previously unsullied natural landscapes on the verge of being lost.[43] With its illustration of Tehran nestled among the foothills of the Alborz Mountains, the cover of a 1978 statistical collection from the Environmental Protection Organization made the point in even starker terms.

In the image, Mount Damavand, a potent symbol of Iranian national identity, stands prominently amidst a robin's egg-blue sky. Lower down, clean air and natural beauty give way to a polluted industrial landscape as unsightly gray clouds collect above the city. In this telling, air pollution stands not just between the residents of Tehran and Damavand, but also their country as a whole and the natural beauty it contained. The harm depicted was an aesthetic one that went beyond the physical danger of smoke and pollution, a point reinforced by the collection's back cover and the blasted, industrial landscapes it presented.[44] Far from exceptional, such representations conjured a vision of Iran defined by industry's hazardous byproducts, an experience that commentators decried again and again as they mourned the loss of Tehran's "pleasant and attractive weather" and the "clear and beautiful sky" that Jahānshāh Sāleh had loved. In their place had come choking fumes, obscuring haze, and rent earth.

The Nature of Air Quality

No Iranian city faced the issue of air pollution as intensely or with a knot of social, economic, topographic, and climatic causes as intractable as Tehran. A small town that was made the country's capital in 1786 by Āqā Muhammad Khan, the first Qājār shah, by the mid-1960s the city was growing at an immense rate. At the time, with land reform and increasing exposure to global markets upending Iran's agrarian economy, large numbers of rural inhabitants were flocking to Tehran in search of work, swelling the city to more than 4.5 million inhabitants by the mid-1970s. It was a trend reinforced by the steady political and economic centralization of Iran over the course of the twentieth century, and in the late Pahlavi period Tehran accounted for approximately half of Iran's GNP and some 40 percent of all national investments, including 60 percent of those made in the industrial sector. But interwoven with those political and economic histories was Tehran's natural setting, and life in the city, sprawling across hundreds of square miles and 3,000 feet of elevation, had long been shaped by the dominating presence of the Alborz mountains. Stretching across northern Iran from Azerbaijan to Khorāsān, the Alborz, capped at more than 18,000 feet by the volcanic Mt. Damavand, separates the lush forests of the Caspian coast from the country's more arid interior. Situated amidst the range's southern slopes between 3,500 and 6,500 feet of elevation, Tehran's topography has been described as a basin or bowl, bounded to the north and east by the mountains and, to a lesser extent, by the Kavir desert to the south. With flows of humid air from the Caspian blocked by the Alborz, Tehran is left with a semi-arid climate marked by dry air, hot summers, and cool winters. Prevailing winds are largely out of the west. Because of this topographic setting, from its historical core, much of the city's expansion has been oriented westward through the Alborz foothills, even as affluent northern districts have climbed the mountain slopes and poorer southern ones have stretched into the arid plains. Indeed, in Tehran, social hierarchy and elevation are strongly linked, with the city's wealthier residents enjoying the lower density, cooler temperatures, cleaner water, and greener surroundings of neighborhoods that sit high up in the mountains. In contrast, the city's older central and southern areas, sitting lower in elevation, are much hotter, much drier, much more densely populated, and by the 1960s, facing significant problems with traffic and pollution. Many of Tehran's new inhabitants, generally with little to their name and few connections, crowded into these latter neighborhoods, creating very dense conditions in much of the city even as state authorities worked to restructure it around open thoroughfares and automobile use.[45]

With a fast-growing population, a burgeoning industrial sector, and an exploding number of private automobiles, by the mid-1960s Tehran's troubles with air pollution were accelerating at a rapid pace. The causes of declining air quality were not entirely anthropogenic, however, as the city's natural topography worked to intensify the issue. Writing in 1966, Jahānshāh Sāleh, basing his conclusions on ten years of observation, reported that the mountains prevented the easy movement of air through Tehran, enabling noxious gases and particulates to accumulate above the city.[46] As a group of hired American consultants moreover noted in 1967, with a large number of industrial facilities having been established in the city's southwestern reaches, the region's westerly prevailing winds often blew their emissions over the urban center and trapped them against the slopes of the Alborz.[47] Sāleh and the American experts were far from alone in connecting Tehran's poor air quality to its natural setting, and in the late Pahlavi period, sustained (if largely disorganized) attention was paid to the subject by Iranian researchers. With little support and scattered across a number of universities, government offices, and the research arm of NIOC, these scientists and engineers often worked alone or in small groups, piggybacking on existing organizations like the Iranian Oil Institute and its journal to share their findings.[48] Many were self-taught or had only limited training in environmental topics, but their work nonetheless illuminated the character of Iran's urban air quality issues. Sponsoring and disseminating much of their research were institutions like NIOC and the Ministry of Petroleum, and the equipment needed to monitor air quality was often owned and operated by such organizations. Indeed, as early as 1948, NIOC had established temporary observation points around Tehran, and by the 1970s, those short-lived systems had been replaced by a larger network of permanent stations operated by the company, government ministries, and universities.[49]

It was with those sources of data that one NIOC engineer found in 1972 that Tehran's air quality was often at its worst in the early morning hours.[50] Writing in the same issue of the Iranian Oil Institute's monthly journal, ʿAlirezā Moshref Razavi, a meteorological engineer and weather forecaster for the Iranian government, used such observations to explain one of the most important natural causes of Tehran's intensifying problem with air pollution: atmospheric inversions.[51] Linked to the city's semi-arid climate and mountainous topography, inversions reflected the powerful environmental circumstances that shaped the lives of Tehran's residents. A layer of air that increases in temperature with altitude, a reversal of the norm, atmospheric inversions produce stable volumes of air that are devoid of convective currents. Without such currents carrying them upward and dispersing them across a wider area, pollutants like smoke and smog can remain trapped near the ground and over time accumulate. Tehran's climate and terrain combined to make it particularly vulnerable to the phenomenon, and between 1967 and 1971 Moshref Razavi found that inversions formed on approximately 70 percent of days. Most were radiational inversions, appearing when the ground quickly radiated away the day's warmth on cool and clear nights. Though many dissipated in the heat of the next day's sun, in the right circumstances such conditions could persist for days or weeks, a possibility made more likely in Tehran by the wind-blocking effect of the Alborz.[52] As contemporary observers like Jahānshāh Sāleh thus knew, the city's growing struggle with poor air quality was not only a product of population growth, the rising use of automobiles, or even the strengthening of industry; it was instead the result of a complicated and strongly hybrid causal web, one that linked together human activity and the uncontrollable forces of topography and nature.

With atmospheric inversions and mountain ranges an immutable fact of life in Tehran, experts like Moshref Razavi argued that a rapid reduction in emissions was the only way to avoid repeating the deadly experiences of the world's industrialized cities.[53] By the 1970s, the environmental conditions of industrialized societies were being treated as ominous specters by many Iranian officials, with one group warning that Tehran risked becoming like the "large cities" of Los Angeles, Chicago, and London, places where air pollution had "put the health of residents . . . in danger." The metropolises of the world's wealthy nations—with their bustling density, busy airports, long ribbons of highway, and prosperous industries—were in this telling refigured into tragic specters of modernity gone wrong. "There is no doubt that in our country a rise in the standard of living of people and the industrialization of the country that is the natural goal of the White Revolution is the wish of all patriotic people," one report stated, "but regard must be given by the authorities . . . because right now we see that the great country of America is captive to the problem of pollution."[54] Or, as one advocate from the Health Society of Tehran bluntly stated in the mid-1960s, oil was "the type of fuel destroying the human race."[55] But such commentators never advocated for a retreat from industrialization, arguing instead for an even tighter embrace of it. "Anywhere you bring technology, humanity has also rushed to bring . . . new problems for the environment," the head of the Oil Society of Iran wrote, before adding that "confronting industrial pollution is not possible except through . . . the latest scientific and technological innovations."[56]

Such comments reflected a widespread belief among Iranian officials in the possibility of a technological fix for declining air quality. Indeed, interventions like exhaust scrubbers for factories and catalytic converters for automobiles had been pursued as soon as air pollution had become a subject of public concern in Iran.[57] But for many, no potential solution was believed to be as effective as widespread natural gas utilization. In 1966, Jahānshāh Sāleh, by that point the president of the University of

Tehran, expressed a desire to have gas replace the industrial use of fuel oil, writing that "in this country that is everywhere full of gas, why do our factories and workshops consume fuel oil and on what principle have they not brought gas to Tehran?"[58] Five years later, in the summer of 1971, Sa'id Shaybāni, head of petrochemical research at NIOC, placed natural gas at the center of the company's efforts to fight air pollution. In a speech reproduced in both the company magazine and as a standalone publication, Shaybāni became the first to systematically address the environmental benefits of gas use.[59] He described Earth's biosphere as a fragile web, one that had been "ruptured" by toxic pollutants that threatened the "body and soul of humanity." Focusing on what he saw as the tight connection between rising energy consumption and declining air quality, Shaybāni argued that natural gas utilization could break the link between industrialization and environmental degradation, thereby enabling economic development without ecological cost. His position reflected an understanding of air pollution as something rooted in tangible social and medical maladies rather than abstract global trends, for he also noted and simultaneously dismissed the significance of the world's fossil fuel-driven rise in atmospheric carbon dioxide levels. In Shaybāni's view, natural gas, by producing much less of the visible smoke and smog that were his primary concern, was thus "one of the cleanest and most desirable" of fuels, a solution that required an extensive program of infrastructure development but was nonetheless "the simplest way of preventing air pollution."[60]

The enthusiasm that Shaybāni and Sāleh expressed for using gas as a tool of environmental preservation found support within NIGC. Indeed, Mohsen Shirāzi, head of the IGAT-1 program, argued in 1974 that the experiences of industrialized societies had shown that the best way of reducing pollution was natural gas utilization.[61] None of the three men suggested that Iran's policies of industrializing development be curtailed, not even in the face of a rapidly worsening urban air quality crisis. Their position was not based on abstract commitments, however, but on the specific availability of large natural gas reserves in the country. NIGC's own experiments backed their convictions. Using the average energy requirements of a hundred homes in Tehran for their tests, NIGC researchers had shown that the use of coal would produce some 10,280 kilograms of sulfur trioxide, 2,320 kilograms of unburned hydrocarbons, 7,720 kilograms of carbon monoxide, and 960 kilograms of "cancerous and dangerous" nitrogen oxides. Oil was generally cleaner, producing 5,670 kilograms of sulfur trioxide, ninety kilograms of unburned hydrocarbons, 530 kilograms of carbon monoxide, and 2,500 kilograms of nitrogen oxides. Natural gas was by far the least polluting, producing only 4.0 kilograms of sulfur trioxide, no unburned hydrocarbons or carbon monoxide, and 680 kilograms of nitrogen oxides.[62] Pahlavi officials' understandings of environmental harm and protection were in this way made part of Iran's emerging sociotechnical imaginary of gas, a resource that seemingly enabled Iran's urban residents to consume vast quantities of cheap energy without also demanding that they live amidst dense clouds of smoke and soot.

In recent years, scholars have noted that sociotechnical imaginaries are often also spatial imaginaries, with technology shaping how place and space are understood.[63] Indeed, Iran's industrializing policies had made the country's cities into sites of smoke, haze, and sickness. But it was a transformation that natural gas infrastructures were seemingly poised to undo. Iranians' memories of cities like Tehran as having once been places of "pleasant and attractive weather" promoted the adoption of natural gas fuel in ways that went above and beyond its potential as a source of energy and economic benefit. In this way, gas and environmental health were made core parts of the imagined futures that motivated Pahlavi officials. The monarchy's turn toward industrializing developmentalism as a source of political legitimation guided their work, but so did notions about the environmental and aesthetic qualities of the urban spaces they wished to create. In their estimations, Tehran would forever be a hybrid environment, ruled as much by nature as by the shah, and serving both masters meant utilizing natural gas, seemingly the only fuel source capable of doing so.

The Smoke of Asphalt and Brick

That many of the solutions proposed for addressing Iran's air quality challenges, including natural gas utilization, were top-down in nature reflected the persistence of approaches to environmental protection that largely prioritized elite expertise and the state.[64] Despite widespread beliefs that environmentalism and industrial interests are necessarily at odds, such adversarial relationships are often contingent upon particular political systems or historical circumstances.[65] In Iran, as elsewhere, environmentalist aims were at times integrated into existing political and economic systems, transforming otherwise unrelated actors and organizations into conduits of environmental policy. Indeed, with natural gas positioned as a key means of mitigating air pollution, in the 1970s, NIGC was functionally put in charge of the country's efforts to improve urban air quality.[66] For that reason, in places like Tehran and Shiraz, where IGAT-1 made large quantities of gas available, the company's efforts to build distribution networks were shaped by both economic and environmental concerns. In the eyes of NIGC officials working in those cities, factories and other industrial facilities were considered to be far better candidates for gas service than residential neighborhoods. It was a view that was rooted not just in the comparative efficiency of providing service for a few large consumers rather than many small ones, but also in the fact that industrial emissions were known to be a significant contributor to urban air pollution.[67] Thus, while Iran's Department of Environment supported policies that promoted natural gas utilization, it had little actual hand in either its planning or execution.[68] Much of that fell to NIGC officials and the local authorities and factory owners with whom they dealt.

There was often a great deal of slippage between local and national perspectives in Iranian discussions of air quality. No locale received more attention than Tehran, a product of the crisis proportions that air pollution had reached in the city, the elite nature of Iran's discourse on the issue, and the city's status as the country's demographic, administrative, cultural, and economic center of gravity. As officials like Shaybāni had made clear, the use of natural gas to mitigate smoke and smog was in part a response to the challenges that the city's topography and climate presented. It was a choice that was moreover rooted in a previously failed effort to reduce emissions from the city's asphalt producers. Invoking the municipality's duty to "curb factories, workshops . . . and all businesses and trades that create noise and disturbances or produce smoke and/or putridity (*'ofunat*)," in the early 1970s Tehran's asphalt companies had been ordered to install emissions scrubbers on their "unsuitable and unprofessional smokestacks."[69] The importance of the producers to the capital region's economy complicated matters, however, even after the shah, witnessing their heavy emissions with his own eyes, had ordered the companies to move. When Tehran's mayor warned that abruptly shutting the plants would "stop asphalt work in and around the city as well as . . . the 100,000-person [Āryāmehr] stadium," factory owners were offered a short grace period to come into compliance.[70] The owners agreed, though they requested a longer delay as none of the necessary equipment was produced in Iran, promising to voluntarily "stop their factories" should the revised deadline be missed.[71] As quickly became apparent, however, the use of exhaust scrubbers at facilities like Tehran's asphalt works was largely ineffective in reducing the amount of visible pollution in the city's skies, an outcome that pushed Pahlavi officials further toward prioritizing the use of natural gas by the city's industrial facilities.

But while Tehran grabbed the lion's share of attention, in the early 1970s Shiraz was also facing challenges from declining air quality. With a comparatively long history of natural gas utilization by the Shiraz fertilizer plant and other nearby facilities, expanding service was in some ways more straightforward than in Tehran. Indeed, it was exactly that history that had enabled Mohsen Qahremāni's private attempt to transition to gas fuel. Responding to Muhammad Reza Shah's displeasure with the area's polluted skies and his subsequent order that Shiraz's brick kilns be converted to gas, in the summer of 1972 the provincial governor began a drive to have all the city's factories make the switch to the new fuel. Qahremāni's earlier failed effort had shown, however, that the conversion process would not be easy, as even the expensive import of expertise and equipment had not prevented catastrophic breakdown and financial loss. But rather than regarding his attempt as a waste, Qahremāni instead offered his story in support of the provincial government's initiative, suggesting that other factory owners learn from his experience.[72] Many of his counterparts—while broadly supportive of the effort to move to gas, with some lamenting that Shiraz was "dense with the smoke of crude oil"—responded by demanding enough government support to avoid the trouble that had befallen Qahremāni.[73] In response, NIGC quickly hired and dispatched an "expert" from Pakistan to evaluate the laying of gas lines, and his "promising" evaluation of the situation led to a request for more specialists to be sent.[74] But at a late July meeting between the mayor of Shiraz and the "majority of the owners of the brickmaking furnaces of Shiraz," the offer of expert analysis was deemed insufficient.[75] Worried about the financial implications of even a successful program to convert their factories, the owners demanded financial support in the form of fee waivers and grants of equipment.[76] While refusing to offer any monetary incentives, NIGC did promise additional expert support, a pledge that amounted to little even as authorities continued to order factories to switch to gas through late 1972 and into 1973.[77]

With the completion of the IGAT-1, Tehran had seemingly become as conducive to industrial natural gas utilization as Shiraz, and early in 1973, a few weeks after the city's asphalt producers had been ordered to curb their emissions, a number of spool-makers were instructed to coordinate with NIGC and begin the process of converting their facilities.[78] As in Shiraz, many of Tehran's industrialists joined the process of mitigating air pollution in an effort to influence its contours and outcomes. In November of that year, factory owners and the city government agreed to the formation of an "expert committee composed of representatives from the health ministries, social affairs and work, the economy, the steel organization, and the capital municipality" to discuss practicable steps in fighting air pollution.[79] Nonetheless, with the shah once again taking a personal interest in the matter, municipal authorities continued to press owners to take quick action, imploring NIGC to prioritize service to industrial plants that sat near existing gas lines.[80] Indeed, though other technical interventions like emissions filters continued to be relevant, they were also increasingly understood as insufficient to the task of improving the city's air quality. In December 1973, Tehran's mayor, for example, proposed yearly inspections of the city's asphalt works in order to build "confidence" in their improving environmental standards. At the same time, he also declared that "in addition to the installation of exhaust filters, at the first possible opportunity they [the asphalt plants] must obtain a gas connection."[81]

With natural gas utilization central to the fight against air pollution, the success of that effort hinged on NIGC's ability to quickly build distribution networks. By the mid-1970s, company officials had gained considerable experience and confidence, at times both overruling their hired consultants and ignoring local authorities who balked at the idea of tearing up streets to lay pipe.[82] With IGAT-1's branch line to Tehran running through the city's industrialized southwest, the area became a natural locus for the kind of fuel conversion programs that officials like Shaybāni called for.[83] For that reason, by the spring of 1974, NIGC was well into the process of pipelaying there, and by the fall the area's "large factories" and some of Tehran's most heavily polluting brick, plaster, and lime kilns were all being prioritized for gas service.[84] Brickmakers in particular drew critical attention, with one high-ranking NIGC official describing the "column of smoke possibly kilometers long on Tehran's southern horizon" that was produced by their indiscriminate use of tires, rubber rings, old plastic, wood, and "anything that burned" as fuel.[85] Their switch to gas thus came under the "direct supervision of NIGC," and by the end of 1974 the company was in the process of laying gas lines to twenty-two brick kilns while another nine had signed agreements to manage their own conversion.[86] By that point, Tehran's factory owners, wary of potential disruptions to their businesses, were largely willing, though not always eager, participants in the effort to protect the city's air quality. In many ways, they were right to be worried, for the conversion process was proving to be far from straightforward, and it took the "studies and efforts of . . . specialists" to resolve the "multiple safety and technical issues" that cropped up. With many owners souring on the project because of mounting costs and technical difficulties, NIGC eventually agreed to a program of financial assistance in order to push the project forward.[87]

Further south, in Shiraz, scant progress was made until early in 1974 when a report highlighting the city's continued problems with air pollution prompted the shah to again order that "all furnaces across the country that consume fuel oil . . . be converted to gas."[88] Flowing through the Ministry of the Interior, the Fārs provincial government, and then to the regional branch of NIGC, the shah's directive spurred a flurry of new activity. Eager to demonstrate the viability of gas utilization, Mohsen Qahremāni again volunteered his facilities for conversion, and a group of NIGC specialists worked through the cold winter months to modify his kilns. With their work complete in March 1974, Qahremāni reported excitedly on the success of the project, writing that he hoped the example would help "liberate the country from air pollution and provide a better natural environment."[89] But while it did spur calls for the remainder of Shiraz's brickmakers to switch to gas, some of Qahremāni's peers continued to drag their feet. As in Tehran, many worried about both the costs of conversion and about NIGC's actual ability to quickly provide reliable gas service.[90] In response, and in the interests of timeliness, NIGC thus accepted a proposal from some owners to import their own equipment. But for the first time the company also threatened to forcibly close any company that did not take action.[91] Delays that were not rooted in the resistance of owners soon cropped up, however, as NIGC found itself unable to fulfill the commitments it had made. Responding to a demand in the summer of 1974 that conversions proceed for the "protection of workers' health and the cleanliness of the natural environment," owners complained that progress was stalled because NIGC had postponed pipelaying work to March 1975.[92] Indeed, NIGC officials had reported that despite the readiness of industrial facilities to receive gas, they could not promise service in less than eight months' time.[93] Some of that delay was rooted in NIGC's failure to account for the actual energy requirements of industrial facilities. In Shiraz, gas service agreements had originally provided for a maximum utilization of 145 cubic meters of gas per hour.[94] In the winter of 1975, however, one owner deemed that amount to be insufficient, asking instead for 350 cubic meters of gas per hour, the minimum needed to maintain his existing production levels.[95] Despite other owners expressing similar discontent, NIGC pushed back, stating that such modifications would not only require "contractual adjustments" but also "renewed plan[ning] for the primary gas network" that would only cause further delays. The dispute continued until May 1976 when it was decided that NIGC would "provisionally increase" delivery volumes to 250 cubic meters per hour; any who desired more could sign new contracts and assume responsibility for converting their facilities.[96] Nonetheless, despite hopes that the compromise would prompt more industrial plants to make the switch to gas, few did so, and gas utilization continued to grow at a slow rate.[97]

Tehran also saw long waits between conversion work and the actual availability of gas, with factory owners and NIGC officials spending many months pointing fingers at each other.[98] In both Tehran and Shiraz, the fundamental issue was that NIGC exercised little control over its own supply of equipment. Any promised start date at best only "approximately corresponded" to when the company expected to take delivery of the European imports upon which it depended.[99] This was not an isolated issue, and throughout the process of converting the two cities' industrial sectors to gas, the promises and deadlines demanded by political and administrative leaders often ran headlong into such barriers. Despite the powerful nationalist claims that had become embedded in natural gas and its technologies, Iran was simply not self-sufficient in meeting the industry's needs. This continued to be true into the spring of 1975, when both the shah and Iran's prime minister once again commanded the "acceleration" of conversion work and an investigation into its slow pace.[100] While the inquiry proved to be more performative than substantive, the heightened attention did result in greater financial support for factory owners.[101] All that again amounted to little, however, and in August 1976 it was reported that while many of Tehran's factory owners had complied with the order to convert their facilities, the "lack of a primary [gas] network" meant that gas was still unavailable. Still, the municipal governments of Tehran and Shiraz were not deterred, and they instructed the factory owners to maintain their readiness in anticipation of service.[102] As of March 1978, most were still waiting.[103]

While NIGC played a dominant role in Iran's natural gas sector, the contours of gas utilization in the country were also shaped by the interests of private actors. It is a history that troubles top-down perspectives of Iranian development that implicitly assume a passive society subject to the whims of the Pahlavi shahs and their technocratic elites.[104] Despite the authoritarian nature of the state, its policies found significant complications as they met the interests and opinions of industrialists around the country. Some factory owners embraced the new fuel with fervor, attempting to convert their furnaces and kilns before any official effort was underway. Others dragged their feet, worried about costs, disruptions in productivity, and the subsequent viability of their businesses. Rarely rising to the level of outright refusal, their reluctance influenced not only the financial terms of gas conversion, but also its technical ones. NIGC's policies were a central expression of Iran's environmentalist impulses in the 1970s, and as such the interests, concerns, and demands of factory owners were as well. The independent pursuit of gas by people like Mohsen Qahremāni was reflective of an interest in environmental protection that existed in relationship with but was nonetheless distinct from official concerns, and one that often pushed ahead of what governmental institutions could provide. In this way, the experience of fuel conversion among the industrialists of Tehran and Shiraz points to a complex interplay between state institutions and potential consumers of natural gas, one that reflected not only commercial and economic concerns, but environmental ones as well.

Three Pieces of Filter Paper

Though industrial air pollution received the lion's share of attention during the 1960s and 1970s, the rapid growth of automobile ownership in Iran prompted similar environmental concerns. Many modernizing Iranian officials saw motor vehicles as an indispensable feature of modern life, but one that threatened the health and beauty of the country. In the eyes of some, however, as was the case with industry, gas seemingly promised a way out of that trap. In early 1975, while reporting that automobiles released some 1,000 tons of "damaging materials" into Tehran's air each day, NIGC's Mohsen Shirāzi proposed a long-term goal of using Iran's "abundant" and "inexpensive" endowment of natural gas to mitigate those emissions. In the meantime, however, he also suggested mixing liquid gas into the diesel that powered Tehran's many city buses, a step he argued would be a straightforward way to reduce their emissions by half.[105]

In actuality, Shirāzi's proposal glossed over what had been a contentious episode in the history of the United Bus Company of Tehran. Founded in the 1950s as a system of public transit serving Tehran and its surrounding areas, in the mid-1960s the company and NIGC had begun experimenting with the use of liquid gas and "torque toppers." Used to mix diesel and liquid gas, officials had hoped that the devices could "curb [the] smoke" of Tehran's public buses and better their ability navigate the city's steep terrain through increased horsepower.[106] At the time, however, modifying the buses was expected to be a difficult and expensive task, and there were significant questions surrounding the reliability of the proposed fuel system. It was thus only over the objections of both Tehran's municipal government and the bus company itself that the experiment went forward under the supervision of NIGC. In an effort to gather representative data from across Tehran, company officials chose the high and steep terrain of the northern Nārmak neighborhood for their study as well as the hot and dusty streets of Nāziābād in the south.[107] While the numerous "technical complications" that arose once the buses were in service proved that concerns about the reliability had been well founded, much more worrying for many observers was the issue of safety. Husayn Razm-Ārā, head of the bus company, described one incident in which a modified bus had scraped across railroad tracks and been damaged, spilling liquid gas onto the street. Had the bus been stopped, Razm-Ārā fretted, something as small as a "spark from a purveyor of [roasted] corn" could have set off a deadly explosion in the middle of Tehran. With the debate raging, the shah undertook an "inspection" of the experimental program, ultimately deciding that the existing plan be scrapped in favor of a much more measured one that only involved a few buses in Tehran's steepest northern neighborhoods.[108]

While that early trial with modified city buses was largely unsuccessful, it did pave the way for similar efforts in the 1970s. Some of them were suggested by private firms, as was the case in 1975 when Jonub Gāz, a small distributor of liquid gas in Shiraz, tried and failed to convince the provincial governor to award it a contract to convert the city's taxis to use gas fuel.[109] A year later, a Dutch firm made a likewise unsuccessful pitch to convert Shiraz's buses and taxis, arguing that it could both increase the performance and lower the emissions of the city's system of public transit.[110] But the largest attempts were again public, particularly a 1974 program to move all of Tehran's municipal buses to the exclusive use of natural gas.[111] Set in motion by royal order and building upon the torque topper experiment of the previous decade, the project was a collaborative effort between NIGC, the United Bus Company, Sherkat-e Butān (a distributor of liquid gas), and several universities.[112] Though inspired by earlier programs in places like Chicago, by the mid-1970s there was no city in the world where any such system was in "practical use."[113] There were thus few established standards or best practices to draw upon, a situation that led Taqi Mosaddeqi, NIGC's chief executive, to express frustration that momentum from the company's earlier experiments with torque toppers had been allowed to dissipate.[114] Complicating things even more was the fact that there was little consensus on either the best type of fuel to use, the expected workload of the buses, or the intended size of the fleet.[115] Some officials worried about the price of liquid gas and the need to create a new network of fueling stations.[116] Others feared the costs of conversion, a serious concern as there were nearly 2,000 buses operating in the city with plans in place to add a thousand more.[117] The project nonetheless moved ahead, with initial trials taking place in May 1974. More strenuous testing took place three years later, when a modified Mercedes-Benz bus was used on one of Tehran's existing routes.[118] Powered by butane and accommodating forty-four seated passengers, for four months the bus wound its way between Tehran's hilly northern neighborhoods of Seyyed Khandān and Darband.[119] Results were promising, with the bus exhibiting good performance and producing relatively low amounts of "smoke." But there were significant downsides as well, particularly the cost of butane and the fact that if the liquid gas system was not tuned properly then it would pollute more than a comparable diesel engine. The tests were nonetheless deemed a success, and a larger pilot program to serve the neighborhood of Tajrish was recommended.[120] At the same time, however, it was also decided that a large quantity of new diesel buses would be purchased for the remainder of Tehran, as a speedy expansion of the city's public transit system was deemed necessary for a "reduction in air pollution" and an "improvement in the state of traffic" in the city."[121]

Even as the project moved ahead, however, serious concerns continued to percolate among municipal officials, with some fretting that liquid gas was dangerous in a city famous for its heavy traffic and "lack of regard for regulations by drivers."[122] Others like Husayn

Razm-Ārā, despite reiterating a commitment to fighting air pollution, balked at the prospect of overhauling Tehran's entire bus fleet, arguing that the company was already having trouble meeting the public's demand for service despite running an unsustainable number of routes.[123] But coming at the end of the 1970s, what ultimately killed the initiative was Iran's revolutionary unrest, the fall of the Pahlavi monarchy, and the onset of the Iran-Iraq War. Despite that outcome, and for all the doubts that some had, the effort to employ liquid gas in Tehran's system of public transit reflected a very real desire among Pahlavi officials to address the city's problems with air quality. As with the conversion of factories and workshops, Tehran's topography played an important role in setting the terms of the problem. Vehicle emissions were subject to the same intensifying effects that had prompted the shah to order the "smoke" of factories to be mitigated. Added was the need for any kind of motor vehicle to labor up the steeply inclined streets of Iran's capital, a task that the introduction of liquid gas promised to make both easier and less polluting. Like Pahlavi officials' contemporaneous efforts to convert industrial facilities to natural gas, the program to use liquid gas in city buses thus expressed a conviction that environmental preservation and the benefits of fossil fuel consumption were not mutually exclusive.

In the 1970s, the belief that expanding systems of public transit were needed to not only help improve air quality but also the "state of traffic" in Tehran reflected the rapid growth of private automobile use in the country. Emerging as a mass consumer item in the early twentieth century in North America, after the Second World War widespread motor vehicle use began to spread to other regions of the world. Often connected to the process of suburbanization and carrying culturally resonant notions of prosperity, industrialized societies like those of Western Europe and Japan began to see rates of automobile ownership rise in the 1950s.[124] While the Global South lagged as a whole, trends there were uneven. In Iran, with its petroleum wealth, developmentalist state, and growing economy, automobile use began to climb rapidly in the 1960s and 1970s. It was a process reinforced by industrializing policies that promoted the domestic manufacture of motor vehicles, and between 1965 and 1975 total production rose in the country from

7,000 units per year to 109,000.[125] Consequently, between 1970 and 1974, the number of private cars in Tehran nearly doubled, going from 347,000 to 660,000, and prompting new worries for those concerned for the city's air quality. As Muhammad 'Ali Mostofizādeh, a petrochemical engineer and head of standards and regulation at NIGC, wrote in early 1976, gasoline and diesel engines had "endangered most of our cities" and "brought forth air that is harmful for the respiration of humanity, animals, and even plants." Using the smog problems of American cities as a warning, he argued that existing programs for industrial natural gas utilization were insufficient for protecting Iran's urban air quality, as motor vehicles threatened to overwhelm any improvements they offered. Thus, Mostofizādeh argued, while technical interventions like catalytic converters could help, the best possible solution would be to make use of gas in motor vehicles. As he wrote,

> [if] 3 pieces of filter paper . . . are placed in front of the exhaust of three different types of vehicles, the first of which would be for an uncontrolled gasoline engine, it would become completely black; the second of which would be for a controlled gasoline engine that has a [catalytic converter] embedded in its exhaust . . . the color [of the filter paper] would be dark gray; and the third one, which is held in front of the exhaust for a gas-burning engine, would remain completely white.

Unlike the liquid gas that had been suggested for Tehran's public buses, however, Mostofizādeh advocated for the use of compressed natural gas. Liquid gas may have had lower upfront costs, but he considered it to be a poor choice because "the amount of production of this gas is not enough in Iran to provide fuel for all vehicles" and "its price . . . is relatively expensive." Compressing natural gas into canisters, on the other hand, could tap into Iran's huge natural gas reserves via the IGAT-1 pipeline system.[126]

But while Mostofizādeh believed compressed natural gas to be the best option for the capital, IGAT-1 did not serve the entire country, making the fuel much less attractive outside of large cities like Tehran, Isfahan, and Shiraz. For places without access to natural gas, he recommended the use of liquid gas, arguing that the continued availability of diesel and gasoline would still allow for intercity transport and offset the inevitable fragmenting of the country's fuel supply.[127] But with his plan,

Mostofizādeh erected an implicit hierarchy that privileged certain urban locales over the rest of Iran, functionally elevating the environmental needs of Tehran over the economic good of other communities. It was a hierarchy that extended to questions of safety as well. In the 1970s, many Iranians worried that installing a canister of compressed natural gas in their automobile was akin to sitting on a bomb. Attempting to disprove that perception, Mostofizādeh presented an argument rooted in engineering schematics, technical data, and vivid anecdotes of gas canisters surviving violent and fiery collisions. But his conclusion that compressed gas was "much safer" than liquid gas was also rooted in the urban fabric of places like Tehran. As Husayn Razm-Ārā had also worried, Mostofizādeh saw the use of liquid gas, heavier than air and liable to pool on the ground and explode, as inherently dangerous in a city known for unruly traffic and frequent accidents.[128] In this way, just as Tehran's topography and explosive growth made gas utilization an important means of fighting air pollution, so too did they influence the forms that such use should take. In Mostofizādeh's view, the streetscapes within which new gas technologies would be used were as important as their technical operation. Liquid gas was more dangerous because it was heavier than air, but also because Tehran's drivers lacked regard for traffic regulations. In other locales, smaller in size and not as overwhelmed by traffic, Mostofizādeh considered liquid gas to be safe, a difference rooted not in any material or technological difference but rather the social conditions that surrounded fuel use.

Mostofizādeh's ideas were built on years of research that NIGC, the Department of Environment, and the University of Tehran had undertaken on the use of compressed natural gas. As part of the state's broader effort to improve urban air quality, researchers from the three institutions had opted to undertake a pilot program to "familiarize public opinion with [compressed] gas fuel." Focusing on Tehran's taxi cabs due to their extensive daily mileage, the project aimed to convert 2,000 automobiles to the use of compressed gas. To support that new fleet, five fueling stations were to be established in locations chosen for topographic and social diversity: two in working-class neighborhoods in the city's southern reaches, one in the west near Mehrābād airport, one in the middle-class suburb of Tehrānpārs to the east, and one in the city's congested center. At each station, gas from the city's expanding distribution network would be compressed into canisters and dispensed to waiting vehicles. Though plans were in place to have the first fueling stations and modified vehicles operational by the summer of 1976, in the end the project did not come to fruition, undone by the realization that converting 2,000 cars to gas was insufficient to make a measurable difference in Tehran's air quality.[129] Nor was Tehran the only city where experiments with compressed natural gas were taking place, as in Shiraz a similar project was under consideration. In October 1975, Manuchehr Piruz, the governor of Fārs, sent a letter to the chief executive of NIGC that praised the company's compressed gas program in Tehran. Expressing confidence that the endeavor would help mitigate air pollution in the capital, he requested that the "good national project" be extended to Shiraz so as to reduce air pollution and protect the monuments and gardens for which the city was becoming internationally known.[130] Once home to revered scholars, mystics, and poets, and near to notable ruins of Iran's ancient empires, Shiraz had by the mid-1970s become a destination for international tourists, a host to numerous cultural festivals, and an important site for Pahlavi nationalist mythmaking.[131] But while many agreed that reducing vehicular emissions would help the city's burgeoning tourism sector, there was much less consensus about whether compressed or liquid gas fuel was the best choice. Initially, in November 1975, it was decided that liquid gas would be used, with one taxi to be modified as a testbed. This approach amounted to little, however, as despite the project's budgetary needs being placed "under the special protection of the Queen of Iran," little money or activity proved to be forthcoming. For that reason, in the summer of 1976, the program was "suspended" despite the "promising" performance of the single converted taxi.[132]

Standing in the way of the Shiraz municipal authority's experiment with liquid gas were NIGC officials and their own effort to modify a thousand taxis in the city to use compressed natural gas.[133] But that initiative foundered too, a victim of unforeseen complexities and substandard equipment. While Shiraz's existing natural gas network could be tapped for supply, the system operated at far too high a pressure for the refueling equipment, necessitating the use of costly and complicated pressure regulators. Even more worrisome was the poor design and excessive weight of the gas fuel system installed in the taxis, a problem of such severity that it ultimately led to the program's cancellation in late 1976.[134] Even the infusion of equipment from Tehran's terminated program failed to revive the project when it was determined in the spring of 1977 that a minimum of 10,000 modified vehicles would be needed for any noticeable improvement in Shiraz's air quality.[135] Thus, in the end, none of the Pahlavi-era efforts to bring gas energy to the country's roads proved successful. But the insistence with which Iranian officials returned again and again to the idea of vehicular gas use expressed a persistent conviction that Iran's urban environments would only be threatened by the burgeoning use of automobiles in the country. But compared to the provisioning of gas for industry, itself far from straightforward, Iranian authorities found it to be much harder to design an effective, safe, and efficient system for using gas in the transportation sector. Compounding the problem was the significant uncertainty surrounding the best form of fuel to use, as both liquid and compressed natural gas were accompanied by a host of both technical and social implications. Nowhere was that more true than in Tehran, the center of Pahlavi-era efforts to combat urban air pollution, and a place where topography acted to both concentrate emissions and influence requirements for vehicular performance. In this way, air quality again acted as a site where gas utilization connected Iranians' sociotechnical imaginaries of gas energy to their spatial imaginaries of Iranian cities. Indeed, as Pahlavi officials spent considerable effort analyzing the movement of gas fuel through busy city streets, the push of topography's intensification of air pollution and the pull of unruly traffic epitomized the hybridity of Iran's urban spaces. Industrializing development had transformed Iranian cities into polluted places, and natural phenomena worked to make it worse. Confronting that dilemma required far-reaching changes to the functioning of Iranian society. For many, the answer was natural gas, a resource that promised a future of accelerating energy consumption, but more than that, by reducing emissions at their source, one that was environmentally healthy as well.

***

Addressing a symposium on air pollution in 1972, Manuchehr Eqbāl, chairman of NIOC, drew an explicit connection between technological advancement and environmental harm. Speaking to a group of academics, ministerial officials, and petroleum industry officials, he warned that

> humanity, who today with the aid of its own knowledge and amazing technology has set foot beyond Earth (*koreh-ye khāk*) and stepped into the sweep of the cosmos, now faces an unfavorable environment that was unsought and unwittingly gained via the same excellent technology.[136]

Eqbāl was not alone in this view in 1970s Iran. Three years later, Shāhpur ʿAbdul Rezā Pahlavi, half brother to the shah and noted proponent of wildlife conservation, remarked to the assembled members of an international petroleum trade group that the "protection of our natural environment and the quality of life of present and future generations" rested upon their willingness to embrace new technologies. Humanity and nature were intertwined, Pahlavi argued, and while industrialization threatened to upset the world's ecological balance, it could also "supply the technology of environmental protection" that was becoming increasingly important in Iran and around the world.[137]

That these two men, elite figures in a self-consciously industrializing state, would so clearly link technology to both environmental loss and preservation was an expression of the environmentalist impulses that were threaded through the developmentalist policies of late Pahlavi Iran. With environmental degradation threatening the fundamental validity of the country's modernizing project in the eyes of many, Iranians faced a choice: prosperous industrialized life or clean air. Intensified by inescapable topographic and climatic factors, the smoke and haze that lay suspended above Iran's cities were both unsightly stains and frightening specters of an unwanted future. But air pollution was also seen as a challenge that could be solved through the application of new technologies. Natural gas, particularly after the completion of the IGAT-1 pipeline, seemingly offered a way to continue industrializing while also avoiding the worst environmental effects of that choice. It was with gas that the "ghoul of pollution" that was banging on the "gates of . . . health," as one NIGC official put it in late 1974, could be vanquished.[138] For such figures, it seemed easier to double down on the fossil fuel use that had brought cities like Tehran to the brink of crisis than to imagine a modern society that consumed less energy or allowed environmental considerations to constrain its growth. The confidence of people like Sa'id Shaybāni that natural gas energy was the "simplest" solution to air pollution, despite the huge infrastructural undertaking its use would require, pointed to the hope that Iranian officials placed in technological solutions. Such convictions demonstrate both the necessity of understanding NIGC as more than a simple supplier of gas fuel and the benefit of broadening our notions of what constitutes environmental action to include people and organizations whose interests at first blush seem antithetical. What many Iranians imagined and hoped for was a future of high energy consumption and the economic growth it would enable. Environmental degradation undermined that vision of prosperity, thereby thrusting Iran's national gas company into the position of providing not just cheap energy but also clean air. NIGC's activities, despite centering on the provisioning of fossil carbon to consumers, thus cannot be fully understood through perspectives that assume ecological disregard. Rather, development and environmentalism were interwoven in Pahlavi Iran, linked by natural gas.

In late Pahlavi Iran, commentators spoke frequently of the country's cities being transformed from places of "pleasant and attractive weather" into ones fouled by the airborne remnants of fuel combustion. The close connections that were drawn between air quality and natural gas utilization were built upon implicit understandings of cities as hybrid spaces. In this way, the desire to return places like Tehran to their former natural beauty, or at least forestall their continued deterioration, embodied an environmentalism that was rooted in a marriage of sociotechnical and spatial imaginaries. In contrast to postrevolutionary perspectives that have often cast Pahlavi officials as being a monolithic bloc of *gharbzadeh* compradors—a term popularized by the revolutionary writer Jalal Al-e Ahmad to refer to a loss of Iranian cultural identity and an unthinking adoption of Western lifestyles, norms, and beliefs—many who worked in the late monarchical state were not wholly blind to the faults of the North Atlantic world.[139] While those within Iran's increasingly powerful and populist revolutionary movement of the 1970s largely articulated critiques of modernity and development through perspectives that emphasized social and economic justice rather than environmental harm, for their part, influential Pahlavi officials like Manuchehr Eqbāl, Sa'id Shaybāni, and Mohsen Shirāzi advocated for the use of gas fuel not in imitation of Euro-American modernity, but in an intentional attempt to improve upon it. They looked at what had befallen the polluted cities of the industrialized world and made specific, informed attempts to build a society that would avoid such a fate. Such modernizing ambitions expressed not only developmentalist aims, but also, implicitly, anticolonial ones rooted in the use of domestic resources to create a society that surpassed those that had long dominated Iran. In an era when Iranian leaders sought to cast their country as neither East nor West but as a "Great Civilization" (*tamaddon-e bozorg*) positioned to lead both, natural gas was to be an agent of both transformation and preservation, a technical solution for reconciling tensions inherent in the country's developmental policies.[140] For many Pahlavi officials, natural gas was the key to finding a better balance between industrialization and the natural world, a way to both enjoy the benefits of intensive energy consumption and avoid the dark clouds of soot and smoke that had swept over much of the world.[141]

Five
National Capillaries

In the summer of 1977, the residents of Shiraz's Shishehgari neighborhood wrote to a long list of government officials demanding to be able to make use of natural gas energy. Lodging angry protests and claiming to represent "over a thousand households" that had been "for years anticipating gas," the letter's seventy-one signatories decried the lack of service for their homes, pointing out that adjacent areas, most notably an industrial district to the south, were already making use of the new fuel. Coming in the late 1970s, their grievances reflected the growing visibility of natural gas use in Iran's urban areas and the increasingly strong connection that Iranians drew between the provision of infrastructural services and their status as citizens. As Shishehgari's residents argued, though the shah himself had promised that "all people [would] benefit from the consumption of gas," the lack of service in their neighborhood meant that all its "hard workers" had been unjustly "cast aside" by the National Iranian Gas Company.[1] Nor were they alone in their complaints. All along the IGAT-1 corridor, Iranians had watched the pipeline become operational, had witnessed it jumpstart gas utilization, and had heard the triumphant words of national leaders. Many of them also saw that they were being left out. Indeed, for all shah's promises of universal gas service, the construction of distribution networks was an uneven affair, built piece by piece, neighborhood by neighborhood, with some areas receiving service years or decades before others. Such irregularity, and the disregard for some that it seemingly represented, would, in the 1970s and after, come to strongly shape the politics of natural gas energy in Iran.

Much of the unevenness of Iran's gas distribution system was rooted in the commitment of Pahlavi officials to particular notions of efficiency. As part of its work on the IGAT-1 project, NIGC had been charged with evaluating Iran's potential gas markets and overseeing the construction of networks to serve them. Much more than had been true for the IMEG-designed main pipeline, the basic technical operation and geographic spread of Iran's gas distribution systems reflected the choices and concerns of NIGC personnel. For many of them, working within a broader developmentalist drive aimed at rapid industrialization, the quick provision of large volumes of natural gas fuel was paramount. Industrial facilities were thus prioritized in many regions, their enormous energy demand (and polluting emissions) offering a rapid and significant return for the government's investment. Similar perceptions created an order of precedence out of Iran's cityscapes, the physical characteristics of urban districts becoming supposedly objective and neutral determinants that, in practice, favored the comparatively wealthy. Gas lines were easiest to install where streets were wide and buildings were made of brick, concrete, and steel, attributes associated with new postwar developments that were mainly inhabited by Iran's educated middle and upper classes. Older quarters, defined by maze-like alleys and densely packed residences, were deemed less suitable for distribution networks, leaving their often lower- and working-class households to look on as affluent areas received gas service sooner. At the end of the line were Iran's rural dwellers, deliberately left waiting as cities were given priority.

Over the past decade, scholars have paid increasing attention to infrastructures as sites where the contours of citizenship and national belonging are negotiated.[2] As they have found, political subjectivities are actively produced in daily interactions with infrastructures, giving rise to notions of "infrastructural citizenship" that reflect how people understand their relationships with the state, make claims upon it, and gain visibility for themselves and their communities.[3] In a similar fashion, in the charged atmosphere of 1970s Iran, natural gas service became a flashpoint for debates about whether or not Iranians actually benefited from the monarchy's prioritization of industrialization and economic growth. For those like the residents of Shishehgari, the desire for gas service was not simply an issue of convenience or cost or even environmental cleanliness; it was rather a question of what was owed to them as citizens. Across the country, in regions both urban and rural, Iranians demanded to join this new era of industrialized convenience and high energy consumption. The exclusion of many from a high profile benefit of the country's developmental programs became symbolic of what they perceived as a broader Pahlavi neglect. Late in the decade, such frustrated expectations were folded into the wave of revolutionary politics that was sweeping the country. To a populist movement accusing the monarchy of endemic corruption and favoring the interests of foreign powers over that of Iranians, the fact that many people could not use gas while large quantities were simultaneously being exported to the Soviet Union was proof that Pahlavi state did not have their interests at heart. Iran's gas resources should be for the use of its citizens, they insisted, a demand for cheap energy that echoed and renewed the deep vein of resource nationalism that ran through the country's political culture. But rather than target foreign governments or international oil firms, it was an impulse now aimed squarely at Iran's own government. Still, whatever fierce opposition to the Pahlavi state and its policies there may have been, evident in this revolutionary charge was the fact that the choice to embrace natural gas utilization, and the industrialized prosperity it augured, was not in question. The problem, as people like the residents of Shishehgari argued, was that Iran's gas infrastructures were expanding too slowly, and that whatever the country's total rate of gas consumption might have been and however fast it was rising, too few Iranians reaped its benefits in their daily lives.

After 1979, under the Islamic Republic, that perspective became ascendant, and the new government put considerable effort into bringing gas service to towns and neighborhoods that had been previously left out. In place of the Pahlavi era's rhetorical emphasis on monumental structures like pipelines and refineries, state officials now prominently celebrated the number of households connected to the network, championed the provisioning of gas fuel to rural areas, and for a time ended exports to the Soviet Union. In doing so, they reinscribed the legitimating power of natural gas even while they drew sharp contrasts with their Pahlavi predecessors. Indeed, despite a shift toward highlighting gas service to small consumers, the gas policies of the Islamic Republic were remarkably similar to those of the prerevolutionary monarchy. Even the discursive construction of gas as a tool of political legitimation remained congruent in important ways. Building seamlessly upon foundations laid in the previous two decades, officials of the Islamic Republic maintained a preoccupation with technical sophistication, scale, the conquest of nature, the sovereign control of natural resources, and, most fundamentally, industrializing development and high energy consumption. Again promised was a future of cheap and bountiful energy, one that asserted the nation's independence, enabled prosperous lifeways, and included more Iranians than ever before.

A Benefit for All

When the residents of Shishehgari complained in the late 1970s that their homes lacked the gas service that a neighboring industrial district already enjoyed, they were adding a new chapter to a debate that was by that point more than two decades old. Predating the IGAT-1 project in its entirety, the question of natural gas in Shiraz went back to the mid-1950s and the construction of the Shiraz Chemical Fertilizer Plant. It was at that time that H.E. Mehdi Farrokh, the Fārs provincial governor, embarked on a sustained campaign to bring natural gas energy to city residents. Home to some 150,000 people, in the 1950s, Shiraz was beginning to change as new neighborhoods were built, new industries were established, and Pahlavi national mythmaking elevated the city as a hub of Persianate culture.[4] Farrokh sought to foster this transformation, placing the provisioning of gas to residents squarely within his developmental priorities for the city. While the governor failed in the summer of 1955 to gain the support of the U.S. Point Four mission for his ambition, in December of that same year he succeeded in pushing the Plan Organization to ask the National Iranian Oil Company to explore the idea as being in the "interest of the majority of the people" of the region.[5] For its part, NIOC was already doing just that, setting aside funds in its 1956 budget to study the possibility of provisioning gas to the city.[6] There was a great deal of enthusiasm among Iranian officials for the suggestion, enough that NIOC, concerned about the "huge investment" such a project would require, was forced to ask the prime minister for help in preventing others from forging ahead.[7] Evaluations were not expected to be complete until early 1957, however, and Farrokh, denouncing the delay, pushed for quicker action, arguing that gas service would accelerate Shiraz's economic growth.[8] The governor had good reason to be anxious, for by that point the city's demand for energy was already outstripping the supplies of coal and charcoal upon which it depended. While oil products could fill the gap, the relative proximity of Shiraz to Iran's primary petroleum producing region made natural gas use an alluring prospect capable of "transforming people's lives" and spurring further industrialization.[9] As would be true in the IGAT-1 program a decade later, industrial and residential gas use were conceptually joined in the minds of many government officials in 1950s-era Shiraz. For them, it made little sense to build a system to provide gas only to industrial facilities when its benefits could be spread much more widely. It was a position that would be oft repeated in the following years, but one that in practice would be realized only in part as Iran's gas infrastructure took shape.

Work on natural gas distribution began in earnest in late 1958 when NIOC committed to supporting the planned chemical fertilizer plant in Marvdasht on the outskirts of Shiraz. To be fed from the Gachsārān oil field via a pipeline more than a hundred miles in length, construction on the facility began in the spring of 1959.[10] By the next year, however, NIOC had begun resisting the possibility of providing service to city residents, arguing that it was "not part of the pipelaying plan" and "not affordable." The company instead sought to exclusively serve "major consumers" like the fertilizer plant, a decision opposed by both provincial and national officials.[11] The prime minister, once having shielded the company from outside pressure, this time quietly pushed NIOC to keep its promises to build a city gas network.[12] The governor of Fārs went further, thundering that the people of Shiraz would accept nothing less than the start of full-scale gas service the moment the pipeline was operational.[13] Nonetheless, when gas began to flow in April 1962 it went only to the chemical plant and a few other nearby facilities.[14] Even the start of full-scale industrial gas service in 1963 did little for the area's residents, as planning work for them had been suspended the previous year on account of insufficient funding.[15] But people like Farrokh were not so easily dissuaded. Writing to the prime minister and praising the Pahlavi state's commitment to "using the country's natural resources for the benefit of [all] classes," he again pressed for financial support for the project from the Plan Organization, proclaiming that the "biggest celebrations" in Shiraz's history would erupt should residents receive gas service.[16] For its part, the Plan Organization refused to help, citing a reduction in its budget and the impending end of the Second Development Plan later that year.[17]

But support for residential gas service in Shiraz was not as unanimous as Farrokh had implied. When first proposed, it had been controversial among the area's civic organizations, with the Agricultural Union of Fārs hailing its ability to bring "prosperity" while the Chamber of Commerce of Bandar Bushehr worried that existing commercial activity would be impeded.[18] The head of the Shiraz Electrical Company opposed it as well, arguing in the summer of 1962 that tearing up roads and alleys to lay new pipe would be costly and disruptive. More important still, in his eyes, was the fact that some two-thirds of city residents were crammed into old, densely packed, and haphazard neighborhoods, potentially dangerous conditions for the use of flammable natural gas. Even wealthier, more spacious districts were not immune from such dangers, he contended, for many households there employed servants that should not be trusted with the "precise technical" task of using gas. Electricity was easier, safer, and already in place, he reasoned, advancing a position that had possibly more than a little self-interest behind it.[19] Nor was the power company head the only one with business interests at stake. In July 1962, the owner of a small liquid gas company offered to bottle natural gas from the Gachsārān-Shiraz pipeline for delivery to residential and commercial customers.[20] Such an approach would avoid the expense of laying gas lines, NIOC noted, but it was also "not practical" as pipeline gas was mostly methane and canister systems predominantly used propane and butane.[21] Thus, while opposition to residential gas service in Shiraz was generally rooted in commercial concerns and at times articulated in insulting statements about the abilities of the city's domestic workers, such framings inadvertently underscored Iranians' perception of gas energy as something modern and sophisticated. It was an understanding that was reinforced by the strong developmentalist impulses that were embedded in Iran's natural gas program, and while doubts (sincere or not) about the ability of ordinary Iranians to make use of gas energy would fade away, questions about which neighborhoods and what kinds of urban spaces were suitable for service would come to hold particular sway in the expansion of Iran's gas infrastructure in the 1960s and 1970s.

Despite the resistance of NIOC and others, Shiraz's local and provincial authorities remained committed to the prospect of piped natural gas energy. Throughout 1962, officials like the mayor of Shiraz pressed to make it a reality in the city, asking that it be included in the country's upcoming Third Development Plan.[22] That persistence paid off, and at an August meeting called to discuss the "demand of the people for the use of gas energy," NIOC declared that while its focus would remain on large industrial consumers, it was willing to reconsider its earlier opposition to residential service. In any case, the inhabitants of Shiraz would not gain access in a uniform way, as it was the opinion of the assembled officials that older neighborhoods—with their narrow, winding alleyways and mud homes—were not suitable candidates for natural gas, exactly as had been argued by the head of Shiraz's electrical company. Service would therefore be restricted to newer areas of the city.[23] Despite shifting its position at the August meeting, NIOC remained unenthusiastic in the following months, pointing not only to the considerable uncertainty surrounding costs but also the fact that the system's overall design remained inchoate.[24] Frustrated, Shiraz's municipal authorities, with little knowledge of the task and less experience, sought to hire their own contractors and forge ahead. Iran's national oil company blocked that move as well, proposing instead to hire a European firm to design and build a small, experimental system that would serve 500 residences.[25] Even with this offer, NIOC continued to drag its feet, and it would not be until 1966, some four years later, that plans for such a test program would be finished, and it would not be until December 1967 that the first residential consumers, some 285 in total, would begin receiving gas. More than two years later, in 1970, that number would stand at only 773.[26]

By the time gas began flowing to residences in Shiraz, the Shiraz Chemical Fertilizer Company in Marvdasht, the original impetus for the transmission of natural gas to the Fārs region, had been in operation for more than four years. The long deferral of residential use and its ultimately restricted implementation were clear expressions of NIOC officials' prioritization of industrial gas service and the limited ability of political authorities on both local and national levels to sway the ostensibly technical decisions they made. While this dynamic would not meaningfully change anywhere in the country in the following years, Shiraz's experience with natural gas was in some ways unique. In contrast to places like Tehran that received gas service as part of the enormous IGAT-1 program and the extensive national planning it embodied, Shiraz's system was built in large part due to the commitment of local authorities to the idea and the willingness of H.E. Mehdi Farrokh, the governor of Fārs, to shepherd the project forward. Shiraz would nonetheless come to be seen as a pioneer of urban natural gas utilization in Iran, and the decisions made there anticipated much of what would take place on a broader scale in the 1970s.[27] Most notable was the use of streetscapes and architecture to determine which neighborhoods would receive gas service and when. In Shiraz, despite rhetoric emphasizing the benefits that natural gas could bring to all Iranians, newer neighborhoods were deemed suitable for service while older districts were not. Rooted primarily in financial considerations and a desire for efficiency, such choices nonetheless also seemingly suggested what Pahlavi officials thought Iran's industrialized future should look like and who could be part of it.

The Spectacle of Modern Empire

From the moment of its conception, the IGAT-1 project had been envisioned as a single system connecting the fossil hydrocarbon pools of Iran's southwest to the cities and towns of the country's central and northern regions. The sociotechnical imaginary that had been developed as part of the Pahlavi state's public discourse on natural gas had moreover established it as a God-given resource that all Iranians had the right to benefit from. As had been true in Shiraz, however, what was built was in the end highly uneven. In the 1970s, distribution networks were constructed for industrial districts long before residential; newer and wealthier city neighborhoods typically enjoyed gas service before older and poorer ones; and rural areas were almost entirely left out. These discrepancies had deep roots, stretching as far back as the National Iranian Gas Company's original planning for IGAT-1. They were further reinforced by technical choices and an enduring commitment by project managers to seeing the project's primary aim as one of accelerating Iran's industrialization and economic growth.

Even before ground was broken on IGAT-1's main pipeline, Iranian officials were already facing difficult and consequential choices about the distribution of gas. First among them, coming in the mid-1960s, was pricing. While export prices were determined in contractual negotiations with the Soviet Union, inside the country they would be set by NIGC. Overseeing the entirety of Iran's domestic petroleum industry, NIOC and NIGC operated as state-owned monopolies, setting prices in accordance with larger policy goals rather than as part of an open market. While aiming to avoid operating at a financial loss, NIGC officials chose to promote the use of gas by setting rates comparable to, or even cheaper than, competing oil products. They further sought to mitigate financial obstacles to gas adoption, using low prices to offset the costs that consumers, particularly industrial facilities, would incur by transitioning to gas. But raw growth in gas utilization was not the only imperative at work, and over the course of the 1960s and 1970s, NIGC would also use gas prices to manage and balance the operations of country's growing gas network. Standardized charges were used to shape demand throughout the day, with officials working to reduce strain on the system during periods of peak usage. Charges were also based on geography, with installation fees for customers rising as distance from the system's primary infrastructure grew. Gas pricing was moreover used to further other policy goals. With Tehran having become the center of gravity for the country's industrial sector, a development that had markedly worsened the city's air quality, government officials used energy charges to push new industrial facilities away from the city. Transport tariffs for natural gas were thus highest in the capital region, while Khuzestān, the center of Iran's petroleum industry, had the lowest. The rest of the country constituted a third pricing region, paying a tariff that sat in between.[28]

NIGC's pricing schemes were a product of natural gas's tight integration with Pahlavi officials' developmentalist ambitions and evidence of how they prioritized those goals. The logics of gas utilization in Iran rested on a desire to substitute the new source of energy for rising domestic oil consumption and thereby free the latter for export. By setting the costs of gas energy to be below that of oil fuels, company officials thus aimed to both spur demand and reduce barriers to its adoption. Such choices had important secondary aims as well, both working to shape the geographic distribution of industry and improve Tehran's urban air quality while also supporting the government's broader industrializing goals. In the eyes of many Pahlavi officials, none of these aspirations were achievable without natural gas, a conviction that points to the co-constitution of Iran's energy infrastructure and the state's modernizing ambitions in the mid-twentieth century. Indeed, in their broad strokes, those ties between national development planning and natural gas were publicly discussed in Iran, often presented as evidence of what the government claimed was the country's increasingly rapid transformation into a prosperous modern society. In late 1971, the audience for such claims became global as part of Muhammad Reza Shah's celebrations of the (ostensible) 2,500th anniversary of the Persian Empire's founding. Culminating in days of festivities near the ruins of Persepolis and attended by hundreds of foreign dignitaries, "the Party," as some derisively came to call the lavish and very expensive celebrations, were aimed at presenting "Iran as a land of stability and prosperity . . . dragged from destitution into modernity by the two Pahlavi monarchs."[29] Marked by the promotion of Iran's pre-Islamic past and Persian cultural heritage—as well as the sweeping of homeless encampments and the hiding of slums—such claims were built most directly on the modernizing policies of Muhammad Reza Shah. Indeed, the months and years preceding the celebration had seen an acceleration of dam, highway, and airport construction as the Pahlavi state used such infrastructures to showcase the country's emergence as a regional power.[30] Iran's expanding gas infrastructure, as an ambitious and sophisticated technological undertaking, fit squarely within that paradigm.

As part of the shah's festivities, officials from NIOC and NIGC offered presentations on Iran's oil and gas sector, four of which were collected in a NIOC publication titled *Naft va Zendegi* ("Petroleum and Life"). In one, NIGC's plans for natural gas distribution were described by a man named Mahmud Peydāyesh. Working as part of Iran's petroleum industry, Peydāyesh's account emphasized urban areas, particularly Tehran, and the technical means of gas transport. As described by Peydāyesh, structuring NIGC's plans were pipelines and their crossing of urban space, an approach that conceptually divided Iran's cities into geographic units based on location and expected customer base. Tehran's residential districts, for example, accounting for 40 percent of the city's total area, were divided into three groups by population density. The first was to the south, between the bazaar and the suburb of Rey, near both Tehran's oil refinery and the terminus for the city's primary thirty-inch natural gas pipeline. The built environment of this area had changed little over the previous century, characterized by the presence of many old buildings and a high population density of 300 to 600 people per hectare. The second was located in the city center, corresponding to areas with a large number of newly constructed three- and four-story buildings and a population density of some 100 to 200 people per hectare. The third region covered Tehran's northern districts, characterized by comparatively wealthy families and large homes. Population density there was only fifty to 100 people per hectare, with an average living space of some 200 square meters per person, far in excess of the average 3.4 square meters per person near the bazaar. The city's outskirts would be served by lines running along major roads to Qum in the south, to Sāveh in the southwest, and to Karaj in the west. A fourth line was possible for the numerous brick kilns in the city's southeast. When Peydāyesh spoke in late 1971, market analyses for Tehran's residential and commercial areas were still ongoing, but driving NIGC's plans for gas distribution was an expectation of rapid population growth in Iran's capital, particularly in southern neighborhoods that were seeing a rapid influx of rural migrants. The company's plans thus called for some 80,000 residential and commercial gas connections to be installed over the following decade—a sizeable figure, but one that would come nowhere near to provisioning gas for a city of millions.[31]

Despite that impending shortfall, clear in Peydāyesh's words was NIGC's prioritization of industry, exactly as had been the case in Shiraz a decade earlier. Unlike the city's residential districts, still under evaluation and not expected to receive gas service for many years, Tehran's industrial sites had been carefully mapped and accounted for, and Peydāyesh's presentation was accompanied by a list of all industrial plants eligible for gas, the volumes they could be expected to use, and maps of their locations. Data was presented on 622 separate facilities, ranging from baking to brickmaking to furniture making and much more. Expected gas consumption climbed from a leatherworker's low of 3,000 cubic meters per year to a high of 99.1 million by the region's primary cement works. Most were expected to consume approximately a million cubic meters per year or less, 6.8 percent would need more than 2.0 million, and only sixteen individual facilities required more than ten million. In total, NIGC expected Tehran's industrial sector to consume nearly 1.0 billion cubic meters of gas per year, or approximately 15 percent of the gas carried by IGAT-1 for use within Iran. As Peydāyesh further reported, a pilot program had already been launched to convert the city's largest consumers of energy, and it was hoped that within five years all of the region's industry would make the transition, particularly several hundred notoriously polluting brick, plaster, and lime kilns.[32] While such hopes would prove to be overly ambitious, as the failure to quickly convert polluting industries would make clear in a few years' time (see chapter 4), they nonetheless reflected the high priority NIGC put on industrial gas consumption. Deemphasizing residential gas service in favor of large industrial users was an approach that reflected the state's broader emphasis on rapid economic growth, one that fit neatly into its use of infrastructure as a politically legitimating spectacle. In this way, Peydāyesh's presentation, and the planning that underpinned it, were explicit expressions of the industrializing developmentalism of the late Pahlavi era, something that was championed as part of the 2,500th anniversary celebrations but would soon come to be seen by many as neglectful of ordinary Iranians.

The basic form of Tehran's urban gas system—major lines serving industrial facilities on the city's outskirts, accompanied by a set of smaller-scale residential and commercial networks within the city limits—was one that would be repeated in other cities along the IGAT-1 corridor. Like Tehran, Isfahan was a rapidly growing city with a significant industrial sector, notably the textile mills that employed some 15 percent of the city's workforce. According to Peydāyesh, NIGC expected the city's factories to eventually take some 264 million cubic meters of gas per year, or approximately one quarter of Tehran's total. While the company's aim was to convert all such facilities to gas within five years, there was no commensurate urgency to provision gas to Isfahan's residential areas. Indeed, when Peydāyesh spoke in late 1971, very little work had been done on the subject, and NIGC's objectives were very modest, expecting that only some 25,000 households would be connected to the gas system, a comparatively small number in a city of more than half a million.[33] Plans for the Alborz Industrial City, founded in 1967 to the southeast of Qazvin, likewise relegated residential gas service to secondary status. Established as a special economic zone in accordance with the "world's most modern standards and principles," natural gas was expected there to provide "clean and inexpensive fuel" that would "prevent the pollution of the air."[34] What was not included in NIGC's plans for the area, however, was the provisioning of gas to the 16,000 residential units that were slated to house the city's workforce. Even Shiraz and its existing city gas network were not expected to see quick growth, as NIGC plans called for the number of household connections there to rise from 1,730 in 1971 to only 12,000 a decade later.[35]

That there would in the end be long delays in the provision of gas service to factories in places like Tehran and Shiraz did not erase the fact that NIGC's planning intentionally de-prioritized the residential districts of those same urban areas. Presented by Peydāyesh as part of the shah's great imperial spectacle, such plans were transmuted from technical accounts into an expression of natural gas's place within Pahlavi developmentalism. Evident was an emphasis on scale that placed Iran's budding natural gas infrastructure in the same company as the massive dams and highways that were already serving as monuments to the shah's modernizing ambitions. Discussions of the system's operational functioning and recitations of technical details like pipeline length and size underlined the state's ability to create sophisticated technological systems on a national scale, evidence of the "quick steps" that Iran was taking under the shah's "wise leadership" to narrow the gap with the world's "advanced countries."[36] But more than that, Peydāyesh made explicit the prioritization of industrial use that underlay NIGC thinking. The sociotechnical imaginary of natural gas that was articulated by organizations like NIGC was built around a claim that the Pahlavi state's developmental policies would enable its benefits to flow to all Iranians. In reality, however, NIGC officials had no intention of quickly supplying gas to most residents; if they were to see any benefits of the country's exploitation of gas resources, it would come indirectly as the company prioritized industrial facilities, seeking to spur economic growth and fight declining urban air quality.

Urban Pressures

Much of the planning presented by Peydāyesh was based on the work of the Iranian Management and Engineering Group, the British consulting company that had served as IGAT-1's primary designer. As a complete system built from the ground up, considerations of demand and distribution had been an integral part of IGAT-1's original planning and requirements. But many of the specifics, including the technical and spatial characteristics of distribution networks, were left incomplete while the challenges of gas gathering, purification, and long-distance transport were addressed. It was only after the main pipeline's completion in October 1970 that questions of distribution were pushed to the fore, and it was then that project officials turned their attention to the networks of small- and medium-diameter pipes that would carry natural gas to consumers. It was also a moment when Iranian officials took a more direct hand in shaping the contours of gas utilization, for despite IMEG's centrality to the IGAT-1 program as a whole, it was NIGC planners and engineers who took control of the issue of distribution. For that reason, while some of IMEG's early work would survive, it would be NIGC's decisions—rooted in its own technical, economic, and political priorities—that would most strongly shape Iran's urban gas networks.

With its thriving industrial sector, sprawling residential districts, and political importance, Tehran quickly became the primary focus of NIGC leaders. It was there that many of the most consequential decisions about Iran's urban natural gas systems would be made. Foremost among them was a technical one: the operating pressure of the city's distribution network. At the time, two prevailing views governed the design of city gas networks, one favoring high gas pressures and the other low. Countries like France and the United States had opted for high-pressure systems, deeming them "more economical" over the long term, while British companies had historically favored less expensive networks designed around much lower pressures. IMEG proposed just such a low-pressure system for Tehran, suggesting the construction of a forty-inch pipe that began at the city's gas gateway in the south, ran along the Old and New Shemirān Roads, and terminated at Tajrish Square in the north. That central spine would in turn feed a number of smaller pipes fanning out across the city, all of which would be operated at 2.0 PSI, or roughly one-seventh of sea level atmospheric pressure. For their part, NIGC officials opposed this plan, fearing that its comparatively low throughput would limit the long-term prospects of gas utilization in Tehran and doom the city to "suffer from shortages of gas." They instead favored a network operating at 60.0 PSI, or one that could move thirty times more gas through a pipe of any given diameter than had been proposed by IMEG. Such a system was potentially much more efficient to build and maintain, requiring fewer or smaller pipes to provision the same amount of gas.[37] This had important implications for wavering perceptions of the project's feasibility. Some in the city, including its mayor, Gholāmrezā Nikpey, had already come out in opposition to widespread natural gas utilization. Instead of gas, Nikpey favored greater electrification—as the Ministry of Water and Electricity had done several years prior—as the running of wires was much less disruptive than all the digging and pipelaying a gas system would require.[38] But the use of a high-pressure system, by postponing the time when additional capacity would be needed, seemingly offered a way to minimize such disruptions. IMEG engineers, on the other hand, defended their choice not on the basis of economic efficiency or disruptive construction, arguments they could not win, but rather on concerns for the system's safety, contending that leaks in high-pressure systems were likely to result in fires that were much larger and more dangerous. NIGC officials, in contrast, drawing on the historical record of systems operating outside Iran, argued that a high-pressure leak would be rapidly apparent, making them in practice safer because gas could not slowly "creep" into surrounding areas and accumulate until explosive conditions were created. In the end, however, the debate was not settled on the grounds of technical or safety considerations at all, but on the politics of sovereign development that underlay Iran's push to utilize natural gas. With no clear way to choose between the two proposals, NIGC leaders ultimately based their decision on the fact that IMEG would one day leave Iran and only their own engineers would remain. Iran's urban gas networks would thus be high-pressure systems.[39]

Despite NIGC's rejection of its low-pressure design, IMEG's early decision to orient Tehran's gas distribution around the "status and location of industrial units" remained the foundation of the system's spatial contours. Largely adhering to the plans described by Mahmud Peydāyesh three years earlier, by 1974 three primary gas lines had been placed in the city's industrial outskirts, one running twenty-three kilometers toward Karaj and the Mehrābād airport in the west, another stretching twenty-seven kilometers east to the suburb of Tehrānpārs, and the last going south to Farahābād power plant. Numerous spurs and branch lines served industry in all three areas. But while some of the region's most voracious energy consumers quickly transitioned to gas, notably the cement works and power plant, broader uptake was much slower than anticipated.[40] For all the developmentalist ambitions that were embedded in the IGAT-1 program, both the need to import equipment and the low delivery volumes offered by NIGC left many of the city's industrialists ambivalent about the prospect of using gas. Indeed, by 1977, only 120 manufacturers had made the switch, far less than the more than 600 facilities that Peydāyesh had predicted.[41] Such slow progress reflected the fact that much of the actual difficulty of building

Iran's gas system lay not in the monumental pipelines and refineries that were celebrated by the Pahlavi state, but in the smaller channels that connected consumers to such infrastructure. Even the industrial sector, prioritized in part because of the comparative efficiency with which its network could be built, was not immune to such challenges. In this, NIGC's work paralleled many of Iran's other developmental programs: fired with grand ambitions that were in practice only partially fulfilled.

Much more important, however, as the lag between industrial and residential gas service grew, was a growing sense among ordinary Iranians that they had been intentionally abandoned by the government and NIGC. In Tehran, in the mid-1970s, even as the basic infrastructure for the city's industrial districts was being completed, the possibility of gas service for residential areas was still being studied. While NIOC had jumpstarted the process by building a small, independent network for the Tehrānsar neighborhood, NIGC's plans for the rest of the city were slated to unfold over decades. Indeed, company officials expected that ten years would pass before the number of households using gas would exceed 100,000, a 25 percent increase over Peydāyesh's prediction but nonetheless a very low figure considering Tehran's enormous and growing population.[42] That slow pace was in part rooted in the choice to build a high-pressure distribution network, a decision that had invalidated much of IMEG's early design work. On the other hand, NIGC officials like Mohsen Shirāzi, head of design at the company, lamented their own reflexive turn toward foreign experts like IMEG, and they saw Tehran's gas distribution system as an opportunity to assert their own independent abilities. Looking to the experiences of neighboring countries like Pakistan that had built gas systems without foreign assistance, figures like Shirāzi repeatedly expressed their commitment to Iranian resource nationalism and sovereign development.[43] For modernizing Pahlavi officials, national independence was not distinct from their work to use natural gas to build a prosperous, industrialized society. These goals were understood as being intertwined and mutually reinforcing, and while they were not averse to employing the expertise of foreign contractors like IMEG, they also aimed to cultivate the country's own capabilities and retain as much control as possible.

With IMEG's work superseded, it fell to NIGC to study Tehran's potential gas markets and devise feasible ways of serving them. This was not a straightforward task, as the city possessed no comprehensive maps of sufficient detail. NIGC planners instead turned to the municipality's master plan and coordinated their efforts with the departments of water and electricity. Drawing upon postwar norms of modernist urban redevelopment, Tehran's 1968 Comprehensive Master Plan was intended to discipline the city's rapid, and largely haphazard, growth. Aimed at reducing density and channeling further expansion into newly created suburban neighborhoods that were connected by a network of highways, the plan forthrightly embedded social class into the city's urban fabric. Under it, the density and built environment of new neighborhoods were purposely tied to wealth, with affluent neighborhoods marked by large homes, gardens, and swimming pools; middle-class areas by spacious apartments, parks, and cafes; and working-class districts by apartment blocks, mosques, and public transit. Social engineering was a core objective, with the visibly better lifestyles of wealthy residents intended as an incentive for productivity among the working classes.[44]

Though less explicit, NIGC's planning for residential gas service in Tehran likewise reflected class-based conceptualizations of the city. Rather than divide the capital on a strictly geographic basis, the company organized its surveys around dwelling size, creating four teams that ranged across the whole of its area. The first group studied residences of 200 square meters or smaller; the second, homes up to 400 square meters in size; and the third, dwellings larger than 400 square meters. A final group surveyed potential large-volume consumers like hotels, restaurants, government offices, public baths, bakeries, hospitals, and educational facilities. Going door to door, the research groups created or updated 500 maps to include all homes and businesses in the city.[45] As had been the case in Shiraz a decade earlier, NIGC's teams quickly determined that "tight and narrow alleys and largely unsuitable residences" made older neighborhoods in large sections of the southern Tehran unsuitable. For the remainder of the city, company planners employed a "mathematical template" that drew upon factors like population density and average energy use to grade areas on their readiness for gas service. Again as had been true in Shiraz in the 1960s, this process did not automatically select Tehran's wealthier districts, but it did push NIGC to prioritize recently built neighborhoods that were often home to middle- and upper-class households. But with many of those newer areas in the north and thus relatively distant from the city's primary gas entrepot, the company instead opted to extend lines first to two lower-income neighborhoods in the south, Nāziābād and Kuy-e Nohom-e Ābān (today Kuy-e Sizdah-ye Ābān). Both were among a number of planned residential communities conceived in the 1940s and were of comparatively recent vintage, Nāziābād having been built in 1941 and Kuy-e Nohom-e Ābān in 1953. Inspired by contemporary urban design in Europe and North America, the two neighborhoods featured wide streets, low-rise apartments, single-family homes, and modern construction materials, characteristics that prompted NIGC analysts to deem them to be excellent candidates.[46] In contrast to persistent rumors that the neighborhoods had been chosen as experimental spaces due to their inhabitants' relative social marginalization, the decision was instead rooted in the nature of the neighborhoods' streetscapes and their locations within the city.[47] Nāziābād and Kuy-e Nohom-e Ābān proved to be the exception, however, and from that point on the ostensibly objective formula that company analysts used strongly selected for the city's more affluent areas. Indeed, in 1974, while construction was ongoing in Nāziābād and Kuy-e Nohom-e Ābān, NIGC chose the wealthy northern neighborhoods of Sahābqarānieh, Niāvarān, and Z'afarānieh to be next in line for gas service.[48]

Once again as Shiraz's experiences had shown in the mid-1960s, the construction of residential distribution networks in late 1970s Tehran was a very slow process, ultimately defined more by how little was done than by how much. Despite the hopes for independent operations that NIGC leaders like Mohsen Shirāzi had articulated, much of the actual engineering work for Iran's urban gas distribution systems was contracted to Sofregaz, an affiliate of France's state-owned firm Gaz de France. But French help did not accelerate the realization of NIGC's plans in Tehran, in part due to Sofregaz leaders' desire to redo the entire process of mapping and market analysis that NIGC had already undertaken. While the French company eventually acquiesced to demands that it accept the existing work, progress was nonetheless even slower than had been the case in the city's industrial areas.[49] Indeed, ambitions were very modest but rarely met. Initial plans for the Kuy-e Nohom-e Ābān neighborhood, for example, called for approximately 3,500 residential and commercial connections to be installed, while in Nāziābād some 10,500 households were expected to receive service.[50] As of 1977, however, despite 460 kilometers of pipe having been laid, a total of only 2,000 connections had been completed. Plans for 54,000 connections in the north were still largely in the design stage, with only Gishā, in north central Tehran, seeing service to a paltry 115 residences. Nor was such sluggishness confined to NIGC's operations, as NIOC's independent work in the Tehrānsar neighborhood had by that point also fallen significantly behind schedule.[51]

Amidst the rising revolutionary turmoil of the 1970s, such delays would come to have significant political repercussions. For many citizens, the extensive failure by NIGC to build residential distribution networks, particularly in a city of the size and importance of Tehran, communicated a seeming lack of urgency on the part of the Pahlavi state to include them in what was celebrated as one of the country's great modernizing projects. Whatever the inherent difficulties of creating gas infrastructure in 1970s Iran might have been—and NIGC's troubles in importing the equipment needed for industrial connections point to some—many Iranians interpreted the long delays as willful indifference to their aspirations for better living standards. They were well aware that natural gas was (slowly) beginning to flow to consumers across the country, and those who went without were confronted by the fact that there seemed to be no real prospect of it appearing in their own lives anytime soon. Even those who lived in areas deemed most suitable for gas service often found themselves waiting far longer than either they or NIGC had anticipated. That disappointing reality quickly began to erode the plausibility of official claims to be harnessing Iran's natural resources for the benefit of all its citizens, particularly in relation to the enormous volumes of gas that people knew were being sent to the Soviet Union. More than the lack of gas itself, painful as that might have been to some, it was the disparity between expectation and reality that angered many Iranians. It was a promissory gap that had been created, in large part, by the Pahlavi state's own efforts to harness natural gas as a tool of political legitimation, and one for which it would soon pay a steep price.

A Hierarchy of Benefit . . .

By the end of the 1970s, NIGC had undertaken market studies and design work in twenty-five Iranian cities.[52] During that time, each year approximately 9.5 billion cubic meters of gas were exported to the Soviet Union and nearly 3.0 billion consumed domestically. While exports had remained steady over the course of the decade, Iran's own utilization had grown dramatically, more than doubling between 1973 and 1977.[53] Tehran accounted for the largest share of that total, at more than 1.0 billion cubic meters in 1977, while Shiraz and Isfahan each took 500 million and 450 million cubic meters, respectively.[54] Reflecting the government's prioritization of industry and its sizeable energy requirements, some 80 percent of that consumption went to replace the use of fuel oil and diesel, two products heavily used in such settings.[55] While many urban districts were slated to receive gas service by the mid-1980s as part of Iran's Sixth Development Plan, the 1978 cancellation of five-year planning by the shah in favor of more open-ended developmentalist programs created significant question marks about when and if such ambitions would be fulfilled.[56] But as limited as progress had been in Iran's cities, rural regions were even more neglected, under the 1973–78 Fifth Development Plan receiving less than 0.21 percent—200 million of 96.7 billion rials—of the government's total investments in gas infrastructure.[57]

Rural gas service had never been a major part of planning for the IGAT-1 project, and in the mid-1970s NIGC officials considered it to be little more than an experimental idea. They nonetheless explored the possibility on a very small scale, undertaking a survey of 640 towns and villages along the IGAT-1 corridor in order to find a site for a trial gas system. Using factors like proximity to the main pipeline, total population, nature of the built environment, and record of engagement with other developmentalist programs, they ultimately chose the village of Guyim, some twenty miles north of Shiraz, as the best candidate. This was not, however, a gift of gas service. While the company would help with the technical aspects, officials expected residents to personally invest in the project by building new kitchens and paying for the installation gas lines. NIGC moreover positioned the use of gas as part of a broader push for modernizing development in the village, promising to also build a school, establish a medical clinic, and provide a small electricity generator.[58] In large part, this process was a success, and having seemingly raised the general standard of living for Guyim's residents, the project inspired NIGC officials to envision the prospect of gas service in hundreds of other towns and villages along the IGAT-1 route. It also spurred them to imagine other potential uses for natural gas energy. In rural Iran, gas could be used not just for cooking in homes that met the requisite "technical qualifications," but could also heat mosques and public baths, provide power for drilling deep wells and pumping water, and fuel for electricity generation.[59] Such ambitions, and the criteria through which Guyim was chosen, demonstrated the modernizing impulses that underlay Pahlavi developmentalism in general and Iranian natural gas utilization in particular. Gas was not simply a means of providing energy, as important as that was for industrializing development, but also an avenue by which the country and the lives of its citizens could be transformed. In the end, however, despite NIGC's growing interest in rural service, such aspirations would not be realized, and Guyim would be the only such community connected to Iran's burgeoning gas system during the Pahlavi era.[60]

NIGC's overt tying of gas service to broader developmentalist objectives in Guyim reflected how tightly interwoven the two projects were. Years of official rhetoric had positioned gas as a major vehicle for the country's advancement and made its infrastructures highly charged symbols of not only Iran's growing prosperity but also the Pahlavi promise to improve the lives of all Iranians. For a time, the state had been generally successful in that goal. In the 1960s the country's economy had expanded rapidly, growing at an average annual rate of 11.4 percent, while per capita GDP had risen nearly 60 percent over the same period. Even as oil production quadrupled from 1.0 million to 4.0 million barrels per day, the country's growth was relatively broad-based, with 23 percent of new jobs coming in the agricultural sector, 44 percent in industry, and 33 percent in services. By the mid-1970s, however, Iran's economy had come to be dominated by the sale of oil and the state's distribution of its rents. Particularly after the 1973–74 oil crisis caused petroleum prices to spike, the benefits of Pahlavi industrializing policies increasingly flowed to the wealthy and well-connected even as a flood of petrodollars spurred rapid inflation and reduced the purchasing power of the country's middle and lower classes.[61] The state's favoring of urban industry moreover undermined Iran's rural agrarian economy and supercharged an already steady flow of people migrating from the countryside into cities. While some of those migrants found employment in the country's expanding industrial sector, most were left impoverished, crowded into inadequate housing in neglected urban neighborhoods and exposed to yawning social gaps that amounted to a seemingly obvious disregard for their welfare.[62] Indeed, the question of whether the Pahlavi state's extensive efforts at industrialization had been beneficial for the majority of Iranians was a highly politicized one in the 1970s. While some enjoyed newfound prosperity in Iran's changing economy and embraced the consumerist lifestyles it permitted, many others objected not only to the weakening of their traditional livelihoods but also the loss of what they deemed to be more authentically Iranian ways of being. Often articulated through culturally resonant Shi'i thought and symbolism, populist issues of economic and social justice came to animate Iran's revolutionary movement, pitting what some described as Iran's *mostaz'afin* (oppressed) classes against the wealthy and powerful *mostakbarin* (oppressors) who tormented them. But whether leftist, nationalist, or Islamist, and whatever language they might have used to describe it, what united many Iranians in the 1970s was the increasing sense that the prosperity promised to them by Pahlavi developmentalism would forever remain out of reach.[63]

As an emblem of Iran's developing modernity, natural gas had by that point become an important means by which people understood their own status, often taking on symbolic import in ways that ran counter to the triumphalist narratives promulgated by state officials. Amidst Iran's growing revolutionary turmoil, the uneven spread of natural gas infrastructure thus became a significant issue for people living in areas that had been left underserved. In response, some gathered signatures and petitioned local officials for gas service; others complained vociferously that surrounding neighborhoods were receiving gas while they were not; still more, invoking the White Revolution's official name of the "Shah and People's Revolution," wrote to NIGC to stridently demand that it provide gas, particularly in places where they claimed the necessary infrastructure already "passed directly in front" of their homes.[64] Indeed, for many of those residing in the towns and villages of the IGAT-1 corridor, a lack of gas service while living in such close proximity to existing gas infrastructure came to be seen as a willful breaking of the Pahlavi state's developmental promises. Such was the case in June 1976 when the administrative council of Sepidān, a rural and mountainous county north of Shiraz, gathered to discuss the region's energy needs. Opening with a panegyric to the shah's "quick and progressive advancements" in the country, the council turned to discussing "difficulties" like poor roads, unstable electricity supplies, and a lack of natural gas service. Believing that a major gas line would cross "within at most twenty kilometers of [the county capital] Ardakān," council members requested that NIGC connect the town and its surrounding villages to the country's growing gas network. Their desire was rooted in practical understandings of gas as a source of convenient and reliable energy, and they hoped that it would help alleviate the "very short supplies" of fuel that residents faced when winters of "heavy snows of six to seven meters" made travel and resupply difficult. While the area had traditionally relied on brush and forest wood for fuel, as had once been true in much of the country, population growth had led to overharvesting and subsequently to government restrictions.[65] By the 1970s, kerosene and other oil products had replaced much of that natural fuel in Iran's rural areas, but deliveries were often difficult for the mountain villages that dotted Sepidān.[66] Unlike kerosene and diesel, natural gas could flow year round, a potent material benefit that underlay the council's call for service. But the nearby passage of gas infrastructure was significant as well, for the assumption of Sepidān's council was that a small branch from the main line to their towns and villages was a relatively insignificant undertaking for NIGC. That belief, whether true or false, was one that would be voiced time and again in Iran's rural areas, a consistent drumbeat that would strengthen the connection between natural gas service and revolutionary politics.

In May 1978, a conflict over the availability of natural gas began to brew in Kāzerun, a county lying west of Shiraz. It was there that the head of the county board, a man by the name of Abdullāh Hushdārān, noted that the IGAT-1 branch line for Shiraz would pass through the area. Owing to that proximity, the board had formally requested from NIGC that the town of Kāzerun be included in its plans for gas distribution in the region.[67] In contrast to the praise that Sepidān's council had heaped on the shah and his leadership two years earlier, by 1978 the mood within Fārs province had begun to change, with local leaders like Hushdārān now openly criticizing the government and the outcomes of its developmental initiatives. Indeed, a month after the meeting, Hushdārān used an article from the national daily *Kayhān* to send a veiled but nonetheless pointed message to the provincial governor, suggesting that while the shah's developmental policies had brought real benefit to the country, little of it was evident to the public at large. Highlighting the argument in an attached copy of the article, Hushdārān noted that he had previously said "the same thing" and went on to "beg" the governor to pay heed to his community's needs, particularly their request for gas service.[68] Unlike in Sepidān in 1976, his remarks went well beyond the practicalities of gas use to embrace its significance as a symbol of Pahlavi developmentalism. While government officials had long trumpeted gas as a new energy source for a prosperous future, the residents of Kāzerun, like many around the country, had yet to see any evidence of it in their daily lives. It was a gap that threatened to discredit the state's entire project of modernizing developmentalism. Indeed, for people like Hushdārān and those he claimed to represent, the end of the 1970s was a moment when the benefits of natural gas infrastructure moved from aspirational hope to unfulfilled promise—and many were beginning to demand that they be given their due.

The ability of provincial authorities to influence NIGC policy was, however, limited. At one meeting of the Kāzerun board in August 1978, a representative insisted that the governor act to have gas infrastructure built in the region, pointing to the fact that NIGC's chief executive had "promised" that natural gas pipes would be laid for Kāzerun "at the proper time."[69] When nothing was done over the next several weeks, frustrations boiled over. One member openly denounced NIGC and its priorities, again noting that "gas pipes have been laid within seven kilometers of Kāzerun" and demanding that the "authorities in the matter" not be "indifferent" to the area's residents. Why should the people of Shiraz have access to gas and not those of Kāzerun, he argued, as

> all Iranians have a right to equally and justly make use of the developmental benefits of this property. Why is piped gas (*gāz-e lulehkeshi*), at a minimum, to be consumed in one city, and in another like Kāzerun, because pipes have not been laid, people must consume liquid gas at an expensive price incomparable to piped? This very issue is, and will be, the cause of people's discontent and they will become burdened by these various discriminations (*tab'aizāt-e gunāgun*). Whatever correspondence they send, entreaties they make, [or] moans they give, no one listens . . . [and] people are upset and unhappy because Kāzerun is not intended to be part of the project of laying gas pipes.[70]

For the residents of Kāzerun, gas use carried political import that went well beyond whatever practical benefits it might have brought. As Iranians, they expected to be able to take full advantage of the "developmental benefits" that Pahlavi gas policies promised. To be denied them was not merely an issue of timing or a misfortune of geography, but an injustice wrought by seemingly willful discrimination on the part of NIGC and the state. As board members asked: "how is it that it is possible for NIGC to lay pipes from the gas production region to Āstārā [some 1,100 kilometers distant], but for a distance of seven kilometers they bring excuses and objections?"[71]

While the people of places like Kāzerun and Sepidān could and did use liquid gas for fuel, generally delivered in canisters by private companies that residents complained provided poor service, they nonetheless saw piped natural gas as a superior source of energy.[72] Liquid gas was more expensive and could not be reliably delivered during snowy winters. In this way, as was true in Pahlavi Iran's official narratives of natural gas more broadly, rural residents' notions of gas were tied to its technologies of use, a sociotechnical imaginary that imbued meaning in gas as both a material resource and a political symbol. Instead of a story about monumental arteries measured in hundreds of miles and billions of cubic feet of gas, their concerns were spoken in the vernacular of the system's capillaries, the threads of pipe that moved gas the final few feet to their homes. For those like the members of the Kāzerun county board, to be denied access to the national gas network was to be cut off from the new lifeblood of the Iranian nation and rendered inferior to those whose communities were vitalized by it. Made clear by figures like Hushdārān was the fact that many Iranians desired to incorporate natural gas into their daily lives, actively wishing to take part in the industrialized and energy-intensive lifeways that the Pahlavi state was fostering in the country. For such people, and many like them all over Iran, gas was at root valuable not just because it was something promised by the government but because it seemingly offered a future of cheap and clean energy. Their anger lay not in a rejection of the Pahlavi state's idealized future of copious energy consumption, but in the mundane reality of being excluded from it.

The residents of Kāzerun and Sepidān never received what they sought from the shah's government. Coming several weeks after the outburst at the August board meeting, NIGC's response acknowledged the considerable efforts that had been made by local authorities to push for gas service, but went on to declare that

> gas delivery (*gāzresāni*) to the various cities and points that exist along the route of the cross-country natural gas pipeline [IGAT-1] must be qualified on technical and economic criteria and principles. The passage of regional gas pipes does not alone justify gas delivery. Therefore, as soon as gas delivery to the county of Kāzerun can be justified from the perspective of conformity with [those] criteria, the necessary action will be taken.[73]

Repeated attempts over the next few months to press the company further only met with frustration, and by January 1979 members of the Kāzerun county board seemed to have acquiesced to the finality of NIGC's decision.[74] Acquiescence did not mean acceptance, however, and Abdullāh Hushdārān vented his anger in front of his peers. After reading NIGC's letter aloud, he angrily denounced the company's entire natural gas program, saying that

> for years in this country the law of the jungle operated and was enforced, and the people were strangled and suffocated by the injustice and oppression of the governments of the time. But the height of unfairness, and counter to human principles, is that the gas that is one of our natural and national resources is for us forbidden (*harām*) . . . but for foreigners overseas (*khārejiān-e māvarā behār*) and others who are thousands of kilometers distant from Iran's borders, it is lawful (*halāl*). With complete indifference [they] lay a pipe within five kilometers of this city in the direction of Gachsārān and Āstārā so they can put cheap gas (*gāz bā qeymat-e nāzel*) at their disposal and provide for their welfare.

In Hushdārān's view, being denied the benefits of Iran's natural gas resources was one of the harshest injustices perpetrated by his government. He considered it the right of Iranians to partake of the "natural and national resources" that lay within their country, believing that their welfare should be prioritized over that of "foreigners." In this way, though both Hushdārān and the officials of the Iranian government wanted "Iran" to benefit from the construction of the IGAT-1 pipeline, they did not fully agree on what that "Iran" was meant to be. In their efforts to find productive uses for Iran's gas, NIGC officials had prioritized large industrial users and a pipeline to the Soviet Union. For them, the overriding imperatives were economic growth and industrialization, made possible by the cheap energy that natural gas could provide. NIGC's "Iran" was thus a monumental one, a nation of advanced technology, huge infrastructures, vast distances, and enormous scales. But such tributes to the expertise of government technocrats meant little to Hushdārān and his fellow residents. For them, all those promises of industrialized prosperity meant little if it could not be felt in the daily lives of Iran's citizens. Their "Iran" encompassed their homes, their communities, and themselves, none of whom were benefiting from the wealth that lay under their country's soil. In Hushdārān's eyes, they, as Iranians, were the true owners of Iran's natural gas, and the fact that foreigners at far remove were able to utilize it while it was still inaccessible to them was a fundamental injustice. As he went on to say,

> we who are the original owners of this God-given wealth plea that the people of this city are ready to render a claim to a branch [of the

> pipeline] and fulfill any other conditions. They answer that the laying of the cross-country gas pipe does not alone justify the laying of pipes to deliver gas to the city of Kāzerun . . . . Consider [that] the workers of the autocratic (*khudkāmeh*) government are prepared for a county of 200,000 people to face hardship (*mazifeh*) and discomfort (*nārāhati*) for the protection of the interests of foreigners and several gas distribution and sale companies . . . . There are no rules and regulations for this extortion and injustice because the plunderers were all companions, but the stretching of one strand of gas pipe five kilometers wants technical and economic principles. But the day of reckoning (*ruz-e hesāb*) approaches and the time has come that the wealth of Iran be for Iranians.[75]

For Hushdārān and those who agreed with him, the priority given to foreign consumers was nothing less than a willful act of plunder. Neither distance nor difficulty seemed to be an obstacle for sending gas to the Soviet Union, but a litany of excuses cloaked in "technical and economic" objectivity was brought forth to deny similar benefits for the residents of Kāzerun. This discrepancy struck at the heart of the Pahlavi state's developmentalist legitimacy, and speaking at a moment when Iran's long and bloody revolution was on the verge of success, Hushdārān predicted that such failures would lead to a "day of reckoning" for the shah. One way or another, he thundered, the vast natural gas wealth that lay beneath the earth of the country's southwest would be for the "Iran" of the provinces, the small towns and rural folk that had been left behind by a system of gas infrastructure built around large cities and foreign exchange.

. . . And A Day of Reckoning

On January 16, 1979, Abdullāh Hushdārān's predicted day of reckoning arrived. Coming after more than a year of bloodshed and escalating protests, Muhammad Reza Shah's flight into exile marked the downfall of the monarchy and the beginning of the end of the Pahlavi government. Within two weeks of Ayatollah Ruhollah Khomeini's February 1st return to the country, Shāhpur Bakhtiār, appointed prime minister by the shah in late December, had also fled, leaving Iran under the formal but limited control of a provisional government headed by Prime Minister Mehdi Bāzargān. Real power lay in the hands of Khomeini, however, exerted through an informal but nonetheless potent network of revolutionary committees in towns and cities across the country. Though a bloody and years-long power struggle between Khomeini's followers and other opposition groups was just beginning, in 1979 the momentum toward the establishment of an Islamic Republic was unstoppable. It would be backed by an overwhelming majority of Iranians in a March referendum and again—though by a smaller margin—in December when the country's new constitution was approved. Aided by the crisis of the September 1980 Iraqi invasion and the onset of what would prove to be the twentieth century's longest conventional war, by the mid-1980s Khomeini had largely asserted his authority over rival revolutionary groups and the Iranian state. By this point, the new Islamic Republic was already extending its influence deep into Iranian society, aiming to implement the intertwined Islamist and populist impulses that had animated its creation.

In its first decade, the Islamic Republic took an aggressive stance toward the management of the country's economy, seeking to excise what it deemed to be the plundering of the country's wealth by a corrupt and westernized elite. But far from tearing down the administrative state built by Muhammad Reza Shah and his father, the Islamic Republic adopted it wholesale, purging only its highest levels. Fueled by oil revenues and seeking to remake a society under the pressures of total war, Iran's bureaucracy was expanded as the remits of existing agencies were increased and six new ministries were created. Much of the state's new effort was directed into a cultural revolution aimed at Islamizing public life—reorienting the judicial system toward the *shariʿa*, prohibiting alcohol, enforcing conservative dress codes, censoring art and media, and expelling large numbers of students and faculty from schools and universities—and eliminating what Khomeini and his followers considered to be unwanted foreign cultural norms. Dreams of autarky led to the rapid nationalization of banks, large industries, and infrastructural systems, particularly those owned by the shah's former courtiers and influential Jewish and Baha'i businessmen who either fled or were executed for their association with the former regime. While Khomeini said little about economic affairs, many who supported the 1979 revolution called for collective benefits like free housing, the outlawing of usury and monopolistic hoarding, universal pensions and disability pay, and the elimination of poverty. While the majority of such goals proved to be well beyond the ability of Iran's new leaders to implement, their underlying populist orientation would in the following decades guide the Islamic Republic's economic policies.[76]

In the 1980s, animated by the revolutionary populism that had propelled it to power, Iran's new government poured considerable effort into improving rural living standards. Though the county board that had provided Abdullāh Hushdārān his public voice had been dissolved upon the shah's fall, his call for the "wealth of Iran [to] be for Iranians" was answered as the Islamic Republic used its developmental programs to signal an aggressive commitment to bettering the lives of Iran's lower classes.[77] Toward that end, and with an eye to the rural regions that had been largely left out of Pahlavi modernizing projects, policies of agricultural pricing that had favored urban consumers over rural producers were eased while extensive programs of education, healthcare, roadbuilding, electricity generation, piped water, and natural gas service were introduced. This general reorientation moreover altered the public expression of narratives surrounding natural gas utilization. As had been true prior to the revolution, public relations materials from state institutions and the national petroleum companies were used to communicate an official sociotechnical imaginary centered on gas exploitation. Aimed at both specialized and general audiences, these books, magazines, and pamphlets articulated a vision of gas energy grounded in both technological systems and an imagined future of national prosperity and independence—much as had been the case in the Pahlavi era. But such strong parallels notwithstanding, the tenor in which those claims were made had shifted. In place of the paeans to monuments of sophisticated technology that had been common in the prerevolutionary era, under the Islamic Republic, gas service for ordinary households was celebrated and bellicose anticolonial politics championed.

"Before the governance of the Islamic Republic," NIGC declared in one colorful 1985 book, "when Iran was the field of invasion of international colonialists and plunderers," the "valuable substance" of natural gas had been "looted [and] put at the disposal of the eastern superpower [the USSR]." Under Iran's new government, however, and the "eviction of eastern and western colonialists," the country's gas resources had been reclaimed so that the "deprived people of our homeland could also enjoy welfare and comfort."[78] As it happened, gas exports to the Soviet Union had indeed been ended in April 1980, a decision rooted in a dispute over pricing but often trumpeted as evidence of the Islamic Republic's anticolonial bona fides. Another 1985 publication from the Ministry of Petroleum, a pamphlet of seven black-and-white pages titled "The National Iranian Gas Company in the Service of the Oppressed" (*Sherkat-e Melli-ye Gāz dar Khedmat-e Mostaz'afin*) reinforced that perspective. Beginning with an ode to God and praise for Khomeini's leadership, the pamphlet proclaimed that a sacred duty had been placed "on the shoulders of the Muslim nation of Iran" to bring "dignity" to the "oppressed of the world." More than a restatement of revolutionary themes, such notions were a key organizing principle for the narratives of natural gas that were constructed in the 1980s. Casting the ministry and NIGC as part of a "government of servants" (*dolat-e khedmatgozār*) working to bring the "blessing of gas" to "many fellow urban and rural countrymen," the pamphlet sought to demonstrate the Islamic Republic's commitment to Iranians' well-being over that of foreigners, stating that the

> National Iranian Gas Company . . . was created under the previous regime with the goal of extortion for the eastern superpower [the Soviet Union] and the protection of global imperialism. During the years [13]49 [1970] to [13]58 [1979], a total of approximately 70 billion cubic meters of natural gas was exported to the Soviet Union at a meager price (*qaymat-e nāchiz*). With the victory of the Islamic Revolution and the disintegration (*gosasteh shodan*) of the binds of tyranny, colonialism, and exploitation, gas—this great blessing of God—and its expansive industry, in both concept and actuality, was placed in the service of the people of our homeland . . . [with] the cutting of the export of gas to the Soviet Union as the first change.[79]

In this view, it was unconscionable that vast infrastructures could be built to transport natural gas hundreds of miles to foreign markets while so many Iranians, like those living in the mountain villages

Sepidān and Kāzerun, were left wanting. In this way, Iran's cessation of gas exports to the USSR, in actuality a result of a Soviet refusal to pay the higher rates that postrevolutionary Iranian negotiators had suddenly demanded, was refigured as a newfound prioritization of Iran's own needs.[80] As early as the mid-1960s, disputes over the price of gas shipments with the Soviet Union had been common, and facing the prospect of losing the project in its entirety, Muhammad Reza Shah had at the time ordered an acceptance of Soviet terms. By the 1980s, however, Iran's new leaders were no longer willing to bend, choosing to accept a loss of export revenue in order to publicly take a stronger anticolonial position. Thus, while resource nationalism had been essential to natural gas's political meaning in the Pahlavi era, under the Islamic Republic its significance was heightened to new rhetorical levels and more strongly linked to the possibility of everyday consumption by Iranians of every social class and standing.

But according to both publications, even more significant than the cessation of exports was the Islamic Republic's efforts to expand gas utilization within the country. As they both argued, cheap and plentiful energy, something that had been wrongly denied by the monarchy, was a right of all Iranians and one that the new government endeavored to fulfill. Toward that end, NIGC and the Ministry of Petroleum worked to quickly expand the prerevolutionary period's "demonstrative and very incomplete" residential gas systems.[81] Their efforts bore fruit, and by 1984, having accelerated prerevolutionary policies of substituting gas for kerosene and diesel in "industries, power plants, cities, and major centers of consumption where the supply of liquid fuel faces difficulties, " they had nearly tripled the 1979 rate of consumption of 9.6 million cubic meters to 26.1 million. Furthermore, even as they sought to draw sharp distinctions between the pre-and postrevolutionary periods, NIGC and the ministry continued to frame their achievements through the legitimating power of statistics and numerical data: five cities connected to the country's gas network before the revolution and twenty-three after; one village—Guyim—provided gas service in the Pahlavi era and forty-eight under the Islamic Republic; 270 industrial facilities using gas at the end of the 1970s and 602 by the middle of the next decade; 50,000 gas connections in 1979 versus 310,000 in 1985; and

27,000 gas users under the shah and 257,000 under the Islamic Republic. Posed side by side, these figures prompted readers to reflect that the new era was 260,000 gas connections, 230,000 customers, eighteen cities, and forty-seven villages better than the old. Promises of future service were delivered in a similar fashion, and it was reported that 255,000 meters of pipe were being laid for 28,000 residential and commercial connections in a further ten cities.[82] As was common under the shah, this was the politics of numbers, a way of giving the imprimatur of fact and objectivity to the political claims the pamphlet was advancing. In this focus on the capillaries of gas infrastructure, the prerevolutionary preoccupation with scale was not eclipsed but redirected, measured now not in billions of cubic feet of natural gas extracted, refined, or transported but rather in the number of people and communities served. This was as much a modernizing vision as anything that had been articulated under the Pahlavi state, but one that emphasized the modest gas systems that Iranians encountered in their daily lives. Rather than foregrounding monumental pipelines and refineries as evidence of technological mastery and dominance over the land, under the Islamic Republic new rhetorical markers were created that gave symbolic priority to the small pipes of residential distribution networks. Resource nationalism and technological mastery remained powerful in this narrative, in some ways were even stronger, but now they would be tied more closely to Iranians' everyday lifeways. The sociotechnical imaginary of natural gas that had first coalesced under the shah thus remained largely intact in the postrevolutionary era, continuing to structure both NIGC's activities and their political salience within Iran.

Competing notions of how to best prioritize Iran's gas infrastructure notwithstanding, unchanged across the revolutionary divide was a fundamental orientation toward industrialization, development, and high energy use. Indeed, NIOC, NIGC, and the Plan Organization, in large part staffed by the same personnel, survived the revolution intact, and many of their existing policies were carried over into the postrevolutionary period. Within that context, natural gas utilization became a source of political legitimacy for the Islamic Republic in ways that remained strikingly similar to those of the Pahlavi era. Evident in the discourse of the 1980s was the fact that the sociotechnical imaginary that put natural gas energy at the center of a 'modern' Iran was never in question, despite the revolutionary change in government. Iranians, both elite and not, monarchist and revolutionary, embraced gas energy as the means by which their country could become sovereign, modern, clean, and energy intensive. The demand for cheap energy by people like the residents of Sepidān and Kāzerun was reflective of both the politics of access and the technical characteristics of gas as a source of energy. Natural gas was seen as superior to both oil products and liquid gas, understood as being more convenient, cheaper, and more modern. To be denied its use, particularly in regions where pipelines already passed close at hand, was therefore, in essence, to be denied one's full rights as an Iranian. Such grievances drove the gas policies of the early Islamic Republic, but the expanded urban and rural distribution networks that the new government used to demonstrate its commitment to economic and social justice—growing nearly ninefold between 1978 and 1984—were in many ways extensions of what had been laid in the Pahlavi era.[83] Nonetheless, the newfound concern for distribution and everyday use remained materially predicated upon the landscape-conquering pipelines and towering refineries that had marked Pahlavi rhetoric, and they continued to be depicted as part of Iran's sociotechnical imaginary of gas.[84] Such choices reinscribed an imagined future of clean, convenient, and abundant energy, one marked by a valorization of technological sophistication, a preoccupation with large quantifiable scale, and a celebration of a domestic source of inexpensive energy. Under the Islamic Republic, natural gas thus remained a crucial pillar of official ambitions to create a prosperous and sovereign Iran, something expressed in both ideational and practical terms through an altered but ultimately harmonious rearticulation of Pahlavi developmentalism.

Indeed, the Islamic Republic's emphasis on expanding distribution to rural and residential consumers did not overturn the fact that the vast majority of natural gas used in Iran—some 97 percent in 1984—continued to be dedicated to electricity generation and the industrial sector.[85] Such continuity, despite the often intense rhetorical exhortations to the opposite, demonstrated that the natural gas policies of the early Islamic Republic in many ways represented an acceleration, expansion, and intensification of those of the Pahlavi era. Over the following three decades, the sustained commitment of Iran's post-revolutionary government to widespread distribution would make natural gas into the primary energy source of Iranian society, a process that necessitated greatly expanded service to both residential and industrial applications alike.[86] In this way, by the early 2010s, Iran's gas system was drawing upon numerous associated and non-associated sources to produce more than 160 billion cubic meters of refined gas each year—some twenty-five times more than had been carried by the original IGAT-1 line. To enable such consumption, the country's gas infrastructure had been expanded to encompass a dozen refineries, eleven major pipelines totaling almost 22,000 miles in length, and more than 146,000 miles of distribution lines serving 16.3 million connections in 920 cities and 12,504 villages. Out of a total nationwide usage of approximately 151 billion cubic meters per year in 2012, electricity generation and residential fuel consumption each represented roughly 27 percent of the total, followed by industry at 17 percent, and petrochemicals at 14 percent. All told, by that year, some fifty-five million Iranians had access to piped natural gas, 96 percent of city residents, and 54 percent of rural.[87] Air quality concerns continued to be relevant even as motor vehicles came by the mid-1990s to account for some 70 percent of Tehran's atmospheric pollutants.[88] Reviving and expanding efforts undertaken in the late Pahlavi era to fight declining air quality, compressed natural gas was again tapped as a new source of fuel for the country's transportation sector, this time with much more success, and by 2015 more than 4.07 million vehicles were making regular use of CNG.[89]

Such continuity was further underscored by the public exaltation of the conquest of nature. First published in the spring of 1991, a monthly magazine from NIGC titled *Nedā-ye Gāz* ("The Call of Gas") was filled with articles and photographs that reported on the "battle with nature" that the company undertook in its effort to expand the country's gas network.[90] Using similar language and much of the same visual grammar that had been evident in Pahlavi-era publications, depictions of NIGC's work were again presented as "proof" of the state's ability to overcome all challenges in providing benefits for the people of Iran. In periodicals like *Nedā-ye Gāz*, in contemporaneous pamphlets, and in later collections of oral histories, the spectacle of technological achievement and the domination of nature was again proclaimed through frequent descriptions of rough terrain and the ingenuity needed to conquer it; through photographs of enormous pipelines stretching along passages of scraped earth and crawling up mountainsides; and through the purported objectivity of statistics representing the expansion of Iran's natural gas infrastructure to new regions.[91] In this way, despite rhetorical attempts by leaders of the Islamic Republic to distance themselves from their monarchical predecessors, the same promises that had structured Muhammad Reza Shah's legitimation strategies continued to operate after his fall. More to the point, whether residential, industrial, or vehicular, the widespread use of natural gas energy in twenty-first-century Iran was built upon foundations that had been first laid in the Pahlavi era. Evident in this history is a sustained and congruent commitment by Iran's pre- and postrevolutionary governments to fostering a new era of bountiful energy use with natural gas infrastructures and the fossil hydrocarbon exploitation they enabled. In the 1990s, coming on the heels of a decade of rationing and privation during the Iran-Iraq war and its immediate aftermath, leaders of the Islamic Republic opted to rebuild Iranian society by both continuing and reinscribing Pahlavi gas policies. Indeed, despite altered postrevolutionary rhetoric, what was articulated and enacted was a set of natural gas policies that maintained and extended both the material systems of natural gas use and the developmentalist orientations that underlay their creation. The quick and publicly celebrated expansion of the country's gas sector reflected that dedication, and as consumption skyrocketed in the first decades of the twenty-first century, the country became ever more reliant on its vast reserves of natural gas. The intense utilization of these resources married the strong anticolonial impulse of Iranian resource nationalism to the industrialized lifeways that both the Pahlavi monarchy and the Islamic Republic aspired to build—and that Iranian citizens had come to demand. The result, an industrialized society ordered around the frenzied consumption of cheap and bountiful fossil hydrocarbon energy, is one that has seemingly fulfilled the dreams that Iranian development planners articulated more than half a century prior.

***

The story of Iranian natural gas utilization is in many ways a top-down one, with national developmental policies and the construction of large infrastructural systems taking center stage. But important influences flowed up from the bottom as well, a result of popular views of natural gas that were allied to, but nonetheless distinct from, those that structured the activities of national institutions. As early as the 1950s, local leaders in places like Shiraz began agitating for the creation of residential gas networks, often aggressively pushing back against authorities in Tehran focused on the provision of gas to industrial consumers. In a pattern that would be repeated time and again in the following decades, officials like those working in NIOC and the Plan Organization resisted the idea of residential gas networks, concerned by their comparative inefficiency and the technical difficulties of serving customers in the haphazard confines of old neighborhoods. It took a years-long campaign on the part of city officials and provincial authorities to push NIOC into building Iran's first residential gas network, and only a small one at that. Despite heady developmentalist rhetoric that promised the benefits of natural gas to all Iranians, that outcome—a small residential network bordering on the experimental—would prove to be the norm through the end of the Pahlavi era, a product of modernizing officials' deliberate and persistent prioritization of the country's industrial sector.

NIOC's position, and similar ones taken by NIGC in the late Pahlavi era, was deeply rooted in the material fabric of Iranian cities. Districts built after the Second World War, characterized by ordered thoroughfares and buildings made of concrete and brick, were considered good candidates for gas service; older quarters, with their narrow, winding streets and closely packed mud homes, were not. There was an inevitable, if not necessarily intended, class distinction operating in such typologies, with middle- and upper-class households more often residing in new neighborhoods and poorer households among the dense alleyways of old. Pahlavi Iran's official sociotechnical imaginary of gas—a national resource that could be harnessed by a sovereign state in support of its ambitions to remake the country as a preeminently modern one—was thus co-constituted with spatial imaginaries about Iran's urban areas. Natural gas, positioned as a modern and technologically sophisticated source of energy, was in this telling best suited for modern urban neighborhoods and their modern, educated, and sophisticated inhabitants. NIGC's experimental system in Guyim reinforced such notions, as the company linked its work in the village to a host of other seemingly unrelated developmental changes. Confirming as they did the modernity of some spaces and the backwardness of others, Pahlavi gas distribution systems symbolically worked to create an Iranian future built around affluent urban areas. Left out of this budding future were the poor and the rural, seemingly consigned to an existence on the margins of a prosperous new society.

While there were very real justifications for favoring industrial consumers over residential and urban areas over rural, among Iranians the explicit and very public lionization of natural gas by state agencies helped create aspirations for cheap and convenient energy that went largely unmet. Officials working for NIOC and NIGC were guided by ostensibly neutral and objective notions of economic and developmental efficiency, but the social and political implications of their choices were not easily controlled by the Pahlavi state. In short, the prosperity that was augured by their country's programs of industrializing development was something that many Iranians chose to embrace and, eventually, demand. Whatever disagreements people might have had with Pahlavi policies—and as the 1970s wore on, such differences would become increasingly apparent—the possibility of a future built upon cheap energy was something that many found to be attractive. Those like Abdullāh Hushdārān, head of the Kāzerun county board, chafed at the injustice of watching Iran's gas resources being transported long distances to foreign lands even as NIGC refused to bridge a gap of a few miles for their own communities. In this way, for many Iranians, the significance of gas went well beyond the material benefits it could bring, as important as those were. What angered people like Hushdārān was not just a lack of gas service, but what that lack seemingly represented about their place within Iranian society. The exclusion of communities like Sepidān and Kāzerun was understood by their residents not as an unfortunate outcome of technical and economic principles, but as a pointed rejection of their status as Iranian citizens equally entitled to draw upon their land's natural bounty.

In the late 1970s, such perceptions found a home within the broader notions of socioeconomic injustice that animated many who joined Iran's cresting revolutionary movement. In response, the new Islamic Republic oriented its natural gas policies around such demands, quickly moving to extend service to residential and rural consumers. Drawing sharp contrasts with its monarchical predecessors, Iran's postrevolutionary government sought to use such accomplishments to bolster its legitimacy through populist displays of prioritizing Iranians' daily needs. But for all that this change in direction entailed for the country's natural gas infrastructure, and even as it strove to assert its superiority over the monarchy it replaced, the Islamic Republic largely extended and accelerated the gas policies of the Pahlavi era. Indeed, despite the newfound emphasis on residential service, well into the twenty-first century the majority of the gas consumed in Iran was used for industrial applications and electricity generation. Moreover, building upon foundations laid in the 1970s, new distribution networks largely adhered to the contours established during the Pahlavi era. Even the discursive framework within which natural gas was discussed was congruent, continuing to emphasize scale, technological sophistication, material prosperity, national independence, and the conquest of nature.

Beginning in the years after the Second World War, Iranians began to imagine a national future that was defined by bountiful energy use and industrialized prosperity. Their hopes would grow deep roots in the following decades, expanding beyond industry to include the everyday lives of Iranian citizens, thereby coupling energy consumption to notions of national belonging and surviving a revolutionary change in government. Focusing on that history reveals deep continuities between Iran's prerevolutionary Pahlavi monarchy and postrevolutionary Islamic Republic, tight connections made not just in the hard stuff of steel and concrete, but also in Iranians' aspirations for the future of their country. These two periods, often seen as opposed, were thus united by a shared preoccupation with industrializing developmentalism and its material benefits—ambitions that would be made possible, it was promised, by the widespread exploitation of cheap natural gas energy.

Conclusion

Since the early 2000s, the world's attention has been largely focused on Iran's controversial nuclear program, a national project embodying sovereign ambition, technological mastery, and the control of energy. For both supporters and critics of that half-century effort to tame the atom, the Islamic Republic's character as either a revolutionary champion of the exploited or a rogue revisionist state has come to be represented by the reactors and centrifuges that now operate in scattered sites across the country. But that focus has in many ways obscured a more fundamental change in Iranian society, one that embodied the emphasis on sovereign technological mastery even more strongly: the rise of natural gas energy. Modern Iranian society was built with and alongside the sophisticated infrastructures used to produce, move, and consume gas. The geopolitical prominence of nuclear technology notwithstanding, the creation of an industrialized society organized around the widespread use of natural gas is amongst the most important stories of Iran's recent past, an overlooked history allied to but nonetheless distinct from oil's role as a lucrative export commodity. This book has studied Iran's quest for gas energy and the developmentalist aspirations, anticolonial politics, and environmental anxieties that animated it. Neither entirely top-down nor bottom-up, those threads were interwoven with the compositions of hydrocarbon deposits, the warp and weft of sedimentary layers, the high passes of mountain ranges, and the available technologies of fossil fuel production and transportation. Sitting at the porous interface between human infrastructure and the natural world, gas energy was moreover generative of powerful moral and political claims about the future and Iranians' relationship with their land.[1] Presented both before and after the revolution as a liberatory conquest of nature, Iran's gas infrastructures were in reality deeply influenced by the materiality of the natural world. Their creation was never a mere layering of human architecture atop passive natural terrain. They were instead more-than-human assemblages encompassing everything from imagined futures to reservoir pressures, and rather than cocooning Iranians through the technological convenience of modern consumer lifeways, gas infrastructures instead deepened their entanglement with the natural history of their land.

If the creation of coal-based fossil fuel economies coincided with the rise of the nation state as the organizing principle of the world's geopolitical system, then it was a trend both propelled and cemented through the use of petroleum.[2] Indeed, for much of the world, state-based nationhood only arrived amidst decolonization and a contemporaneous rise in oil and gas consumption. That period, most pronounced in the three decades after the end of the Second World War, was one that saw dozens of new states created from the remains of Europe's colonial empires. But for the elites of many such states, "[political] independence was not the end of decolonization;" rather, "development was."[3] As Sukarno, president of newly independent Indonesia, said (and was famously quoted by Marshall McLuhan): "a refrigerator can be a revolutionary symbol—to a people which has no refrigerators."[4] Such perspectives put the world's unequal distribution of wealth and divergent living standards at the center of politics across the Global South, especially as regional leaders and competing Cold War powers sought legitimacy and influence through extensive programs of socioeconomic reform. Unravelling the implications of that sentiment for one corner of the developing world—Iran—has been the aim of this book. There, a combined impulse toward resource sovereignty, national independence, rapid economic growth, and the creation of a consumer society became intertwined with the exploitation of natural gas.

Iranian leaders entered the postwar era with an eye toward modernizing development. For them, industrialization was the key to building an independent and prosperous nation, and they invested considerable amounts of money and effort in its pursuit. Such aspirations were made possible by the country's enormous petroleum reserves, but while oil and its export revenues funded Iran's developmental programs, it was natural gas that would become Iran's most important source of energy. As Iranian industrialization accelerated, gas enabled Iranian modernizers to sidestep a looming crisis: that the very oil that paid for development was also what was needed to power it. With no existing market within reach, natural gas became the foundation upon which Iran's industrial ambitions were conceived, and through the 1970s Pahlavi officials consistently prioritized such consumers for gas service. All along the route of IGAT-1, facilities like sugar refineries, a steel mill, cement producers, chemical plants, brick kilns, and more were connected to the growing gas network, generally well in advance of residential neighborhoods.[5] Such was true for regions outside the IGAT-1 corridor as well. Indeed, though typically smaller in scope, the gas systems that served cities like Mashhad and Ahvāz largely drew upon the same principles that structured Tehran's distribution grid, extending service to industrial consumers years or decades before city residents.[6] Whatever their rhetorical gestures toward a future of widespread national gas use, Iranian officials consistently chose to provision large-scale users like factories and power plants first. It was a decision grounded in notions of efficiency—how to use as much gas as possible as quickly as possible as cheaply as possible—but also one that reflected broader developmental logics that were oriented toward industrialization and intense energy consumption.

At the same time, natural gas became important means by which the Pahlavi state, and later the Islamic Republic, articulated their understandings of what Iranian society could and should be. Expressed was a politics of development rooted in a claimed ability to conquer nature with enormous and technologically sophisticated pipelines and refineries. The reality, however, was something less than total domination.

Those same systems were strongly shaped by natural materialities, as gas compositions, topography, and reservoir pressures placed limits on what was physically possible. For Iranians to use gas energy was thus to enter into a relationship with the natural world that was mediated by the infrastructures that connected them to the hydrocarbon pools of southwest Iran. For Pahlavi-era urban residents, that relationship extended to natural gas's comparatively clean combustion, and they imagined a city free of the oppressive air pollution that had come to mark their daily lives. Alongside triumphant assertions of technological mastery and the politically legitimating spread of consumer lifeways, Iranians' imaginaries of gas energy encompassed a desire to live in an industrialized society free of the environmental harm that they saw blanketing the developed world. As a technical fix to the spiraling problem of air pollution, natural gas promised the possibility of an environmentalist modernity, a pathway to the prosperity of high energy consumption without the noxious clouds that had already begun to burn Iranians' lungs.

But such environmentalist imaginaries proved to be fragile, succumbing within a few short years to the circumstances of revolution and war. Pahlavi gas policies had prioritized urban industry, a choice that dovetailed with growing concerns about air quality. But for supporters of the revolution like Abdullāh Hushdārān, that decision to favor urban areas had been an unjust one, and given the concentration of factories near Iran's major cities, unravelling it would necessarily slow gas-based air quality fixes. In the early 1980s, moreover, though more than a decade of work had gone into building distribution networks, gas was still unavailable in many parts of the country. As a result, factories were often forced to continue some of their most polluting ways, especially after the destruction of the Ābādān refinery in the early days of the Iran-Iraq War caused widespread shortages of oil products. Air quality regulations were thus less welcome than they had been a few years prior, and many were undermined in court, opposed by interested parties, or simply ignored amidst the turmoil of the early Islamic Republic.[7] Throughout the 1980s and well into the 1990s, environmental restrictions thus went largely unenforced, at times with the help of local authorities.[8] Indeed, while officials of the Islamic Republic's

Department of Environment frequently attempted to prevent the use of non-gas fuels due to their "irredeemable destruction of the environment," for many years such efforts had little enduring effect.[9]

While regulators working under the Islamic Republic found only limited success, their consistent work was indicative of a steady if subdued commitment to environmental protection within at least some government circles. In the mid- and late 1990s, that work began to again coalesce into a more concerted attempt to address the country's intensifying problems with pollution. Reviving and extending policies first formulated in the Pahlavi era, in 1993 Tehran's poor air quality was labeled a "high priority environmental and health issue" by the Iranian government.[10] In response, in that same year, Tehran's municipality established the Air Quality Control Company to monitor air pollution and oversee programs for mitigation. Two years later, Iran's parliament enacted a new Clean Air Act aimed at addressing the country's declining urban air quality through a wide-ranging master plan.[11] Unlike in the 1970s when industrial emissions were the highest priority, by the mid-1990s, Iran's transportation sector had become the primary source of air pollution in places like Tehran.[12] In response, draconian traffic regulations were instituted, major infrastructure projects like the expansion of Tehran's metro system were undertaken, and the use of compressed natural gas in motor vehicles was mandated. Such efforts have largely failed to stop a continuing decline in air quality, however, overcome by rapid population growth, poorly planned urban development, intensifying energy use, and even technical deficiencies in vehicular gas fuel systems. Seeking to foster domestic industry, Iran's postrevolutionary officials prioritized the local manufacture of such systems wherever possible, in the process reinforcing a sociotechnical imaginary of gas that continued to lionize economic sovereignty and technological mastery. But Iranian-made CNG systems were often obsolete and inefficient, and thus notably more polluting than those that could be obtained abroad.[13] In the end, whatever theoretical benefits CNG might have had, they were undermined by technical weakness and an overwhelmingly rapid growth in motor vehicle use in the country. While natural gas had for decades married economic and environmental concerns in Iran, that link was not an inevitable outgrowth of methane's comparatively clean combustion profile. It was instead a product of such materiality combined with Iranians' social and political prioritization of the environment. When that prioritization waned, so did the environmental benefit that natural gas brought.

In the end, fuel conversion programs proved insufficient in the face of much stronger developmental commitments to local industry and the voracious consumption of fossil hydrocarbons. Indeed, the legitimation strategies of Iran's pre- and postrevolutionary governments both extolled industrialization and positioned cheap energy as a core benefit of national belonging. Such was the opportunity that natural gas seemed to offer: a pathway to the widespread enjoyment of prosperous, energy-intensive lifeways, one built on sovereign technological mastery, and without the environmental degradation so frequently seen around the world. But such was gas's danger as well, for try as they might, Iranians could not just substitute natural gas for other fuels and thereby improve their urban air. This history suggests that technical solutions to environmental challenges are often deeply fraught and possibly do not exist. There are complex reasons that widespread gas use failed to improve Iran's urban air quality—population growth, geopolitics, topography, contradictory and mutually-exclusive policy demands—and that is exactly the point. While rooted in the particulars of Iranian history, the ways that natural gas was made into Iran's promised but ultimately failed environmental savior are equally visible in discourses surrounding "clean" energy in the region and around the world. As various technologies for carbon capture and green energy production push their way to the fore as potential solutions to climate change, countries like Iran that have already undertaken widespread experiments in deliberate energy transition offer a cautionary lesson. Simply put, any attempt to address environmental crises cannot begin and end with technology.

A Great Acceleration

The story of natural gas energy in Iran is a deeply national one. It manifested the modernizing ambitions of two regimes and embodied notions of national belonging and the right of Iranians to share their country's hydrocarbon wealth. It is also a tale that reflects the entanglement of Iran's human and natural histories, an intertwined and contingent specificity that gave rise to an energy system upon which Iranians now depend. Natural gas was a tremendous opportunity for the country, offering a way toward a seemingly prosperous, industrialized, and independent society free of air pollution. Many Iranians saw in gas the doorway to a modern lifestyle of cleanliness, convenience, and profligate energy use. Many more understood southwest Iran's enormous natural gas reserves as presenting an opportunity to sidestep foreign control of their country's petroleum resources, a potential outcome supercharged by the country's failed history of oil nationalization. For that reason, and aiming to bolster their legitimacy, leaders of both the prerevolutionary Pahlavi monarchy and the postrevolutionary Islamic Republic made gas a central pillar of their developmentalist agendas. Officials charged with implementing Iran's pre- and postrevolutionary gas policies knew the resource as a means of enabling lucrative oil exports while also feeding the country's rapidly growing appetite for energy. They and urban residents alike, besieged by smoke and haze, also hoped (in vain) that it could reduce air pollution and bring back clear skies. But while the deterioration of Iran's urban air quality was rooted in the specific topographies and industrializing policies of the country, factors that this book attends to, the wider developmental dynamics within which it took place point to connections between such local histories and global currents of anthropogenic environmental change. Often headlined by climate crisis but encompassing a long list of outcomes ranging from dammed rivers to reckless synthetic fertilizer and pesticide use to urban sprawl, the effects of human activity on the world's ecosystems have grown at an accelerating rate in the decades since the Second World War. It is a trend that is driven in large part by rising populations in the Global South and the determination of many leaders there to develop their societies in ways that were predicated on industrialization and intense energy consumption.

Seeking to understand global climate change and the world's accelerating ecological crises, scholars have increasingly turned toward the concept of the Anthropocene to explain, for lack of a better phrase, "how we got here." First proposed by Paul J. Crutzen and Eugene F.

Stoermer in the early 2000s, the term aimed to capture the broad and increasingly obvious changes that humanity had wrought in the Earth's climate and biosphere. Seeing wide-ranging anthropogenic shifts like deforestation, biodiversity loss, and rising atmospheric CO_2 levels as significant enough to constitute a new epoch in Earth's history, the two argued that humanity had become a geological force capable of driving planetary transformations. While the notion of the Anthropocene has in the following years been accepted widely (though by no means universally) within both academic circles and society at large, considerable debate nonetheless remains about its possible periodization. Crutzen and Stoermer themselves suggested the late eighteenth century and the dawn of the industrial age as its start, rooting their proposal in the utilization of fossil fuels and the consequent rise of atmospheric carbon dioxide levels.[14] This perspective finds support in the literature on the rise of coal use and its relationship with both European colonialism and the advent of industrial capitalism.[15] In such accounts, the ability to unlock the caloric potential of coal was key to Britain's ability to surpass the seemingly Malthusian limits imposed by land and population. Andreas Malm's work on the history of industrial capitalism has, however, done much to deconstruct naturalized assumptions about coal and steam power's inherent advantage over other means of production in the period.[16] The history he charts, where the prime driver of steam power's adoption was not obvious energetic superiority but the ability of capitalists to situate factories near preexisting pools of human labor, points to the centrality of social and political concerns in the history of energy and climate.

Underlying the idea of the Anthropocene's birth in the Industrial Revolution is, however, a prioritization of carbon emissions as the most salient measure of people's influence on the global environment, a compelling choice in light of the world's climate crises but one that is not universally agreed upon. Other suggested periodizations include the controlled use of fire by humans (some 2.0 million years ago); the extinction of megafauna at human hands (between 50,000 and 10,000 years ago); the development of settled agriculture (8,000 years ago); contact between Afro-Eurasia and the Americas in the fifteenth-century and the Columbian exchange; and the 1945 dawn of the nuclear age and its worldwide depositing of radioactive residues. While disagreement has surrounded each of these proposals and their compatibility with the definitional requirements of a formal geological epoch, each reflects significant human influence on the earth's eco- and planetary systems.[17] But many such definitions are also at root backward projections of developments that took place much later, particularly the rise of atmospheric carbon dioxide levels, and ignore the fact that the Anthropocene's most intense effects have occurred in the decades after the Second World War. In other words, the effects of human action on global environments have been strongest in exactly the era when much of the Global South undertook sprawling programs of industrializing development.

Termed the Great Acceleration by J.R. McNeill and Peter Engelke, the eight decades that followed the Second World War have been in large part defined by a worldwide surge in population and resource consumption that is actively refiguring the earth's surface and atmosphere. The global population more than tripled in that time, from some 2.5 billion to over 8.0 billion people, while societies have become increasingly urbanized. Alongside such explosive growth came a voracious demand for food, materials, and energy, demand driven not just by a raw increase in population but also a general rise in living standards and expectations thereof. While energy consumption had doubled between 1900 and 1950—mostly on the back of coal, the great fuel of the Industrial Revolution—the postwar decades saw it quintuple again, an increase this time enabled in large part by oil and natural gas. The ability to utilize enormous quantities of energy refigured societies and politics alike, as over the previous century the combination of industrialization and extractive colonial empire had concentrated energy-intensive lifeways in Europe and North America. The ambitions of many newly decolonized countries in the Global South to follow a similar path of high energy use and intensive resource consumption would in the postwar decades begin to undermine the supremacy of the world's industrialized North, a process that is still ongoing but has nonetheless driven much of the Great Acceleration's sweeping changes. Indeed, after 1965, countries like China and India saw their energy use increase by sixteen and eleven times, respectively, while over the same period that of the United States rose by only 40 percent, dropping the U.S.'s share of the world's total energy consumption from one-third in the 1960s to one-fifth in 2009. Though different regions have grown at faster or slower rates or industrialized differently in that time, on the whole the second half of the twentieth century was marked by "spectacular growth" in both the human population and its levels of consumption, a set of paired trends at the center of which sat the countries of the developing world.[18]

Understanding how and why that acceleration took place is a goal that lurks behind the national history this book has recounted. While the Great Acceleration and the environmental crises it has spawned can be understood on the level of (human) species and planet, they are at the same time the product of a collection of events and choices made by a large number of people over decades of time in communities all over the world. Some scholars have sought the foundations of those choices in the century prior and the state and empire building that coal enabled. For Victor Seow, coal was the linchpin around which the modern technocratic states were created in East Asia, a "mutual production of calorific and political power" that reflected both the inter-state rivalries of the period and a political culture that employed comparative energy use as a proxy for civilizational worth.[19] But "energy" itself is a historically situated and contested notion, one that came to be overdetermined by an energy-as-work paradigm that framed European and, later, American, imperial projects around the world.[20] Refiguring Dipesh Chakrabarty's call to provincialize Europe by rejecting the presumed universality of its social theories, On Barak has sought to provincialize energy by centering the coal-powered trade networks that connected metropole and periphery in the British Empire—making the history of coal and climate change inherently one of empire and colony as well.[21] Working to excavate the roots of our "addictions" to fossil hydrocarbons and consumption patterns they enable, both Barak and Seow complicate a geographic emphasis on Europe and North America in the history of fossil fuel use, offering important revisionist accounts for how hydrocarbons have become central to our social (and ultimately climatic) worlds.[22] By focusing exclusively on coal, however, they also reproduce a broader analytical emphasis on societies that industrialized comparatively early, ignoring the rapid rise in consumption and emissions of those that did so after World War II. Indeed, centering other forms of energy prompts different Anthropocene periodizations and illuminates important discontinuities between earlier histories and later ones, particularly the significance of nationalism and developmentalist thinking.[23] As this book has traced in postwar Iranian gas utilization, the Great Acceleration was made in such connections, a blending of industrialization, decolonization, and fossil hydrocarbons that catapulted societies of the Global South into voracious energy consumption.

Informed by specific local, national, and regional conditions, officials in places like Iran pursued policies aimed at transforming their societies in ways that promoted economic growth, expanded resource extraction, and intensified energy consumption—policies often understood and grouped under the headings of "development" and "modernization." Those policies were imbricated with other distinct political goals that not only shaped their implementation but also the social and economic changes that came together across the world to give rise to the Great Acceleration. The pursuit of energy-intensive industrialization was not undertaken for its own sake or as an expression of some historical force favoring growth and greater consumption; rather, leaders in the developing world pursued such policies because they served their own political and social ends. Even accounting for their universalizing aspirations and assumptions, the developmentalist paradigms pushed by the United States and the Soviet Union during the Cold War—a period encompassing much of the Great Acceleration's early decades—largely served narrower national purposes. They moreover empowered the state in much of the decolonizing world. For many newly postcolonial elites, development was a way to legitimize their rule, while the competing world powers promoted it to secure strategic resources and grow markets. "Development" and "modernization" were thus multiple, and it was only in the ideological precommitments of their boosters and in their post hoc aggregation by scholars that such parochial interests amounted to a seemingly broad-based and inevitable trend.[24]

The sharply diverging interests of developing and industrialized nations moreover extended to the budding movement toward environmental governance in the 1970s and after, particularly in relation to the extent to which environmental considerations should be allowed to curb economic growth. While environmental challenges like acid rain and ozone depletion were understood as being transnational or even global in scope, policies to address them were largely channeled through national governments that had significant competing demands placed upon them. Having fought in the postwar years to assert sovereignty over their own resources and political futures, at international forums like the 1972 "Only One Earth" conference in Stockholm, leaders from the Global South resisted efforts to slow economic growth or develop in less energy- or resource-intensive ways, arguing that wealthy nations were hypocritical in reaping the benefits of industrialization while demanding that their people go without. In the following years, environmental regulation would thus be increasingly wedded to economic growth under the rubric of "sustainable development," a framing that in most cases did little to slow the growth of energy consumption and resource extraction.[25] Rather than treating the Great Acceleration from a singular global perspective, excavating the particular, if overlapping, histories that comprise it is essential for understanding how and why societies in the Global South became organized around higher energy use and resource consumption. It was a trend that was not a teleological inevitability, but rather one that reflected changing technologies, changing expectations, and a changing world order.

By focusing on postwar Iranian developmentalism and its entanglement with natural gas, this book situates the Great Acceleration within the particulars of history, or at least one corner of it, with the goal of uncovering the choices and aspirations from which it arose. It thus aims to complicate what has been heretofore rendered artificially uniform. In recent years, scholars have begun to critique the analytical and moral flattening that the Anthropocene's species-level viewpoint produces, turning the focus of inquiry from a misleadingly homogenous "humanity" to the particular and often profoundly unequal histories that have come together to produce widespread environmental change. Notions like the Capitalocene and the Plantationocene center processes of capitalist accumulation and their rendering of nature as a source of cheap material inputs and as a dumping ground for waste, in the process tracing the extractivist logics that connect the histories of the exploitation of nature to the exploitation of people.[26] Such understandings nonetheless reproduce the human-centeredness that structures the idea of the Anthropocene and the environmental violence it seeks to describe, thereby overlooking the more-than-human assemblages that arise from the entanglements of society and the natural world.[27] Objecting to the implicit teleology of increasing human mastery over nature present in the idea of the Anthropocene and its derivatives, Anna L. Tsing instead suggests centering notions of precarity, "the condition of being vulnerable to others," and contingency, the "unpredictable encounters [that] transform us," in our analyses of more-than-human interactions.[28] While Tsing is focused on uncovering the alternatives that exist at the margins of capitalist modernity and its alienation of nature, the point remains relevant even at the heart of capitalism, helping us see the contingency and indeterminacy that exist within spaces like those of Iran's petroleum industries. Rather than focusing only on the ways Iranian planners and officials overcame or mastered the challenges of the natural world, this book attends to how those same natural forces shaped the creation of Iran's natural gas infrastructure, expanding the frame of human-nature interactions to include nonliving factors—topography, the fracture systems of limestone rock, the compositions and pressures of underground petroleum reserves—as crucial influences on the construction of a society dependent on vast hydrocarbon abundance. Viewing the past through universalizing perspectives like the Anthropocene is insufficient for fully comprehending our world's widespread environmental changes, requiring instead deeper examinations of local and national histories. Iran's pathway to high energy consumption was not identical to others, not only because of the country's human history but also because of its natural one. Paying heed to their connections is necessary to see that Iranians not only eagerly joined in Cold War-era developmentalism for their own political and economic reasons, but that they did so in ways that reflected the natural specificities of their country. The abstract global aggregates of the Anthropocene and Great Acceleration have little to say on such a subject, their analytical framing unsuited to understanding why and how Iranians—or any other polity—came to rely on fossilized carbon.

***

The industrialized society that Iranians built in the late twentieth century was from the beginning stalked by the dangers of environmental violence. Despite the hopes that Iranian officials placed on it, natural gas utilization largely failed to produce clean air in Iran's urban areas. Far from unique, the environmental promises that Iranians projected on natural gas are mirrored in its positioning as a transitional fuel in the fight against global climate change.[29] But despite its comparative cleanliness, gas utilization could simply not overcome the raw growth of energy demand in the country, ultimately failing to stem the decline of air quality and leaving places like Tehran some of the most polluted in the world. This too mirrors the story of natural gas on the world stage, as carbon emissions have continued to rise and maintain high levels even as gas consumption has risen.[30] The story of Iranian encounters with gas energy demonstrates a longer braiding of hydrocarbon use and environmental concern than has been previously understood—a history that complicates a straightforward equating of fossil fuel combustion with environmental ignorance and disregard. At the same time, its failures and insufficiencies are perhaps an ominous foreshadowing of our own inability to effect environmental change through technological solutions. Uncomplicated ideas of replacing fossil fuels with renewables like wind and solar overlook the fact that historical energy transitions have been far from smooth, being not only lengthy and uneven processes, but also intensifying the use of existing fuels in the meantime.[31] They also ignore the historical circumstances that have pushed developing societies like Iran to embrace intense hydrocarbon usage as a politically charged expression of national sovereignty.

For Iranians, natural gas utilization was rooted in the politics of resource nationalism and a desire to enjoy the material benefits of industrialized prosperity. The country's very high per-capita energy consumption is an outcome of those twin impulses. It was also a trajectory paved in part by environmentalist hopes. Such tangled economic, political, and environmental motivations reflect the centrality of energy infrastructures to contemporary societies and the sometimes divergent, sometimes congruent, but always seemingly better futures that Iranians imagined with natural gas. But natural gas has been a double-edged sword, one with which Iranians asserted their country's independence while simultaneously embedding themselves deeper into natural processes that threaten their destruction. While state officials largely succeeded in using gas to make their developmental ambitions a reality, that carbon-intensive path has rebounded heavily on the country, with climate change helping turn Iran into one of the most water-stressed and drought-prone regions on earth.[32] For Iranians, the path to an energy-intensive society was forged through an overlapping set of interests and incentives animated by a powerful anticolonial push for economic sovereignty. Tracing that history, as this book has done, is key to understanding not only how modern Iranian society was built, but also how and why the promise of development has curdled into the terrifying reality of climate crisis. That on some level Iranians succeeded in their aims while contributing to the conditions of civilizational ruin is neither the point nor beside it; rather, it is that our world's unfolding climate crises are the outcome of histories like it, each reflecting distinct combinations of socioeconomic conditions, political circumstances, and the natural world. This is perhaps the broadest lesson of Iran's history with natural gas: that the global phenomenon of the Great Acceleration was both multiple and deeply historical, created by overlapping and contingent entanglements of human and natural histories. It is also the story of our world's unfolding ecological disasters, not of a heedless *Homo sapiens* burning the past to feed the present at the cost of the future, but of political opportunities seized, powerful national aspirations fulfilled, and catastrophic environmental dangers overlooked.

Notes

Introduction

1. *Energy Institute Statistical Review of World Energy 2024*, "Gas: Consumption – Bcf (from 1965)," available at https://www.energyinst.org/statistical-review (accessed May 9, 2025).

2. Daniel Yergin, *The Prize: The Epic Quest for Oil, Money & Power* (New York: Free Press, 1991); Daniel Yergin, *The Quest: Energy, Security, and the Remaking of the Modern World* (New York: The Penguin Press, 2011).

3. H. Mahdavy, "The Patterns and Problems of Economic Development in Rentier States: The Case of Iran," in *Studies in the Economic History of the Middle East*, ed. M.A. Cook (London: Routledge, 1970). See also Michael L. Ross, *The Oil Curse: How Petroleum Wealth Shapes the Development of Nations* (Princeton, NJ: Princeton University Press, 2012) for a synthesis of the large body of literature on this topic.

4. Mostafa Elm, *Oil, Power, and Principle: Iran's Oil Nationalization and its Aftermath* (Syracuse, NY: Syracuse University Press, 1992); Ervand Abrahamian, *Oil Crisis in Iran: From Nationalism to Coup d'Etat* (Cambridge, UK: Cambridge University Press, 2021); David S. Painter and Gregory Brew, *The Struggle for Iran: Oil, Autocracy, and the Cold War, 1951–1954* (Chapel Hill, NC: The University of North Carolina Press, 2022).

5. Mattin Biglari, "Iranian Oil Nationalisation as Decolonisation: Historiographical Reflections, Global History, and Postcolonial Theory," in *Iran*

and Global Decolonisation: Politics and Resistance After Empire, eds. Firoozeh Kashani-Sabet and Robert Steele (London: Gingko, 2023), 101–133. See also Hamid Dabashi, *Iran: A People Interrupted* (New York: The New Press, 2007).

6. James M. Gustafson, *The Lion and the Sun: Environmental History and the Formation of Modern Iran* (London: I.B. Tauris, 2025), 175–193.

7. Homa Katouzian, "Riza Shah's Political Legitimacy and Social Base, 1921–1941," in *The Making of Modern Iran: State and Society Under Riza Shah, 1921–1941*, ed. Stephanie Cronin (London: Routledge, 2003), 15–36.

8. Shahbaz Shahnavaz, *Britain and the Opening Up of South-West Persia, 1880–1914: A Study in Imperialism and Economic Dependence* (London: RoutledgeCurzon, 2005).

9. Arash Khazeni, *Tribes and Empire on the Margins of Nineteenth-Century Iran* (Seattle, WA: University of Washington Press, 2009).

10. Leonardo Davoudi, *Persian Petroleum: Oil, Empire and Revolution in Late Qajar Iran* (London: I.B. Tauris, 2021).

11. Touraj Atabaki, "Indian Migrant Workers in the Iranian Oil Industry," in *Working for Oil: Comparative Social Histories of Labor in the Global Oil Industry*," eds. Touraj Atabaki et al. (Cham, Switzerland: Palgrave Macmillan, 2018), 189–226; Kaveh Ehsani, "Social Engineering and the Contradictions of Modernization in Khuzestan's Company Towns: A Look at Abadan and Masjed-Soleyman," *International Review of Social History* 48, no. 3 (2003), 361–399; Robert Vitalis, *America's Kingdom: Mythmaking on the Saudi Oil Frontier* (Stanford, CA: Stanford University Press, 2007).

12. Valérie Marcel and John V. Mitchell, *Oil Titans: National Oil Companies in the Middle East* (Baltimore, MD: Brookings Institution Press, 2006).

13. Christopher R.W. Dietrich, *Oil Revolution: Anticolonial Elites, Sovereign Rights, and the Economic Culture of Decolonization* (Cambridge, UK: Cambridge University Press, 2017).

14. Timothy Mitchell, *Carbon Democracy: Political Power in the Age of Oil* (London: Verso, 2011).

15. Mandana E. Limbert, *In the Time of Oil: Piety, Memory, and Social Life in an Omani Town* (Stanford, CA: Stanford University Press, 2010); Matthew T. Huber, *Lifeblood: Oil, Freedom, and the Forces of Capital* (Minneapolis, MN: University of Minnesota Press, 2013); Bob Johnson, *Carbon Nation: Fossil Fuels in the Making of American Culture* (Lawrence, KA: University Press of Kansas, 2014); Arbella Bet-Shlimon, *City of Black Gold: Oil, Ethnicity, and the Making of Modern Kirkuk* (Stanford, CA: Stanford University Press, 2019). See also Sheena Wilson, Adam Carlson, and Imre Szeman, eds., *Petrocultures: Oil, Politics, Culture* (Montreal, Canada: McGill-Queen's University Press, 2017).

16. David G. Victor, Amy M. Jaffe, and Mark H. Hayes, eds., *Natural Gas and Geopolitics: From 1970 to 2040* (Cambridge, UK: Cambridge University Press, 2006); Per Högselius, *Red Gas: Russia and the Origins of European Energy Dependence* (New York: Palgrave Macmillan, 2013); Jeronim Perović, ed., *Cold War Energy: A Transnational History of Soviet Oil and Gas* (Cham, Switzerland: Palgrave Macmillan, 2016).

17. Zsuzsa Gille, *From the Cult of Waste to the Trash Heap of History: The Politics of Waste in Socialist and Postsocialist Hungary* (Bloomington, IN: Indiana University Press, 2007). See also Oran R. Young, *Resource Regimes: Natural Resources and Social Institutions* (Berkeley, CA: University of California Press, 1982).

18. Mary Douglas, *Purity and Danger: An Analysis of Concepts of Pollution and Taboo* (London: Routledge, 1966, 2002).

19. Vittoria Di Palma, *Wasteland: A History* (New Haven, CT: Yale University Press, 2014).

20. Brian Larkin, "The Politics and Poetics of Infrastructure," *Annual Review of Anthropology* 42 (2013), 328.

21. Langdon Winner, "Do Artifacts Have Politics?," *Daedalus* 109, no. 1 (1980), 121–136; Timothy Mitchell, *Rule of Experts: Egypt, Techno-Politics, Modernity* (Berkeley, CA: University of California Press, 2002), especially the first chapter "Can the Mosquito Speak?;" On Barak, *On Time: Technology and Temporality in Modern Egypt* (Berkeley, CA: University of California Press, 2013); Ronen Shamir, *Current Flow: The Electrification of Palestine* (Stanford, CA: Stanford University Press, 2013); Jessica Barnes, *Cultivating the Nile: The Everyday Politics of Water in Egypt* (Durham, NC: Duke University Press, 2014); Jennifer L. Derr, *The Lived Nile: Environment, Disease, and Material Colonial Economy in Egypt* (Stanford, CA: Stanford University Press, 2019).

22. Bruno Latour, *The Pasteurization of France*, trans. Alan Sheridan and John Law (Cambridge, MA: Harvard University Press, 1988); Bruno Latour, *We Have Never Been Modern*, trans. Catherine Porter (Cambridge, MA: Harvard University Press, 1993); Bruno Latour, *Aramis, or the Love of Technology*, trans. Catherine Porter (Cambridge, MA: Harvard University Press, 1996).

23. Donna Haraway, "A Cyborg Manifesto: Science, Technology, and Socialist-Feminism in the Late Twentieth Century," *Socialist Review* 15 (1985), 65–108; Zoe Todd, "An Indigenous Feminist's Take on the Ontological Turn: 'Ontology' is Just Another Word for Colonialism," *Journal of Historical Sociology* 29, no. 1 (2016), 4–22.

24. Langdon Winner, "Upon Opening the Black Box and Finding it Empty: Social Constructivism and the Philosophy of Technology," *Science, Technology,*

& Human Values 18, no. 3 (1993), 362–378; Andreas Malm, *The Progress of this Storm: Nature and Society in a Warming World* (London: Verso, 2018).

25. Jane Bennett, "The Agency of Assemblages and the North American Blackout," *Public Culture* 17, no. 3 (2005), 445, n. 2.

26. Jane Bennett, *Vibrant Matter: A Political Ecology of Things* (Durham, NC: Duke University Press, 2010), 31.

27. Ciruce A. Movahedi-Lankarani, "Precarious Petroleum: Volatile Reservoirs, Varied Natural Gas Compositions, and Development in 1960s Iran," *Comparative Studies of South Asia, Africa and the Middle East* 44, no.1 (2024), 3–17.

28. Gabrielle Hecht, *The Radiance of France: Nuclear Power and National Identity after World War II* (Cambridge, MA: MIT Press, 1998), 15.

29. Katayoun Shafiee, *Machineries of Oil: An Infrastructural History of BP in Iran* (Cambridge, MA: MIT Press, 2018), 7.

30. Geoffrey C. Bowker, *Science on the Run: Information Management and Industrial Geophysics at Schlumberger, 1920–1940* (Cambridge, MA: MIT Press, 1994).

31. Christopher F. Jones, *Routes of Power: Energy and Modern America* (Cambridge, MA: Harvard University Press, 2014).

32. Penny Harvey and Hannah Knox, *Roads: An Anthropology of Infrastructure and Expertise* (Ithaca, NY: Cornell University Press, 2015).

33. Patrick Clawson, "Knitting Iran Together: The Land Transport Revolution, 1920–1940," *Iranian Studies* 26, no. 3–4 (Summer/Fall 1993), 235–250.

34. Ervand Abrahamian, *A History of Modern Iran* (Cambridge, UK: Cambridge University Press, 2008), 92–93.

35. Amin Banani, *The Modernization of Iran, 1921–1941* (Stanford, CA: Stanford University Press, 1961); Stephanie Cronin, *The Making of Modern Iran: State and Society Under Riza Shah, 1921–1941* (New York: Routledge, 2003); Touraj Atabaki and Erik J. Zürcher, *Men of Order: Authoritarian Modernization Under Atatürk and Reza Shah* (London: I.B. Tauris & Co. Ltd, 2004).

36. Ali Gheissari, *Iranian Intellectuals in the 20th Century* (Austin, TX: University of Texas Press, 1998); Ali Mirsepassi, *Intellectual Discourse and the Politics of Modernization: Negotiating Modernity in Iran* (Cambridge, UK: Cambridge University Press, 2000); Farzin Vahdat, *God and Juggernaut: Iran's Intellectual Encounter with Modernity* (Syracuse, NY: Syracuse University Press, 2002); Farhang Rajaee, *Islamism and Modernism: The Changing Discourse in Iran* (Austin, TX: University of Texas Press, 2007); Afshin Marashi, *Nationalizing Iran: Culture, Power, and the State, 1870–1940* (Seattle, WA: University of Washington Press, 2008); Ali Mirsepassi, *Political Islam, Iran, and the Enlightenment: Philosophies of Hope and Despair* (Cambridge, UK: Cambridge University Press, 2011); Ali M. Ansari, *The*

Politics of Nationalism in Modern Iran (Cambridge, UK: Cambridge University Press, 2012); Reza Zia-Ebrahimi, "'Arab Invasion' and Decline, Or the Import of European Racial Thought by Iranian Nationalists," *Ethnic and Racial Studies* 37, no. 6 (2014), 1043–1061; Ali Mirsepassi, *Iran's Troubled Modernity: Debating Ahmad Fardid's Legacy* (Cambridge, UK: Cambridge University Press, 2019).

37. Cyrus Schayegh, *Who Is Knowledgeable Is Strong: Science, Class, and the Formation of Modern Iranian Society* (Berkeley, CA: University of California Press, 2009); Firoozeh Kashani-Sabet, *Conceiving Citizens: Women and the Politics of Motherhood in Iran* (Oxford, UK: Oxford University Press, 2011); Firoozeh Kashani-Sabet, "Dressing Up (or Down): Veils, Hats, and Consumer Fashions in Interwar Iran," in *Anti-Veiling Campaigns in the Muslim World: Gender, Modernism and the Politics of Dress*, ed. by Stephanie Cronin, 149–162 (New York: Routledge, 2014).

38. Camron Michael Amin, *The Making of the Modern Iranian Woman: Gender, State Policy, and Popular Culture, 1865–1946* (Gainesville, FL: University of Florida Press, 2002); Afsaneh Najmabadi, *Women with Mustaches and Men Without Beards: Gender and Sexual Anxieties of Iranian Modernity* (Berkeley, CA: University of California Press, 2005); Janet Afary, *Sexual Politics in Modern Iran* (Cambridge, UK: Cambridge University Press, 2009).

39. Ervand Abrahamian, *Iran: Between Two Revolutions* (Princeton: Princeton University Press, 1982), 146–147.

40. Firoozeh Kashani-Sabet, *Heroes to Hostages: America and Iran, 1800–1988* (Cambridge, UK: Cambridge University Press, 2023), 221–238.

41. Mahdavy, "Patterns and Problems;" Farhad Daftary, "Development Planning in Iran: A Historical Survey," *Iranian Studies* 6, no. 4 (1973), 176–228; Homa Katouzian, *The Political Economy of Modern Iran: Despotism and Pseudo-Modernism, 1926–1979* (New York: New York University Press, 1981), 213–331; Hooshang Amirahmadi and Farhad Atash, "Dynamics of Provincial Development and Disparity in Iran, 1956–1984," *Third World Planning Review* 9, no. 2 (1987), 155–185; Kamran Mofid, *Development Planning in Iran: From Monarchy to Islamic Republic* (Cambridgeshire, UK: Middle East and North Africa Press, Ltd., 1987); Hadi Salehi Esfahani and M. Hashem Pesaran, "The Iranian Economy in the Twentieth Century: A Global Perspective," *Iranian Studies* 42, no. 2 (2009), 177–211; Pooya Azadi, Mohsen B. Mesgaran, and Matin Mirramezani, *The Struggle for Development in Iran: The Evolution of Governance, Economy, and Society* (Stanford, CA: Stanford University Press, 2022).

42. Hossein Razavi and Firouz Vakil, *The Political Environment of Economic Planning in Iran, 1971–1983: From Monarchy to Islamic Republic* (Boulder, CO: Westview Press, 1984); Frances Bostock and Geoffrey Jones, *Planning and Power*

in Iran: Ebtehaj and Economic Development Under the Shah (London: Frank Cass and Company Limited, 1989); Azadeh Mashayekhi, "The 1968 Tehran Master Plan and the Politics of Planning Development in Iran (1945–1979)," *Planning Perspectives* 34, no. 5 (2019), 849–876.

43. Daniel Lerner, *The Passing of Traditional Society: Modernizing the Middle East* (New York: The Free Press, 1958); W.W. Rostow, *The Stages of Economic Growth: A Non-Communist Manifesto* (Cambridge, UK: Cambridge University Press, 1960); David E. Apter, *The Politics of Modernization* (Chicago: The University of Chicago Press, 1965).

44. Ramin Nassehi, "Domesticating Cold War Economic Ideas: The Rise of Iranian Developmentalism in the 1950s and 1960s," in *The Age of Aryamehr: Late Pahlavi Iran and Its Global Entanglements*, ed. Roham Alvandi (London: Gingko Library, 2018), 35–69.

45. Michael Adas, "Modernization Theory and the American Revival of the Scientific and Technological Standards of Social Achievement and Human Worth," in *Staging Growth: Modernization, Development, and the Global Cold War*, eds. David C. Engerman, Nils Gilman, Mark H. Haefele, and Michael E. Latham (Amherst, MA: University of Massachusetts Press, 2003), 39.

46. Matías Dewey and Kedron Thomas, "Futurity Beyond the State: Illegal Markets and Imagined Futures in Latin America," *Latin American Politics and Society* 64, no. 4 (2022), 1–23.

47. Arjun Appadurai, *The Future as Cultural Fact: Essays on the Global Condition* (London: Verso, 2013), 265. See also Wendell Bell, "What Do We Mean by Futures Studies?," in *New Thinking for A New Millennium*, ed. Richard A. Slaughter (London: Routledge, 1996).

48. Daniel Rosenberg and Susan Harding, "Introduction: Histories of the Future," in *Histories of the Future*, eds Daniel Rosenberg and Susan Harding, 3–18 (Durham, NC: Duke University Press, 2005).

49. Victoria De Grazia, *Irresistible Empire: America's Advance Through Twentieth-Century Europe* (Cambridge, MA: The Belknap Press of Harvard University Press, 2005), 1–4.

50. Michael E. Latham, *Modernization as Ideology: American Social Science and "Nation Building" in the Kennedy Era* (Chapel Hill, NC: University of North Carolina Press, 2000); Nils Gilman, *Mandarins of the Future: Modernization Theory in Cold War America* (Baltimore, MD: Johns Hopkins University Press, 2003); David Ekbladh, *The Great American Mission: Modernization and the Construction of an American World Order* (Princeton, NJ: Princeton University Press, 2010).

51. Thomas M. Ricks, "U.S. Military Missions To Iran, 1943–1978: The Political Economy of Military Assistance," *Iranian Studies* 12, no. 3–4 (1979), 163–193;

April R. Summitt, "For a White Revolution: John F. Kennedy and the Shah of Iran," *Middle East Journal* 58, no. 4 (2004), 560–575; Victor V. Nemchenok, "'That So Fair a Thing Should Be So Frail:' The Ford Foundation and the Failure of Rural Development in Iran, 1953–1964," *Middle East Journal* 63, no. 2 (2009), 261–284; Jasamin Rostam-Kolayi, "The New Frontier Meets the White Revolution: The Peace Corps in Iran, 1962–76," *Iranian Studies* 51, no. 4 (2018), 587–612; Pamela Karimi, *Domesticity and Consumer Culture in Iran: Interior Revolutions of the Modern Era* (New York: Routledge, 2013), 84–119.

52. Gregory Brew, *Petroleum and Progress in Iran: Oil, Development, and the Cold War* (Cambridge, UK: Cambridge University Press, 2022).

53. The literatures on dependency theory and world-systems analysis are vast and contested. See, for example, Paul A. Baran, *The Political Economy of Growth* (New York: Monthly Review Press, 1957) and Immanuel Wallerstein, *The Modern World-System I: Capitalist Agriculture and the Origins of the European World-Economy in the Sixteenth Century* (New York: Academic Press, 1974).

54. Fred Halliday, *Iran: Dictatorship and Development* (New York: Penguin Books, Ltd., 1979).

55. J. Jasper Deuten and Arie Rip, "The Narrative Shaping of a Production Creation Process," in *Contested Futures: A Sociology of Prospective Techno-Science*, eds. Nik Brown, Brian Rappert, and Andrew Webster (London: Routledge, 2000).

56. Harro van Lente, "Forceful Futures: From Promise to Requirement," in *Contested Futures: A Sociology of Prospective Techno-Science*, ed. Nik Brown, Brian Rappert, and Andrew Webster (London: Routledge, 2000).

57. Cyrus Schayegh, "Iran's Global Long 1970s: An Empire Project, Civilisational Developmentalism, and the Crisis of the Global North," in *The Age of Aryamehr: Late Pahlavi Iran and Its Global Entanglements*, ed. Roham Alvandi (London: Gingko, 2018); Robert Steele, *The Shah's Imperial Celebrations of 1971: Nationalism, Culture and Politics in Late Pahlavi Iran* (London: I.B. Tauris, 2021); Edward F. Fischer, *The Good Life: Aspiration, Dignity, and the Anthropology of Wellbeing* (Stanford, CA: Stanford University Press, 2014).

58. Ali M. Ansari, "The Myth of the White Revolution: Mohammad Reza Shah, 'Modernization' and the Consolidation of Power," *Middle Eastern Studies* 37, no. 3 (July 2001), 1–24.

59. Cyrus Schayegh, "'Seeing Like a State': An Essay on the Historiography of Modern Iran," *International Journal of Middle East Studies* 42, no. 1 (2010), 37–61.

60. Cyrus Schayegh, "Iran's Karaj Dam Affair: Emerging Mass Consumerism, the Politics of Promise, and the Cold War in the Third World," *Comparative Studies in Society and History* 54, no. 3 (2012), 626.

61. Antina von Schnitzler, *Democracy's Infrastructure: Techno-Politics and Protest after Apartheid* (Princeton, NJ: Princeton University Press, 2016); Charlotte Lemanski, "Infrastructural Citizenship: The Everyday Citizenships of Adapting and/or Destroying Public Infrastructure in Cape Town, South Africa," *Transactions of the Institute of British Geographers* 45, no. 3 (2019), 1–17.

62. Mikiya Koyagi, *Iran in Motion: Mobility, Space, and the Trans-Iranian Railway* (Stanford, CA: Stanford University Press, 2021), 190. See also Nikhil Anand, *Hydraulic City: Water and the Infrastructures of Citizenship in Mumbai* (Durham, NC: Duke University Press, 2017).

63. Brian Larkin, *Signal and Noise: Media, Infrastructure, and Urban Culture in Nigeria* (Durham, NC: Duke University Press, 2008), 245.

64. Arturo Escobar, *Encountering Development: The Making and Unmaking of the Third World* (Princeton, NJ: Princeton University Press, 1995, 2012), 10.

65. James Ferguson, *The Anti-Politics Machine: "Development," Depoliticization, and Bureaucratic Power in Lesotho* (Minneapolis, MN: University of Minnesota Press, 1990, 1994).

66. Escobar, *Encountering Development*, 6.

67. James C. Scott, *Seeing Like a State: How Certain Schemes to Improve the Human Condition Have Failed* (New Haven, CT: Yale University Press, 1998).

68. Sara Lorenzini, *Global Development: A Cold War History* (Princeton, NJ: Princeton University Press, 2019).

69. David Blackbourn, *The Conquest of Nature: Water, Landscape, and the Making of Modern Germany* (New York: W.W. Norton & Company, 2006); Toby Craig Jones, *Desert Kingdom: How Oil and Water Forged Modern Saudi Arabia* (Cambridge, MA: Harvard University Press, 2010); Erik Swyngedouw, *Liquid Power: Contested Hydro-Modernities in Twentieth-Century Spain* (Cambridge, MA: Harvard University Press, 2015); Harry Verhoeven, *Water, Civilisation and Power in Sudan: The Political Economy of Military-Islamist State Building* (Cambridge, UK: Cambridge University Press, 2015).

70. Sheila Jasanoff, "Future Imperfect: Science, Technology, and the Imaginations of Modernity," in *Dreamscapes of Modernity: Sociotechnical Imaginaries and the Fabrication of Power*, eds. Sheila Jasanoff and Sang-Hyun Kim, pp. 1–33 (Chicago: University of Chicago Press, 2015), 4. This is a revision of the definition of a sociotechnical imaginary first proposed by Jasanoff and Kim in "Containing the Atom: Sociotechnical Imaginaries and Nuclear Power in the United States and South Korea," *Minerva* 47, no. 2 (June 2009): 119–146.

71. M.M. Bakhtin, "Discourse in the Novel," in *The Dialogic Imagination: Four Essays*, trans. Caryl Emerson and Michael Holquist, ed. Michael Holquist, 259–422 (Austin, TX: University of Texas Press, 1981, 2014).

72. Shaden M. Tageldin, *Disarming Words: Empire and the Seductions of Translation in Egypt* (Berkeley, CA: University of California Press, 2011), 7.

73. Sarah Babb, *Managing Mexico: Economists from Nationalism to Neoliberalism* (Princeton, NJ: Princeton University Press, 2001).

74. Mana Kia, *Persianate Selves: Memories of Place and Origin Before Nationalism* (Stanford, CA: Stanford University Press, 2020).

75. Firoozeh Kashani-Sabet, *Frontier Fictions: Shaping the Iranian Nation, 1804–1946* (Princeton, NJ: Princeton University Press, 1999).

76. David E. Nye, *American Technological Sublime* (Cambridge, MA: MIT Press, 1994); Graeme Macdonald, "Containing Oil: The Pipeline in Petroculture" in *Petrocultures: Oil, Politics, Culture*, eds. Sheena Wilson, Adam Carlson, and Imre Szeman, 36–77 (Montreal, Canada: McGill-Queen's University Press, 2017).

77. Farhad Atash, "The Deterioration of Urban Environments in Developing Countries: Mitigating the Air Pollution Crisis in Tehran, Iran," *Cities* 24, no. 6 (2007), 399–409; Tahereh Saheb, "Air Pollution Governance in Iran: Inhibiting Factors" (PhD diss., Rensselaer Polytechnic Institute, 2015); Vahid Hosseini and Hossein Shahbazi, "Urban Air Pollution in Iran," *Iranian Studies* 49, no. 6 (2016), 1029–1046. For more on Iran's environmentalist movements, see Kaveh L. Afrasiabi, "The Environmental Movement in Iran: Perspectives from Below and Above," *The Middle East Journal* 57, no. 3 (2003), 432–448 and Simin Fadaee, *Social Movements in Iran: Environmentalism and Civil Society* (New York: Routledge, 2012).

78. Emily Brownell, *Gone to Ground: A History of Environment and Infrastructure in Dar es Salaam* (Pittsburgh, PA: University of Pittsburgh Press, 2020).

79. Brett L. Walker, *Toxic Archipelago: A History of Industrial Disease in Japan* (Seattle, WA: University of Washington Press, 2010), 16.

80. Zoé Chateau et al., "Integrating Sociotechnical and Spatial Imaginaries in Researching Energy Futures," *Energy Research & Social Science* 80 (2021), 102–207.

81. Carola Hein and Mohamad Sedighi, "Iran's Global Petroleumscape: The Role of Oil in Shaping Khuzestan and Tehran," *Architectural Theory Review* 21, no. 3 (2017), 349–374.

82. Mattin Biglari, "Toxic Standards: Pollution, 'Slow Violence', and the Environmental History of the Abadan Oil Refinery, Iran," *Journal of Energy History* 12, no. 1 (2024), 1–17.

83. Maarten A. Hajer, *The Politics of Environmental Discourse: Ecological Modernization and the Policy Process* (Oxford, UK: Oxford University Press, 1995), 3.

84. Muhammad Reza Shah Pahlavi, quoted in Schayegh, "Iran's Global Long 1970s," 275.

85. Guiliano Garavini, *The Rise and Fall of OPEC in the Twentieth Century* (Oxford, UK: Oxford University Press, 2019), 210–215.

86. Laleh Khalili, "Crude Knowledge: Petro-Periodicals and Resource Sovereignty," *International Journal of Middle East Studies* 56, no. 3 (2024), 446–464.

Chapter 1

1. National Iranian Oil Company and Iranian Oil Operating Companies, "Present Status of Natural Gas in Iran," in *Proceedings of the Seminar on the Development and Utilization of Natural Gas Resources: with Special Reference to the ECAFE Region,* p. 64–82 (New York: United Nations, 1965).

2. Laurence Lockhart, "Iranian Petroleum in Ancient and Medieval Times," *Journal of the Institute of Petroleum* 25, no. 183 (1939), 1–18.

3. G.M. Lees, "Persia," in *The Science of Petroleum: A Comprehensive Treatise of the Principles and Practice of the Production, Refining, Transport and Distribution of Mineral Oil, Vol. 6, Part 1, The World's Oil Fields: The Eastern Hemisphere,* ed. V.C. Illing (London: Oxford University Press, 1953), 78.

4. R.J. Forbes and D.R. O'Beirne, *The Technical Development of the Royal Dutch/Shell, 1890–1940* (Leiden, Netherlands: E.J. Brill, 1957), 60–66.

5. R.W. Ferrier, *The History of the British Petroleum Company, vol. 1: The Developing Years, 1901–1932* (Cambridge, UK: Cambridge University Press, 1982), 24–27.

6. A.S. Alsharhan and A.E.M. Nairn, *Sedimentary Basins and Petroleum Geology of the Middle East* (Amsterdam, Netherlands: Elsevier, 1997), 467–470.

7. Ferrier, *History of the British Petroleum Company,* vol. 1, 27–28.

8. Alexander Melamid, "The Geographical Pattern of Iranian Oil Development," *Economic Geography* 35, no. 3 (1959), 200; Ferrier, *History of the British Petroleum Company,* vol. 1, 60–62.

9. Daniel Yergin, *The Prize: The Epic Quest for Oil, Money & Power* (New York: Free Press, 1991), 123–127.

10. Arash Khazeni, *Tribes and Empire on the Margins of Nineteenth-Century Iran* (Seattle, WA: University of Washington Press, 2009).

11. Melamid, "Geographical Pattern of Iranian Oil Development," 201–202; Shahbaz Shahnavaz, *Britain and the Opening Up of South-West Persia, 1880–1914: A Study in Imperialism and Economic Dependence* (London: RoutledgeCurzon, 2005).

12. Alsharhan and Nairn, *Sedimentary Basins and Petroleum Geology of the Middle East,* 732–733.

13. Melamid, "Geographical Pattern of Iranian Oil Development," 205–206.

14. Ferrier, *History of the British Petroleum Company,* vol. 1, 262–294 and Appendix 11, tables 11.1–11.8.

15. Yergin, *The Prize*, 130–131.

16. Memorandum from H.W. Lane to L.C. Rice, "Government Enquiry," HWL/370, June 4, 1948, *Production 15th January 1946 to 15th September 1948* (58787), BP Archive, University of Warwick [hereafter *Production 15th January 1946 to 15th September 1948*].

17. Bijan Mossavar-Rahmani, *Energy Policy in Iran: Domestic Choices and International Implications* (New York: Pergamon Press, 1981), 70–71.

18. For more, see chapter two of Katayoun Shafiee, *Machineries of Oil: An Infrastructural History of BP in Iran* (Cambridge, MA: MIT Press, 2018), 57–85. See also Geoffrey C. Bowker, *Science on the Run: Information Management and Industrial Geophysics at Schlumberger, 1920–1940* (Cambridge, MA: MIT Press, 1994).

19. Alsharhan and Nairn, *Sedimentary Basins and Petroleum Geology of the Middle East*, 726.

20. Ciruce Movahedi-Lankarani, "Precarious Petroleum: Volatile Reservoirs, Varied Natural Gas Compositions, and Development in 1960s Iran," *Comparative Studies of South Asia, Africa, and the Middle East* 44, no. 1 (2024), 3–17.

21. National Iranian Oil Company and Iranian Oil Operating Companies, "Present Status of Natural Gas in Iran," 73.

22. Memorandum from G.M. Lees to Deputy Director, Production, "Utilisation of Natural Gas in Iran," no document number, November 11, 1937, p. 1, *Utilisation of Natural Gas* (44113), BP Archive, University of Warwick [hereafter *Utilisation of Natural Gas*].

23. Memorandum Comins to Jameson, Attachment to "Repressuring and Reservoir Gas," no document number, February 16, 1938, *Visit to Iran 1938: Miscellaneous Papers and Notes* (67572), BP Archive, University of Warwick [hereafter *Visit to Iran 1938*].

24. 1 billion cubic feet natural gas to 1 million tonnes oil equivalent conversion ratio taken from *bp Statistical Review of World Energy 2020*, p. 64, accessed June 30, 2021, https://www.bp.com/en/global/corporate/energy-economics/statistical-review-of-world-energy.html; oil production figures from J.H. Bamberg, *The History of the British Petroleum Company, vol. 2: The Anglo-Iranian Years, 1928–1954* (Cambridge, UK: Cambridge University Press, 1994), 69.

25. Ferrier, *History of the British Petroleum Company*, vol. 1, 420–421, 455–457; Bamberg, *The History of the British Petroleum Company*, vol. 2, 192–194; "Meeting Held at Britannic House on Tuesday 30th November, 1937, to Discuss the Problem of the Utilisation of Natural Gas in Iran," no document number, December 3, 1937, p. 1, *Utilisation of Natural Gas*.

26. "Meeting Held at Britannic House on Tuesday 30th November 1937, to Discuss the Problem of the Utilisation of Natural Gas in Iran," no document number, December 3, 1937, p. 1, *Utilisation of Natural Gas.*

27. "Extract from Letter Dated January 19th, 1938, Received from Professor Egerton," no document number, January 19, 1938, *Utilisation of Natural Gas.*

28. Ferrier, *History of the British Petroleum Company*, vol. 1, 456–457.

29. "Meeting Held at Britannic House on Tuesday 30th November 1937, to Discuss the Problem of the Utilisation of Natural Gas in Iran," no document number, December 3, 1937, p. 1–4, *Utilisation of Natural Gas.*

30. Ferrier, *History of the British Petroleum Company*, vol. 1, 397–460; Shafiee, *Machineries of Oil*, 57–85.

31. Anglo-Iranian Oil Company, Ltd., Sunbury-on-Thames Research Station, "Gas Research – M.i.S., Visit of Dr. J.E. Carruthers to Professor R. Robinson at Oxford – November 1st, 1938," no document number, November 1, 1938; Anglo-Iranian Oil Company, Ltd., Sunbury-on-Thames Research Station, "Research Advisory Committee, Meeting to be Held at Britannic House, London, E.C.2, on Friday December 2nd at 2.30 P.M.," no document number, November 25, 1938, p. 1, *Utilisation of Natural Gas.*

32. Anglo-Iranian Oil Company, Ltd., Sunbury-on-Thames Research Station, "Research Advisory Committee, Minutes of Meeting Held on Friday, December 2nd, 1938 at Britannic House," no document number, December 6, 1938, p. 2, *Utilisation of Natural Gas.*

33. Memorandum D. Comins to Jameson, "Utilisation of Fields Gas," no document number, December 16, 1937, p. 1–2; Memorandum from B.K.N. Wyllie, "Is the Manufacture of Magnesium from Dolomite a Possible Mode of Utilising Some of the Surplus Gas in Southern Iran?," no document number, November 10, 1937, attached to Memorandum from G.M. Lees to Deputy Director, Production, November 11, 1937, *Utilisation of Natural Gas.*

34. Memorandum D. Comins to Jameson, "Utilisation of Fields Gas," no document number, December 16, 1937, p. 2.

35. "Meeting held at Britannic House on the 20th January, 1938 to discuss the Problems of the Utilisation of Natural Gas in Iran," no document number, February 14, 1938, p. 1; "Extract from Letter Dated 19th January 1938, Received from Professor Egerton," no document number, January 19, 1938; Letter W. Mitchell to D. Comins, no document number, December 20, 1937 and Letter D. Comins to W.C. Mitchell, no document number, January 13, 1938; Memorandum W.H. Cadman to D. Comins, "New Electro-Thermal Process," no document number, January 18, 1938; Memorandum W.H. Cadman to D. Comins, "Magnesium Production Method," no document number, January 19, 1938, *Utilisation of Natural Gas.*

36. Letter from J.M. Pattinson to E.H.O. Elkington, P/E/155, February 1, 1938; Letter from E.H.O. Elkington to J.M. Pattinson, No. 160, February 10, 1938; Memorandum from D. Comins to Dr. Dunstan, "Utilisation of Iranian Gas," no document number, February 21, 1938; Letter from D. Comins to T.T. McCreath, no document number, March 18, 1938, *Utilisation of Natural Gas.*

37. "Meeting held at Britannic House on the 20th January, 1938 to discuss the Problems of the Utilisation of Natural Gas in Iran," no document number, January 20, 1938, p. 4, *Utilisation of Natural Gas.*

38. Memorandum from D.G. Smith, "Carbon Black," no document number, February 16, 1944, p. 1; Telegram to Jackson, No. 64, July 17 & 18, 1944, p. 2; Memorandum from D.G. Smith, "Carbon Black," no document number, April 3, 1944; Telegram from Jameson to B.R. Jackson, No. 2, May 3, 1944; Telegram to Jackson, No. 64, July 17 & 18, 1944, p. 1–3, *Utilisation of Natural Gas.*

39. Telegram from B.R. Jackson to Jameson, No. 90, May 18, 1944, *Utilisation of Natural Gas.*

40. See, for example, the data reported in Oil Service Company of Iran (Private Company), "Estimated Wet Gas Availability and Utilization, 1974–1978," March 1974 (60103), *Estimated Wet Gas Availability and Utilization, 1974–1978,* BP Archive, University of Warwick.

41. Memorandum from D. Comins to C.A. Harrison, "Potential Gas Supplies in Iran for Carbon Black Manufacture," DC/619, May 12, 1944, *Utilisation of Natural Gas.*

42. Telegram from AIOC Abadan, No. 1085, June 20 & 21, 1945, *Utilisation of Natural Gas.*

43. Telegram from Wright to Jackson, No. 35, June 24, 1944, *Utilisation of Natural Gas.*

44. Telegram from B.R. Jackson to Jameson, No. 26, June 10, 1944, *Utilisation of Natural Gas.*

45. Alsharhan and Nairn, *Sedimentary Basins and Petroleum Geology of the Middle East*, 730.

46. Memorandum from J.M. Pattinson to F.G.C. Morris, "Carbon Black," no document number, June 16, 1947, *Utilisation of Natural Gas.*

47. Letter from F.E. Smith to K.L. Stock, no document number, March 15, 1949, *Iran – Cement Manufacture and Distribution and Utilisation of Gas* (59791), BP Archive, University of Warwick.

48. Anglo-Iranian Oil Company, Ltd., Sunbury-on-Thames Research Station, "Research Advisory Committee, Minutes of Meeting Held on Friday, December 2nd, 1938 at Britannic House," no document number, December 6, 1938, p. 2, *Utilisation of Natural Gas.*

49. Memorandum from P. Meyer to D. Comins, "Requirement for Data on Gas. (Desulphurization of Agha Jari Gas for Carbon Black Production," PRO/131, May 17, 1946, *Utilisation of Natural Gas.*

50. Letter from Moosa Sheybani of Ministry of Finance to A.I.O.C. Ltd., "Utilisation of Oil Gas," 32670, February 1, 1946, *Utilisation of Natural Gas.*

51. Ervand Abrahamian, *Iran: Between Two Revolutions* (Princeton, NJ: Princeton University Press, 1982), 135–149.

52. Shafiee, *Machineries of Oil*, 87–119.

53. International Engineering Company, Morrison-Knudsen International Company, Inc., *Report on Program for the Development of Iran* (San Francisco: n.p., 1947), 3.

54. F. Daftary, "Barnāma-Rīzī, *Encyclopaedia Iranica*, III/8, p. 809–814, available online at http://www.iranicaonline.org/articles/barnama-rizi-planning (accessed January 26, 2020).

55. Morrison-Knudsen, *Report on Program for the Development of Iran*, 6, 244–247.

56. Ibid., 244–248.

57. Letter from J.M. Pattinson to I.M. Jones, "Supply of Gas to Ahwaz Town," D.O. No. 18, September 26, 1947, p. 1, *Iran – Cement Manufacture and Distribution and Utilisation of Gas.*

58. Ibid., 1.

59. Note to Mr. Rice, Mr. Elkington, and Mr. Gass, no document number, September 24, 1947; Letter from J.M. Pattinson to L.C. Rice, "Cement Manufacture and Gas Distribution in Iran," no document number, November 23, 1950, *Iran – Cement Manufacture and Distribution and Utilisation of Gas.*

60. Memorandum from H.W. Lane to L.C. Rice, "Government Enquiry," HWL/370, June 4, 1948 and Memorandum from Waters to Lane, "Natural Gas-Iranian Government Enquiry," DC/JFW/605, June 1, 1948 and attached flow sheets, *Production 15th January 1946 to 15th September 1948.*

61. Shafiee, *Machineries of Oil*, 57–85.

62. Letter from Pir Nia for the Minister of Finance to A.I.O.C. Ltd, [Translated by AIOC], No. 3/29328, August 22, 1948, p. 1–2, *Iran – Cement Manufacture and Distribution and Utilisation of Gas*; "Convention Concluded Between the Imperial Government of Persia and the Anglo-Persian Oil Company, Limited, at Tehran on 29th April 1933," Article 12, paragraph A.

63. Letter from J.M. Pattinson to Chairman, "Gas Supply for Abadan Refinery," no document number, October 4, 1948, attached "Abadan Fuel Consumption" chart, October 1, 1948, *Iran – Cement Manufacture and Distribution and Utilisation of Gas.*

64. Letter from Lane to J.M. Pattinson, "Gas Dehydration," HWL/316, March 31, 1948, *Production 15th January 1946 to 15th September 1948*; Letter from L.C. Rice to J.M. Pattinson, no document number, November 28, 1950; Letter from J.M. Pattinson to Chairman, "Gas Supply for Abadan Refinery;" Memorandum from J.M. Pattinson to N.R. Seddon, "Utilisation of Natural Gas from Agha Jari," JMP/OMS, September 23, 1948; Letter from H.M. Lane[?] to Pattinson, no document number, August 1, 1950; Memorandum from J.M. Pattinson to G.H. Coxon, no document number, December 12, 1950, *Iran – Cement Manufacture and Distribution and Utilisation of Gas.*

65. Abrahamian, *Iran*, 419–420.

66. Parviz Mina, "Oil Agreements in Iran," *Encyclopaedia Iranica*, online edition, 2004, available at http://www.iranicaonline.org/articles/oil-agreements-in-iran (accessed online on April 30, 2020).

67. Memorandum from J.M. Pattinson to L.C. Rice, "Surplus Fields Gas," JMP/OMS, July 30, 1948, *Iran – Cement Manufacture and Distribution and Utilisation of Gas.*

68. Excerpted in Chairman at the First World Petroleum Congress, "Science in the Petroleum Industry," p. 566; excerpted in Memorandum from D. Comins, Field Branch Southwell to Fields Branch Sunbury, "Dossier of Information on Recycling of Products," no document number, April 19, 1940, p. 4–5, *Gas Recycling, Correspondence* (42378), BP Archive, University of Warwick [hereafter *Gas Recycling, Correspondence*].

69. Letter from D. Comins to Jameson, "Repressuring and Reservoir Gas Storage," no document number, February 16, 1938, p. 2, *Visit to Iran 1938, Miscellaneous Papers and Notes* (67572), BP Archive, University of Warwick.

70. L.A. Pym, "Iranian Recoverable Crude Reserves in Proved Fields, March 7, 1945, *Iranian Recoverable Crude Reserves in Proven Fields* (53811), BP Archive, University of Warwick.

71. Letter from N.A. Gass to Arthur fforde [sic], February 6, 1946, *Utilisation of Natural Gas.*

72. Letter from Comins to Elkington, "Ref. Draft Letter to I.G. Enclosed in Mr. Rices D/O Letter of 15.5.46 to Mr. Northcroft re Utilisation of Gas in Iran," p. 1, June 5, 1946; Memorandum from ___ to Comins, "Ref. DC/163 of 5th June, 1946," *Utilisation of Natural Gas.*

73. Memorandum from H.W. Lane, "Fields Production Conferences 1948. Decisions and Recommendations.," 11561, July 29, 1948, *Production 15th January 1946 to 15th September 1948.*

74. Memorandum from H.W. Lane to L.C. Rice, "Government Enquiry," 3.

75. H.S. Gibson, "Multistage Stabilization of Crude," *Transactions of the AIME* 136, no. 1 (December 1940): 25–36.

76. Letter from J.F. Waters to Gass, no document number, August 25, 1948, *Utilisation of Natural Gas.*

77. Letter from D. Comins to Gass, "Iran Gas Position," DC/676, August 6, 1948, p. 1, *Iran – Cement Manufacture and Distribution and Utilisation of Gas.*

78. Memorandum from H.W. Lane to L.C. Rice, "Government Enquiry," 2–3.

79. Memorandum from D. Comins to Jameson, "Your Enquiries re Iran Gas and Karun Irrigation," DC/636, June 22, 1948, *Production 15th January 1946 to 15th September 1948.*

80. "IRAN: Recycling of Gas to Reservoirs," no document number, undated, *Gas Recycling, Correspondence.*

81. Memorandum from D. Comins, Fields Branch Southwell to Fields Branch Sunbury, "Dossier of Information on Recycling of Products," no document number, April 19, 1940, *Gas Recycling, Correspondence*; Letter from J.M. Pattinson to H.S. Gibson, D.O. No. 399, December 22, 1947, *Iran – Cement Manufacture and Distribution and Utilisation of Gas.*

82. Letter from Comins to Elkington, "Ref. Draft Letter to I.G. Enclosed in Mr. Rice D/O Letter of 15.5.46 to Mr. Northcroft re Utilisation of Gas in Iran," p. 2, June 1946, *Utilisation of Natural Gas.*

83. National Iranian Oil Company and Iranian Oil Operating Companies, "Present Status of Natural Gas in Iran," 70; Alsharhan and Nairn, *Sedimentary Basins and Petroleum Geology of the Middle East*, 720–722.

84. National Iranian Oil Company and Iranian Oil Operating Companies, "Present Status of Natural Gas in Iran," 73–74.

85. Gas Study Committee, "Report on the Availability and Utilization of Surplus Natural Gas from Agreement Area in Iran, Tehran, 23 January 1962," p. 1–2, 5, 16–18, *National Iranian Oil Corporation, Report on the Availability and Utilization of Natural Gas from the Agreement Area in Iran* (23383), BP Archive, University of Warwick.

86. Ibid., 1–2, 21–23, 26–27.

87. Ibid., 18, 23, 27–29, 30–35.

88. Ibid., 15, 38–42.

89. Elham Hassanzadeh, *Iran's Natural Gas Industry in the Post-Revolutionary Period: Optimism, Scepticism, and Potential* (Oxford, UK: Oxford University Press, 2014), 34.

90. National Iranian Oil Company and Iranian Oil Operating Companies, "Present Status of Natural Gas in Iran," 73.

Chapter 2

1. United Nations, Economic Commission for Asia and the Far East, United Nations Bureau of Technical Assistance Operations, "Report on the Seminar on the Development and Utilization of Natural Gas Resources," in *Proceedings of the Seminar on the Development and Utilization of Natural Gas Resources: with Special Reference to the ECAFE Region*, 3–24 (New York: United Nations, 1965) 3–4.

2. Iranian Delegation, "Discussion Paper on Utilization of Natural Gas and Its Allied Products as Raw Materials for Fertilizer and Other Chemical Industries," in *Proceedings of the Seminar on the Development and Utilization of Natural Gas Resources: with Special Reference to the ECAFE Region*, 346–354 (New York: United Nations, 1965), 350.

3. United Nations, Economic Commission for Asia and the Far East, "Report on the Second Symposium on the Development of Petroleum Resources of Asia and the Far East," in *Proceedings of the Second Symposium on the Development of Petroleum Resources of Asia and the Far East*, 3–28 (New York: United Nations, 1963), 16.

4. ECAFE, "Report on the Second Symposium on the Development of Petroleum Resources of Asia and the Far East," 4–18.

5. Joe Barnes et al., "Introduction to the Study," *Natural Gas and Geopolitics: From 1970 to 2040*, eds. David G. Victor, Amy M. Jaffe, and Mark H. Hayes, 3–24 (Cambridge, UK: Cambridge University Press, 2006), 5–7.

6. Viacheslav Nekrasov, "Decision-Making in the Soviet Energy Sector in Post-Stalinist Times: The Failure of Khrushchev's Economic Modernization Strategy," in *Cold War Energy: A Transnational History of Soviet Oil and Gas*, ed. Jeronim Perović (Cham, Switzerland: Palgrave Macmillan, 2016), 165–199.

7. Astrid Kander, Paolo Malanima, and Paul Warde, *Power to the People: Energy in Europe over the Last Five Centuries* (Princeton, NJ: Princeton University Press, 2013), 251–332. See also Timothy Mitchell, *Carbon Democracy: Political Power in the Age of Oil* (London: Verso, 2011).

8. Bechtel International Corporation, *Natural Gas Supply for Europe: Preliminary Study for Natural Gas Pipe Line* (San Francisco: Bechtel International Corporation, 1951), 2.

9. Ibid., preface, 2–3.

10. Ibid., 5–7, 14.

11. Letter from Estimating & Technical Expenditure Branch to J.M. Pattinson, "Natural Gas to Europe," no document number, June 6, 1951, p. 1, *Iran – Cement Manufacture and Distribution and Utilisation of Gas* (59791), BP Archive, University of Warwick.

12. Letter from Estimating & Technical Expenditure Branch to Pattinson, "Natural Gas to Europe," no document number, June 11, 1951, p. 2, *Iran – Cement Manufacture and Distribution and Utilisation of Gas.*

13. "Natural Gas to Europe: Economic Survey Estimate," no document number, May 30, 1951, p. 2, *Iran – Cement Manufacture and Distribution and Utilisation of Gas.*

14. Ballou, George T., "Natural Gas in the Eastern Hemisphere—Recent Developments," in *Proceedings of the Seminar on the Development and Utilization of Natural Gas Resources: with Special Reference to the ECAFE Region,* pp. 27–30 (New York: United Nations, 1965), 28.

15. Letter from J.M. Pattinson to Chairman, "Natural Gas to Europe," no document number, June 19, 1951, *Iran – Cement Manufacture and Distribution and Utilisation of Gas.*

16. Letter from Estimating & Technical Expenditure Branch to J.M. Pattinson, "Natural Gas to Europe," no document number, June 6, 1951, p. 2; "Natural Gas to Europe: Economic Survey Estimate," no document number, May 30, 1951, p. 3, *Iran – Cement Manufacture and Distribution and Utilisation of Gas.*

17. ECAFE, "Report on the Second Symposium of Petroleum Resources," 16–17; ECAFE, "Report on the Seminar on Natural Gas Resources," 6.

18. Fred Aftalion, *A History of the International Chemical Industry*, trans. Otto Theodor Benfey (Philadelphia, PA: University of Pennsylvania Press, 1991), 214–242, 262–263, and 275–276.

19. Louis Galambos, Takashi Hikino, and Vera Zamagni, eds., *The Global Chemical Industry in the Age of the Petrochemical Revolution* (Cambridge, UK: Cambridge University Press, 2007), 3–8 and Peter H. Spitz, *Petrochemicals: The Rise of an Industry* (New York: John Wiley & Sons, 1988), 486–488.

20. International Engineering Company, Morrison-Knudsen International Company, Inc., *Report on Program for the Development of Iran* (San Francisco: n.p., 1947), 22–31.

21. David Ekbladh, *The Great American Mission: Modernization and the Construction of an American World Order* (Princeton, NJ: Princeton University Press, 2010), 4.

22. Michael E. Latham, *Modernization as Ideology: American Social Science and "Nation Building" in the Kennedy Era* (Chapel Hill, NC: The University of North Carolina Press, 2000).

23. Nils Gilman, *Mandarins of the Future: Modernization Theory in Cold War America* (Baltimore, MD: The Johns Hopkins University Press, 2003).

24. Eric Helleiner, *Forgotten Foundations of Bretton Woods: International Development and the Making of the Postwar Order* (Ithaca, NY: Cornell University Press, 2014).

25. Amy L.S. Staples, *The Birth of Development: How the World Bank, Food and Agriculture Organization, and World Health Organization Changed the World, 1945–1965* (Kent, OH: The Kent State University Press, 2006).

26. Nick Cullather, *The Hungry World: America's Cold War Battle against Poverty in Asia* (Cambridge, MA: Harvard University Press, 2010).

27. Jacob Shively, "'Good Deeds Aren't Enough': Point Four in Iran, 1949–1953," *Diplomacy & Statecraft* 29 no. 3 (2018), 421–424.

28. Firoozeh Kashani-Sabet, "American Crosses, Persian Crescents: Religion and the Diplomacy of US-Iranian Relations, 1834–1911," *Iranian Studies* 44, no. 5 (2011), 607–625.

29. Bruce R. Kuniholm, *The Origins of the Cold War in the Near East: Great Power Conflict and Diplomacy in Iran, Turkey, and Greece* (Princeton, NJ: Princeton University Press, 1980, 1994).

30. Thomas M. Ricks, "U.S. Military Missions to Iran, 1943–1978: The Political Economy of Military Assistance," *Iranian Studies* 12, no. 3–4 (1979), 163–193.

31. Kamran Mofid, *Development Planning in Iran: From Monarchy to Islamic Republic* (Cambridgeshire, UK: Middle East and North Africa Press, Ltd. 1987), 37–38.

32. Ramin Nassehi, "Domesticating Cold War Economic Ideas: The Rise of Iranian Developmentalism in the 1950s and 1960s," in *The Age of Aryamehr: Late Pahlavi Iran and Its Global Entanglements*, ed. Roham Alvandi (London: Gingko Library, 2018), 42–49.

33. Ervand Abrahamian, *Iran: Between Two Revolutions* (Princeton, NJ: Princeton University Press, 1982), 420.

34. Frances Bostock and Geoffrey Jones, *Planning and Power in Iran: Ebtehaj and Economic Development Under the Shah* (London: Frank Cass and Company Limited, 1989), 125–127.

35. Mofid, *Development Planning in Iran*, 38–47.

36. Nassehi, "Domesticating Cold War Economic Ideas," 50–53.

37. Mofid, *Development Planning in Iran*, 44.

38. Morrison-Knudsen International Company, *Report on the Program for the Development of Iran*, 21.

39. Afsaneh Najmabadi, *Land Reform and Social Change in Iran* (Salt Lake City, UT: University of Utah Press, 1987), 45–47.

40. Abrahamian, *Iran*, 264–265, 281–325, 375–382.

41. Eric J. Hooglund, *Land and Revolution in Iran, 1960–1980* (Austin, TX: University of Texas Press, 1982), 10–35.

42. Najmabadi, *Land Reform and Social Change in Iran*, 51–54.

43. Ibid., 66–75.

44. April R. Summitt, "For a White Revolution: John F. Kennedy and the Shah of Iran," *Middle East Journal* 58, no. 4 (2004), 560–575.

45. Hooglund, *Land and Revolution in Iran*, 49–71.

46. Najmabadi, *Land Reform and Social Change in Iran*, 99–168.

47. Abrahamian, *Iran*, 428.

48. Najmabadi, *Land Reform and Social Change in Iran*, 139–142.

49. Morrison-Knudsen International Company, *Report on the Program for the Development of Iran*, 25, 140–144.

50. Montecatini Preliminary Investigation, "Summary of Ideas and Recommendations Concerning the Most Promising Possibilities for the Utilization of Gas in Iran," April 1957, p. 7–8, 10, 37–44, Development and Resources Corporation Records, MC014, Public Policy Papers, Department of Special Collections, Princeton University Library.

51. "Kārkhāneh-ye Kudshimiāi-ye Shiraz," *Nāmeh-ye San'at-e Naft-e Iran* 5, no. 7 (Mehr 1347), 11.

52. Husayn Mahbubi Ardekāni, *Tārikh-e Moassesāt-e Tamaddoni-ye Jadid dar Iran*, vol. 3, 2nd ed. (Tehran: Enteshārāt-e Dāneshgāh-e Tehran, 1376), 306–307.

53. Ruhollah Sayah, "Kārkhāneh -ye Kudshimiyāi -ye Shiraz," *Nāmeh-ye San'at-e Naft-e Iran* 1, no. 12 (Farvardin[?] 1342), 20–21; Sherkat-e Melli-ye Gāz-e Iran, *San'at-e Gāz-e Iran* (Tehran[?]: Enteshārāt-e Ravābet-e Omumi-ye San'at-e Naft-e Iran, 1352), 36.

54. Manuchehr Eqbāl, foreword to *Kārkhānejāt-e Shimiāi-ye Shirāz* by Sherkat-e Sahāmi-ye Kudshimiāi-ye Iran (Tehran: Enteshārāt-e Omur-e Ravābat-e 'Omumi-ye San'at-e Naft-e Irān, 1353), 1; International Development Association, International Bank for Reconstruction and Development, "The Economic Development of Iran," vol. II (Part I), Sectoral Analyses (Parts I and II) (October 1974), 130.

55. Sherkat-e Sahāmi-ye Kudshimiāi-ye Iran, *Kārkhānejāt-e Shimiāi-ye Shirāz*, (Tehran: Enteshārāt-e Omur-e Ravābat-e 'Omumi-ye San'at-e Naft-e Irān, 1353), 3.

56. "Kārkhāneh-ye Kudshimiāi-ye Shiraz," *Nāmeh-ye San'at-e Naft-e Iran*, 12.

57. Montecatini, "Summary of Ideas and Recommendations," 40–41.

58. Mofid, *Development Planning in Iran*, 37–38, 44.

59. Iranian Delegation, "Discussion Paper on Prospects of Regional Arrangements in Natural Gas Development and in the Establishment of Industries Based on Natural Gas," in *Proceedings of the Seminar on the Development and Utilization of Natural Gas Resources: with Special Reference to the ECAFE Region*, 429–432 (New York: United Nations, 1965), 430–431; Iranian Delegation, "Discussion Paper on the Manufacture and Distribution of Fertilizers in the ECAFE Area," in *Proceedings of the Seminar on the Development and Utilization of Natural Gas Resources: with Special Reference to the ECAFE Region*, 432–436 (New York: United Nations, 1965), 433–436.

60. Regina Lee Blaszczyk, "Synthetics for the Shah: DuPont and the Challenges to Multinationals in 1970s Iran," *Enterprise & Society* 9, no. 4 (2008), 671–672.

61. Fuad Rouhani, *San'at-e Naft-e Iran: Bist Sāl pas az Melli Shodan* (Tehran: Sherkat-e Sahāmi-ye Ketāb-hā-ye Jibi, 2536), 227.

62. R.P. Liddiard, "A Review of Iran's Expanding Gas Industry," August 1975, p. 15–16, *A Review of Iran's Expanding Gas Industry* (31971), BP Archive, University of Warwick.

63. Ciruce Movahedi-Lankarani, "Precarious Petroleum: Volatile Resources, Varied Natural Gas Compositions, and Development in 1960s Iran," *Comparative Studies of South Asia, Africa and the Middle East* 44, no. 1 (2024), 3–17.

64. Morrison-Knudsen International Company, *Report on the Program for the Development of Iran*, 25; Montecatini, "Summary of Ideas and Recommendations," 11.

65. Saghar Sadeghian, "The Caspian Forests of Northern Iran during the Qajar and Pahlavi Periods: Deforestation, Regulation, and Reforestation," *Iranian Studies* 49, no. 6 (2016), 977.

66. Farshad Amiraslani and Deirdre Dragovich, "Forest Management Policies and Oil Wealth in Iran over the Last Century: A Review," *Natural Resources Forum* 37, no. 3 (2013), 170–174.

67. Sherkat-e Melli-ye Naft-e Iran, *Naft va Zendegi* (Tehran: Enteshārāt-e Ravābat-e Omumi-ye San'at-e Naft-e Iran, 1352), 28, figure 6.

68. "Pālāyeshgāh-ye Tehran," *Nāmeh-ye San'at-e Naft-e Iran*, no. 10 (Bahman 1341), 8–9, 45.

69. I. Abaie et al. "History and Development of the Alborz and Sarajeh Fields of Central Iran," paper presented at the *6th World Petroleum Congress, Frankfurt am Mein, Germany, June 1963*, OnePetro.

70. Ardekāni, *Tārikh-e Moassesāt-e Tamaddoni-ye Jadid dar Iran*, 326–331.

71. Sherkat-e Melli-ye Naft-e Iran, *Naft va Zendegi*, 33, figure 11.

72. Montecatini, "Summary of Ideas and Recommendations," 12, 36.

73. Ibid., 49–50.

74. Seyyed Gholāmhusayn Hasantāsh and Mikāyil 'Azimi, *Tārikh-e San'at-e Gāz-e Māy'a-ye Iran* (Tehran: Kavir, 1394), 50–51.

75. Hasantāsh and 'Azimi, *Tārikh-e San'at-e Gāz-e Māy'a-ye Iran*, 52–86.

76. Pamela Karimi, *Domesticity and Consumer Culture in Iran: Interior Revolutions of the Modern Era* (New York: Routledge, 2013), 100–107.

77. Pirouz Ashraf, "Natural Gas Industry in Iran," *Encyclopaedia Iranica*, online edition, 2016, available at http://www.iranicaonline.org/articles/natural-gas-industry-in-iran (accessed on May 3, 2020).

78. Hasantāsh and 'Azimi, *Tārikh-e San'at-e Gāz-e Māy'a-ye Iran*, 87–90, 91–127, 130.

79. Montecatini, "Summary of Ideas and Recommendations," 36.

80. William A. Gamson and Andre Modigliani, "Media Discourse and Public Opinion on Nuclear Power: A Constructionist Approach," *American Journal of Sociology* 95, no. 1 (July 1989), 2–3.

81. Sheila Jasanoff, "Future Imperfect: Science, Technology, and the Imaginations of Modernity," in *Dreamscapes of Modernity: Sociotechnical Imaginaries and the Fabrication of Power*, eds. Sheila Jasanoff and Sang-Hyun Kim (Chicago: University of Chicago Press, 2015), 10.

82. Cyrus Schayegh, "Iran's Karaj Dam Affair: Emerging Mass Consumerism, the Politics of Promise, and the Cold War in the Third World," *Comparative Studies in Society and History* 54, no. 3 (2012), 612–643.

83. Ali M. Ansari, "The Myth of the White Revolution: Mohammad Reza Shah, 'Modernization' and the Consolidation of Power," *Middle Eastern Studies* 37, no. 3 (July 2001), 1–24.

84. Abrahamian, *Iran*, 424, 430.

85. Mofid, *Development Planning in Iran*, 57–59.

86. Mohammad Reza Pahlavi, *Mission for My Country* (New York: McGraw-Hill Book Company, Inc., 1961), 132–160.

87. Title page of *Nāmeh-ye San'at-e Naft-e Iran* 1, no. 1 (Ordibehesht 1341).

88. "Pālāyeshgāh-ye Tehran," 9, 45.

89. "Vas'at-e Dāmāneh-ye Masraf-e Gāz-e Tabi'i," *Nāmeh-ye San'at-e Naft-e Iran* 3, no. 4 (Shahrivar 1343), 8–9.

90. David E. Nye, *American Technological Sublime* (Cambridge, MA: MIT Press, 1994), xi-xiv.

91. Eugene Levy, "The Aesthetics of Power: High-Voltage Transmission Systems and the American Landscape," *Technology and Culture* 38, no. 3 (1997), 575–607.

92. Graeme Macdonald, "Containing Oil: The Pipeline in Petroculture" in *Petrocultures: Oil, Politics, Culture*, eds. Sheena Wilson, Adam Carlson, and Imre Szeman (Montreal, Canada: McGill-Queen's University Press, 2017), 39 and 52.

93. "Vas'at-e Dāmāneh-ye Masraf-e Gāz-e Tabi'i," 8–9.

94. Roland Barthes, *Camera Lucida: Reflections on Photography*, trans. Richard Howard (New York: Hill and Wang, 1981), 25–28, 55, 106–107.

95. Susan Sontag, *On Photography* (New York: Farrar, Straus and Giroux, 1973. Reprint, New York: Picador, 1977), 174.

96. T. Jack Thompson, *Light on Darkness?: Missionary Photography of Africa in the Nineteenth and Early Twentieth Centuries* (Grand Rapids, MI: William B. Eerdmans Publishing Company, 2012), 249.

97. Sontag, *On Photography*, 106.

98. "Vas'at-e Dāmāneh-ye Masraf-e Gāz-e Tabi'i," 8–9.

99. Amitav Ghosh, "Petrofiction: The Oil Encounter and the Novel," *The New Republic* (March 2, 1992), 29–34; Bob Johnson, *Carbon Nation: Fossil Fuels in the*

Making of American Culture (Lawrence, KA: University Press of Kansas, 2014); Macdonald, "Containing Oil," 37–39.

100. Camron Michael Amin, *The Making of the Modern Iranian Woman: Gender, State Policy, and Popular Culture, 1865–1946* (Gainesville, FL: University of Florida Press, 2002).

101. Firoozeh Kashani-Sabet, *Conceiving Citizens: Women and the Politics of Motherhood in Iran* (Oxford, UK: Oxford University Press, 2011).

102. Karimi, *Domesticity and Consumer Culture in Iran*, 84–119.

103. "Cherāgh-e Gāz va Re'āyat-e Nokāt-e Imani dar Khāneh," *Nāmeh-ye San'at-e Naft-e Iran* 4, no. 9 (Bahman 1344), 34–35.

104. "Ta'sisāt-e Pālāyeshgāh-ye Tehran," *Nāmeh-ye San'at-e Naft-e Iran* 4, no. 6 (Ābān 1344), 6–7. The same drawing was used for a similar purpose three years later in "B'azi az Farāvardeh-ha-ye Pālāyeshgah-ye Tehran," *Nāmeh-ye San'at-e Naft-e Iran* 7, no. 6 (Ābān 1347), 18–19.

105. Pahlavi, *Mission for My Country*, 289.

Chapter 3

1. "Mosāhebeh-ye Matbu'āti," *Nāmeh-ye San'at-e Naft-e Iran*, 4, no. 2 (Tir 1344), 3–4.

2. Ibid., 47.

3. Ahmad Ashraf, "Eqbāl, Manuchehr," *Encyclopedia Iranica*, vol. VIII, fasc. 5, pp. 515–517, available online at https://www.iranicaonline.org/articles/eqbal-manucehr/ (accessed September 2, 2025).

4. "Goshāyesh-e Khat-e Dovvom-e Luleh-ye Naft-e Ahvaz-Tehran," *Nāmeh-ye San'at-e Naft-e Iran* 4, no. 8 (Dey 1344), 3.

5. "Chairman's Visit to Iran, October – December 1945," no document number, January 17, 1946, p. 7, *Chairman's Visit to Iran, October to December 1945 (Director's Folder No 23)* (71140), BP Archive, University of Warwick.

6. International Engineering Company, Morrison-Knudsen International Company, Inc., *Report on Program for the Development of Iran* (San Francisco: n.p., 1947), 244–245.

7. Telegram B.R. Jackson to Sir Hubert Heath-Eves, 186, April 3, 1947, p. 2, *Iran: Pipelines: b)Khuzistan/Teheran (1947–1948)* (44015), BP Archive, University of Warwick [hereafter *Iran: Pipelines*]

8. Morrison-Knudsen International Company, *Report on the Program for the Development of Iran*, 245–247.

9. Memorandum to the Deputy Chairman, no document number, April 22, 1947; Memorandum J.M. Pattinson to F.G.C. Morris, "Tehran Fuel Supply," no document number, April 23, 1947, *Iran: Pipelines*.

10. Husayn Mahbubi Ardekāni, *Tārikh-e Moassesāt-e Tamaddoni-ye Jadid dar Iran*, vol. 3, 2nd ed. (Tehran: Enteshārāt-e Dāneshgāh-e Tehran, 1376), 326–331.

11. Letter D.J. Bleifuss to B.R. Jackson, no document number, July 23, 1947, *Iran: Pipelines.*

12. Morrison-Knudsen International Company, *Report on the Program for the Development of Iran*, 247–252.

13. Telegram B.R. Jackson to Sir Hubert Heath-Eves, 186, April 3, 1947, p. 2; Memorandum E.H.O. Elkington to I.M. Jones, 40, June 12, 1947, p. 2, *Iran: Pipelines.*

14. Memorandum from J.M. Pattinson to L.C. Rice, "Notes on Messrs. Morrison Knudsen's Report to the Persian Government on 'Provision of Fuel – Natural Gas," no document number, September 23, 1947, p. 1–3, *Iran: Pipelines.*

15. Memorandum J.M. Pattinson, "Tehran Fuel and Gas Lines," no document number, April 21, 1947, p. 2–3, *Iran: Pipelines.*

16. Memorandum E.H.O. Elkington to I.M. Jones, 40, June 12, 1947, p. 1–2, *Iran: Pipelines.*

17. "Vas'at-e Dāmāneh-ye Masraf-e Gāz-e Tabi'i," *Nāmeh-ye San'at-e Naft-e Iran* 3, no. 4 (Shahrivar 1343), 9.

18. "Goshāyesh-e Khat-e Dovvom-e Luleh-ye Naft-e Ahvaz-Tehran," 3.

19. Christopher R.W. Dietrich, *Oil Revolution: Anticolonial Elites, Sovereign Rights, and the Economic Culture of Decolonization* (Cambridge, UK: Cambridge University Press, 2017); Guiliano Garavini, *The Rise and Fall of OPEC in the Twentieth Century* (Oxford, UK: Oxford University Press, 2019).

20. "Mosāhebeh-ye Matbu'āti," *Nāmeh-ye San'at-e Naft-e Iran* 5, no. 8 (Dey 1345), 4–6

21. "Goshāyesh-e Khat-e Dovvom-e Luleh-ye Naft-e Ahvaz-Tehran," 3.

22. American Embassy Tehran to Department of State, Airgram A-553, February 10, 1966; Iran's Steel Mill/Natural Gas Agreements with the Soviet Union, 3; file AID 6 Iran, 1964–1966 Subject Numeric File, RG 59: General Records of the Department of State, U.S. National Archives [hereafter NARA].

23. Roham Alvandi, "Flirting with Neutrality: The Shah, Khrushchev, and the Failed 1959 Soviet-Iranian Negotiations," *Iranian Studies* 47, no. 3 (2014), 419–440; Roham Alvandi, "The Shah's Détente with Khrushchev: Iran's 1962 Missile Base Pledge to the Soviet Union," *Cold War History* 14, no. 3 (2014), 423–444.

24. U.S. Embassy to Department of State, Telegram 66, July 28, 1966, p. 1–2; file INCO-IRON IRAN, 1964–66 Subject-Numeric File, NARA.

25. Letter to Harold F. Linder from Thomas C. Mann, August 4, 1965, p. 2; file INCO-IRON IRAN, 1964–66 Subject-Numeric File, NARA.

26. Stuart W. Rockwell, Charge d'Affaires ad interim U.S. Embassy Tehran to Department of State, Airgram A-208, November 2, 1964; Soviet Interest in Iranian Steel Mill, file INCO-IRON IRAN USSR; U.S. Embassy Tehran to

Department of State, Telegram 1447, June 18, 1965; file INCO-IRON IRAN, 1964–66 Subject-Numeric File, NARA.

27. U.S. Embassy Tehran to Department of State, Telegram 1447, June 18, 1965; file INCO-IRON IRAN, 1964–66 Subject-Numeric File, NARA.

28. Stuart W. Rockwell, Charge d'Affaires ad interim, U.S. Embassy Tehran to Department of State, Airgram A-27, July 8, 1965; An Official Iranian Attitude Toward Possibility of Soviet-Built Steel Mill; file AID 6 IRAN, 1964–66 Subject-Numeric File, NARA.

29. American Embassy Tehran to Department of State, Airgram A-553, February 10, 1966, 11–12; American Embassy Tehran to Department of State, Airgram A-565, February 14, 1966, p. 2; file AID 6 Iran, 1964–1966 Subject Numeric File, NARA.

30. Ibrāhim Razāqi, *Eqtesād-e Iran* (Tehran: Nashr-e Ney, 1367), 408–409.

31. American Embassy Tehran to Department of State, Airgram A-553, February 10, 1966, p. 3–4.

32. Sherkat-e Melli-ye Gāz-e Iran, "F'āliat-hā va 'Amalkard-e Sherkat-e Melli-ye Gāz-e Iran tay Durān-e Barnāmeh-hā-ye 'Omrāni-ye Chahārom va Panjom," 16 Bahman 2535, p. 1–2; Documents and Publication Center of the Plan and Budget Organization of the Islamic Republic of Iran.

33. Pirooz Ashraf, "Natural Gas Industry in Iran," *Encyclopaedia Iranica*, online edition, 2016, available at http://www.iranicaonline.org/articles/natural-gas-industry-in-iran (accessed on May 6, 2020).

34. Mohsen Shirāzi, *San'at-e Gāz-e Iran: Az Āghāz tā Āstāneh-ye Enqelāb*, Part 2, interview by Golāmrezā Afkhami (Bethesda, MD: Foundation for Iranian Studies, 1999), available online at https://fis-iran.org/fa/ebook/gas/ (accessed September 2, 2025).

35. Report from Sa'id Naqavi for the NIOC to Mehdi Sami'i of the Plan Organization, 2/gl/2372, 14 Bahman 1347, p. 4, *Tarh-e Ehdās Shāhluleh-ye Gāz va Enteqāl-e Ān beh Shoravi* Riāsat-e Jomhuri, National Archives of Iran, Tehran [hereafter *Tarh-e Ehdās Shāhluleh-ye Gāz*].

36. American Embassy Tehran to Department of State, Airgram A-553, February 10, 1966, 1, 3–5, 8–9.

37. Bowler, "IGAT – Note on Procedure to be adopted for the letting of construction and supply contracts," no document number, undated, p. 3–4; Attached to Document 593, May 22, 1966; Attached to Letter July 4, 1966, *Tarh-e Ehdās Shāhluleh-ye Gāz*.

38. "IGAT: Note on possible sources," p. 1–2.

39. Arash Khazeni, *Tribes and Empire on the Margins of Nineteenth-Century Iran* (Seattle, WA: University of Washington Press, 2009), 124–129; Touraj Atabaki, "From *'Amaleh* (Labor) to *Kargar* (Worker): Recruitment, Work

Discipline and Making of the Working Class in the Persian/Iranian Oil Industry," *International Labor and Working-Class History* 84 (2013), 159–175; Touraj Atabaki, "Far From Home, But at Home: Indian Migrant Workers in the Iranian Oil Industry," *Studies in History* 31, no. 1 (2015), 85–114; Carola Hein and Mohamad Sedighi, "Iran's Global Petroleumscape: The Role of Oil in Shaping Khuzestan and Tehran," *Architectural Theory Review* 21, no. 3 (2017), 354–362; Kaveh Ehsani, "Social Engineering and the Contradictions of Modernization in Khuzestan's Company Towns: A Look at Abadan and Masjed-Soleyman," *International Review of Social History* 48, no. 3 (2003): 361–399; Katayoun Shafiee, *Machineries of Oil: An Infrastructural History of BP in Iran* (Cambridge, MA: MIT Press, 2018), 111–155, 189–193. See also Robert Vitalis, *America's Kingdom: Mythmaking on the Saudi Oil Frontier* (Stanford, CA: Stanford University Press, 2007) for more on the ubiquity of such arrangements in the global oil industry.

40. For more on labor movements in Iran's oil industry in the 1960s and prior, see Stephanie Cronin, *Soldiers, Shahs and Subalterns in Iran: Opposition, Protest and Revolt, 1921–1941* (New York: Palgrave Macmillan, 2010), 201–237 and Maral Jefroudi, "Revisiting 'the Long Night' of Iranian Workers: Labor Activism in the Iranian Oil Industry in the 1960s," *International Labor and Working-Class History* 84 (2013), 176–194.

41. Bowler, "IGAT," p. 1; Meeting Minutes, no document number, 12 Tir 1345, p. 1, *Tarh-e Ehdās Shāhluleh-ye Gāz*

42. IMEG, "Iranian Gas Trunkline for N.I.O.C.," no document number, undated, p. 1–2, Attached to letter I.J. Bowler to Nikpey, no document number, July 4, 1966, *Tarh-e Ehdās Shāhluleh-ye Gāz.*

43. "IGAT: Note on possible sources of materials and construction Contractors discussed by Project Stations," no document number, June 30, 1966, p. 1–2; Attached to letter I.J. Bowler to Nikpey, no document number, July 4, 1966, *Tarh-e Ehdās Shāhluleh-ye Gāz.*

44. Bowler, "IGAT," p. 1.

45. "IGAT: Note on possible sources," 4.

46. "IGAT: Note on Sources of Materials and Contractors by Countries," no document number, June 30, 1966, p. 1, *Tarh-e Ehdās Shāhluleh-ye Gāz.*

47. American Embassy Tehran to Department of State, Airgram A-553, February 10, 1966, 5–6.

48. Ministry of the Economy, "Report Pertaining to the Establishment of a Pipe Mill in Iran," no document number, 8 Tir 1345[?], p. 3; *Tarh-e Ehdās Shāhluleh-ye Gāz.*

49. Letter from Eqbāl at NIOC to Prime Minister, no document number, 25 Khordād 1345, p. 3–7, *Tarh-e Ehdās Shāhluleh-ye Gāz.*

50. Kamran Mofid, *Development Planning in Iran: from Monarchy to Islamic Republic* (Cambridgeshire, UK: Middle East and North Africa Press, Ltd., 1987), 58–80.

51. Ashraf, "Natural Gas Industry in Iran;" Report from Sa'id Naqavi for the NIOC to Mehdi Sami'i of the Plan Organization, 2/gl/2372, 14 Bahman 1347, 5, *Tarh-e Ehdās Shāhluleh-ye Gāz.*

52. For more on the history of the Ahvāz pipe mill, see Ciruce Movahedi-Lankarani, "A Firm Policy Decision: Infrastructural Form and Pahlavi Developmentalism at the Ahvāz Pipe Mill," *Iranian Studies* 58, no. 2 (2025), 1–17.

53. S. Naghavi and N. Manoochehri, "Iranian Gas Trunkline Project," paper presented at *7th World Petroleum Conference, Mexico City, Mexico, April 1967*, OnePetro, 255–258.

54. Report from Saljuqi to Prime Minister, Document 2675, 27 Farvardin 1346, p. 1, 4–5, *Tarh-e Ehdās Shāhluleh-ye Gāz.*

55. Percentages calculated from figures provided by *Energy Institute Statistical Review of World Energy 2024*, available at https://www.energyinst.org/statistical-review (accessed May 9, 2025).

56. Vezārat-e Niru, Daftar-e Barnāmehrizi-yi Enerzhi, *Tarāznāmeh-ye Enerzhi-ye Keshvar, 1346–1365* (Tehran[?]: Vezārat-e Niru, 1366), 46, table 6, table 12, 80 table 29; *Energy Institute Statistical Review of World Energy 2024.*

57. S. Naghavi and N. Manoochehri, "Iranian Gas Trunkline Project," paper presented at *7th World Petroleum Conference, Mexico City, Mexico, April 1967*, OnePetro, 257–261, table 5.

58. Shirāzi, *San'at-e Gāz-e Iran*, Part 2.

59. Ibid.

60. Thos H. McLeod, *National Planning in Iran: A Report Based on the Experiences of the Harvard Advisory Group in Iran* (Regina, SK, 1964), 240.

61. Shirāzi, *San'at-e Gāz-e Iran*, Part 2.

62. American Embassy Tehran to Department of State, Telegram 66, July 28, 1966, p. 1–2; file INCO-IRON IRAN, 1964–1966 Subject-Numeric File, NARA.

63. IOEPC, "A Comparison of Three Proposals to Meet the IGAT Demand in 1970, Masjed-e-Suleyman, 14th March 1967," no document number, March 14, 1967, p. 1–2, figure 1; *A Comparison of Three Proposals to Meet the IGAT Demand in 1970, Masjed-e-Suleyman, 14th March 1967* (65067), BP Archive, University of Warwick.

64. Meeting Minutes, no document number, 12 Tir 1345, p. 2, *Tarh-e Ehdās Shāhluleh-ye Gāz.*

65. See J. Addison, "Export of Natural Gas and Gas Liquids," no document number, December 28, 1966 and its attachments; *Iran, Gas/LPG Scheme* (42960), BP Archive, University of Warwick.

66. Naghavi and Manoochehri, "Iranian Gas Trunkline Project," 261; figure from "Review of Gas Matters," p. 4, figure 2; *Natural Gas and LPG, Iran (1965–1966)* (23312), BP Archive, University of Warwick.

67. Naghavi and Manoochehri, "Iranian Gas Trunkline Project," 267–268.

68. IOEPC, "A Comparison of Three Proposals to Meet the IGAT Demand in 1970, Masjed-e-Suleyman, 14th March 1967," no document number, March 14, 1967, *A Comparison of Three Proposals to Meet the IGAT Demand in 1970, Masjed-e-Suleyman, 14th March 1967* (65067); R.P. Liddard, "A Review of Iran's Expanding Gas Industry," August 1975, p. 5, *A Review of Iran's Expanding Gas Industry (1975)* (31971), BP Archive, University of Warwick.

69. Shirāzi, *San'at-e Gāz-e Iran*, Part 2; Gas specifications from Naghavi and Manoochehri, "Iranian Gas Trunkline Project," 262.

70. IOEPC & IORC, "Natural Gas Liquids Project," May 1968, p. 7–11, figures 2.2, 2.3, 2.4, *Supply Aspects of Abadan Gas Oil Desulphurization (1968)* (35368_001), BP Archive, University of Warwick.

71. Naghavi and Manoochehri, "Iranian Gas Trunkline Project," 261–262, 267–268.

72. Shirāzi, *San'at-e Gāz-e Iran*, Part 2.

73. "Shāhānshāh Āryāmehr Ta'sisāt-e Nowin-e San'at-e Naft ra dar Khuzestān Eftetāh Kard," *Nāmeh-ye San'at-e Naft-e Iran* 6, no. 8 (Dey 1346), 7.

74. "Summary of Activities Accomplished of the IGAT-1 Pipeline through 2/9/1346," no document number, undated, p. 1–4, *Tarh-e Ehdās Shāhluleh-ye Gāz.*

75. Meeting Minutes, Document 4/4646, 1 Bahman 1346, *Tarh-e Ehdās Shāhluleh-ye Gāz.*

76. Letter from Sa'id Naqavi to Eqbāl, no document number, 5 Āzar 1346, p. 1–5, *Tarh-e Ehdās Shāhluleh-ye Gāz.*

77. Meeting Minutes, no document number, 3 Esfand 1346, p. 1–3, *Tarh-e Ehdās Shāhluleh-ye Gāz.*

78. "Estimates of Total Installed Cost Up to IGAT Onstream, Early 1970," no document number, July 2, 1966, p. 2, *Tarh-e Ehdās Shāhluleh-ye Gāz*; American Embassy Tehran to Department of State, Airgram A-553, 10 February 1966, 4–5.

79. Meeting Minutes, no document number, 12 Tir 1345, p. 2–3; Memorandum from Manuchehri, no document number, undated, p. 2; Attached to memorandum 41/1/5-m/29, *Tarh-e Ehdās Shāhluleh-ye Gāz.*

80. Shafiee, *Machineries of Oil*, 61–62.

81. R.W. Ferrier, *The History of the British Petroleum Company, vol. 1: The Developing Years, 1901–1932* (Cambridge, UK: Cambridge University Press, 1982), 420–421.

82. Naghavi and Manoochehri, "Iranian Gas Trunkline Project," 261; Shirāzi, *San'at-e Gāz-e Iran*, Part 2.

83. Shirāzi, *San'at-e Gāz-e Iran*, Part 2.

84. Naghavi and Manoochehri, "Iranian Gas Trunkline Project," 262–264.

85. Sherkat-e Melli-ye Naft-e Iran, *Shāhluleh-ye Gāz Iran* (Tehran[?]: Enteshārāt-e Ravābat-e 'Omumi-ye San'at-e Naft-e Iran, n.d.), 10.

86. Ibid., 10.

87. "Masir-e Shāhluleh-ye Gāz," *Nāmeh-ye San'at-e Naft-e Iran* (Mordad, 1349), 9, 44, 50.

88. Naghavi and Manoochehri, "Iranian Gas Trunkline Project," 262.

89. Shirāzi, *San'at-e Gāz-e Iran*, Part 2.

90. Ibid.

91. "Masir-e Shāhluleh-ye Gāz," 9, 50.

92. Report Naqavi to Sami'i, 2/gl/2372, 14 Bahman 1347, 1–3, 14.

93. Report from Sa'id Naqavi for the NIOC to Mehdi Sami'i of the Plan Organization, 2; Ashraf, "Natural Gas Industry in Iran."

94. Ashraf, "Natural Gas Industry in Iran."

95. Theodore M. Porter, *Trust in Numbers: The Pursuit of Objectivity in Science and Public Life* (Princeton, NJ: Princeton University Press, 1995), 8; See also James Ferguson, *The Anti-Politics Machine: "Development," Depoliticization, and Bureaucratic Power in Lesotho* (Minneapolis, MN: University of Minnesota Press, 1990, 1994).

96. "Tarh-e Shāh Luleh-ye Gāz-e Sartāsari-ye Iran," *Nāmeh-ye San'at-e Naft-e Iran* 6, no. 8 (Dey 1346), 13, 42.

97. Ibid., 12.

98. Bob Johnson, *Carbon Nation: Fossil Fuels in the Making of American Culture* (Lawrence, KA: University Press of Kansas, 2014); Kaveh Ehsani, "Disappearing the Workers: How Labor in the Oil Complex has been Made Invisible," in *Working for Oil: Comparative Social Histories of Labor in the Global Oil Industry*, eds. Touraj Atabaki et al., 11–34 (Cham, Switzerland: Palgrave Macmillan, 2018).

99. "Pishraft-e Sari-ye Sākhtemān-e Tarh-e Shāhluleh-ye Gāz" in *Nāmeh-ye San'at-e Naft-e Iran* 7, no. 2 (Tir 1347), 5.

100. Graeme Macdonald, "Containing Oil: The Pipeline in Petroculture," in *Petrocultures: Oil, Politics, Culture*, eds. Sheena Wilson et al., 36–77 (Montreal, Canada: McGill-Queen's University Press, 2017), 60–61.

101. Eugene Levy, "The Aesthetics of Power: High-Voltage Transmission Systems and the American Landscape," *Technology and Culture* 38, no. 3 (1997): 575–607; David E. Nye, *American Technological Sublime* (Cambridge, MA: MIT Press, 1994).

102. Sherkat-e Melli-ye Gāz-e Iran, *San'at-e Gāz-e Iran* (Tehran[?]: Enteshārāt-e Ravābat-e Omumi-ye San'at-e Naft-e Iran, 1352), 13.

103. "Āghāz-e Sākhtemān-e Tasfiyehkhāneh-ye 'Azim-e Gāz-e Bid Boland," *Nāmeh-ye San'at-e Naft-e Iran* 7, no. 9 (Bahman 1347), 14–15.

104. Sherkat-e Melli-ye Gāz-e Iran, *San'at-e Gāz-e Iran*, 1, 3.

105. Vezārat-e Āmuzesh va Parvaresh, *Gāz; Kitāb-hā-ye Khāndani barāye Dāneshāmuzān-e Dabestān* (Tehran[?]: Vezārat-e Āmuzesh va Parvaresh, 1351), 7–13.

106. Sherkat-e Melli-ye Naft-e Iran, *Naft va Zendegi* (Tehran[?]: Enteshārāt-e Ravābat-e Omumi-ye San'at-e Naft-e Iran, 1352), foreword, introductory poem.

107. Victor Seow, *Carbon Technocracy: Energy Regimes in Modern East Asia* (Chicago: The University of Chicago Press, 2021).

Chapter 4

1. For a list of factories in Shiraz, see list from gendarmerie attached to letter t401-77-8-52-12, 20 Khordād 1353; *Tarh-e Gāzresāni beh Kureh-hā va Kucheh-hā va Khiābān-hā va Hazineh-hā-ye Tarh-e Mazbur* (98-293-4335), Ostāndāri-ye Fārs, National Archives of Iran, Fārs Document Center, Shiraz. For more on the fuel sources of brick kilns, see "Naqsh-e Gāz dar Tabdil-e Sukht-e Kureh-ha-ye Ājorsāzi," *Nāmeh-ye San'at-e Naft-e Iran* 14, no. 3 (1354/1975), 11–19.

2. Letter from Manuchehr Piruz to 'Atārad, 7847, 12 Tir 1351; *Mekātebāt-e Sherkat-e Melli-ye Gāz*, NLAI.

3. Memorandum from Lt. Gen. Mohsen Hāshemi-Nezhād to Javād Shahrestāni, 508-14-21-10, 6 Āzar 1352, attached to letter m/2080, 7 Āzar 1352; *Tabdil-e No'-e Sukht-e Kureh-hā-ye Ājorpazi-ye Atrāf-e Tehran beh Gāz* (340–72), Shahrdāri-ye Tehran, National Archives of Iran, Tehran [hereafter *Tabdil-e No'-e Sukht*, NLAI].

4. Letter from Gholāmrezā Nikpey to Lt. Gen. Mohsen Hāshemi-Nezhād, m/2080, 7 Āzar 1352; Letter from the Mayor of Tehran to Mosaddeqi, m/2108, 20 Āzar 1352; *Tabdil-e No'-e Sukht*, NLAI.

5. Hozeh-ye Shomāreh-ye 9, *Hoshdāri beh Maghāmāt-e Mas'ul dar Movārrad-e Āludegi-ye Havā-ye Tehran* (Tehran: n.p., 1346), 7–9.

6. *Energy Institute Statistical Review of World Energy 2024*, available at https://www.energyinst.org/statistical-review (accessed May 9, 2025); J.R. McNeill and Peter Engelke, *The Great Acceleration: An Environmental History of the Anthropocene since 1945* (Cambridge, MA: Harvard University Press, 2014), 9–11.

7. *Energy Institute Statistical Review of World Energy 2024*.

8. Pooya Azadi, Mohsen B. Mesgaran, and Matin Mirramezani, *The Struggle for Development in Iran: The Evolution of Governance, Economy, and Society* (Stanford, CA: Stanford University Press, 2022), 98–102.

9. "Eghdāmāt-e Sherkat-e Melli-ye Naft-e Iran barāyeh Jelugiri az Āludegi-ye Havā," *Nāmeh-ye San'at-e Naft-e Iran* 10, no. 2 (1350), 14.

10. Amit Sadan, "Masters of the Garden: An Environmental History of Nation-Building in Pahlavi Iran," PhD diss. (University of Michigan, 2023).

11. Maarten A. Hajer, *The Politics of Environmental Discourse: Ecological Modernization and the Policy Process* (Oxford, UK: Oxford University Press, 1995), 3.

12. For more on those postrevolutionary movements, see chapter four of Simin Fadaee's *Social Movements in Iran: Environmentalism and Civil Society* (New York: Routledge, 2012).

13. Brett L. Walker, *Toxic Archipelago: A History of Industrial Disease in Japan* (Seattle, WA: University of Washington Press, 2010).

14. Letter from Sherkat-e Sahāmi-ye Ājorfeshāri-ye Shiraz to the Provincial Government of Fārs, 24, 17 Khordād 1351; *Mekātebāt-e Sherkat-e Melli-ye Gāz-e Mantaqeh-ye Shiraz* (293–32994), Ostāndari-ye Fārs, National Archives of Iran, Tehran.

15. Perrin Selcer, *The Postwar Origins of the Global Environment: How the United Nations Built Spaceship Earth* (New York: Columbia University Press, 2018), 62–96.

16. Bryan Campbell Sitzes, "Wildlife and Beyond: The Evolution of Modern Iranian Environmentalism," PhD diss. (University of Texas at Austin, 2024), 214–231.

17. Eskandar Firouz, "National Parks of Iran," *Encyclopedia Iranica*, available online at https://www.iranicaonline.org/articles/national-parks-iran/ (accessed online May 20, 2025).

18. Eskandar Firouz and Daniel Balland, "Environmental Protection," *Encyclopedia Iranica*, available online at http://www.iranicaonline.org/articles/environmental-protection (accessed online May 20, 2025).

19. Barbara Ward and René Dubos, *Only One Earth: The Care and Maintenance of a Small Planet* (New York: W.W. Norton & Company Inc., 1972). Quotation from Selcer, *Postwar Origins*, 201.

20. Selcer, *Postwar Origins*, 201–244.

21. Sitzes, "Wildlife and Beyond," 238–245.

22. Sāzmān-e Hefāzat-e Mohit-e Zist, *Barnāmeh-ye Hefāzat-e Mohit-e Zist dar Durān-e Barnāmeh-ye 'Omrāni-ye Panjom-e Keshvar* (n.p.: Sāzman-e Hefāzat-e Mohit-e Zist, n.d.), 2–3, 6, 15, 18–30, 42–70.

23. "Tehran dar Dud Khafeh Mishavad!," *Ettelā'āt* (20 Tir 1350); Hozeh-ye Shomāreh-ye 9, *Hoshdāri-ye beh Maghāmāt-e Mas'ul*, 2–3.

24. R. Zerbonia and B. Soraya, "Air Pollution Control in Iran," *Journal of the Air Pollution Control Association* 28, no. 4 (1978), 335.

25. Firouz, "Environmental Protection."

26. R. Zerbonia and B. Soraya, "Air Pollution Control in Iran," 335–336.

27. "Az Khatarāt-e Ojāgh-ha-ye Gāz Cheh Midānid?," *Nāmeh-ye San'at-e Naft-e Iran* (Mordād 1341), 20–21; Sherkat-e Melli-ye Naft-e Iran, *Naft va Zendegi* (Tehran: Enteshārāt-e Ravābat-e Omumi-ye San'at-e Naft-e Iran, 1352); Sherkat-e Melli-ye Gāz-e Iran, *San'at-e Gāz-e Iran* (Tehran[?]: Enteshārāt-e Omur-i Ravābat-e Omumi-ye San'at-e Naft-e Iran, 1352).

28. "Pālāyeshgāh-ye Tehran," *Nāmeh-ye San'at-e Naft-e Iran* 1, no. 10 (Bahman 1341), 45.

29. "Dud-e Tehran Tahdid Mikonad," *Ettelā'āt* (15 Āzar 1345), 13.

30. "Tehran dar Dud Khafeh Mishavad!," *Ettelā'āt.*

31. "Dud-e Tehran Tahdid Mikonad," *Ettelā'āt* 1, 13.

32. "Tehran dar Dud Khafeh Mishavad!," *Ettelā'āt.*

33. "Havā-ye Tehran Qābel-e Tahamol Nist," *Ettelā'āt* (4 Bahman 1345/1967); "Dud-e Tehran Tahdid Mikonad," *Ettelā'āt.*

34. "Mobārezeh bā Āludegi-ye Havā," *Nāmeh-ye San'at-e Naft-e Iran* 6, no. 9 (Bahman 1346/1968), 14–16.

35. "Sukht-i Tehran Bāyad Taghir Konad," *Ettelā'āt* (17 Āzar 1345/1966).

36. "Havā-ye Tehran Qābel-e Tahamol Nist," *Ettelā'āt.*

37. Hozeh-ye Shomāreh-ye 9, *Hoshdāri-ye beh Maghāmāt-e Mas'ul dar Movārrad-e Āludegi-ye Havā-ye Tehran* (Tehran: n.p., 1346/1967–68), 11.

38. "Havā-ye Tehran Qābel-e Tahamol Nist," *Ettelā'āt.*

39. Sherkat-e Sahāmi-yi Ājorfeshāri-ye Shiraz to the Provincial Government of Fārs, 24, 17 Khordād 1351; *Mekātebāt-e Sherkat-e Melli-ye Gāz-e Mantaqeh-ye Shiraz* (293-32994), Ostāndāri-ye Fārs, National Archives of Iran, Tehran [hereafter *Mekātebāt-e Sherkat-e Melli-ye Gāz*, NLAI].

40. "Tehran dar Dud Khafeh Mishavad!," *Ettelā'āt.*

41. "Eghdāmāt-e Sherkat-e Melli-ye Naft-e Iran barāyeh Jelugiri az Āludegi-ye Havā," *Nāmeh-ye San'at-e Naft-e Iran* 10, no. 2 (1350), 31.

42. "Paykar bā Āludegi-ye Havā," *Nāmeh-ye San'at-e Naft-e Iran* 10, no. 11 (1351), 35.

43. Ibid., 35.

44. Sāzmān-e Hefāzat-e Mohit-e Zist, *Āmār-e Āludegi-ye Havā-ye Tehran, Seh Māheh-ye Sevvom-e Sāl-e 1357* (Tehran[?]: *Enteshārāt-e Sāzmān-e Hefāzat-e Mohit-e Zist*, 1357[?]), cover and rear cover.

45. Ali Madanipour, *Tehran: The Making of a Metropolis* (Chichester, UK: John Wiley & Sons, Ltd, 1998), 20, 102–127.

46. "Dud-e Tehran Tahdid Mikonad," *Ettelā'āt*, 13.

47. Hozeh-ye Shomāreh-ye 9, *Hoshdāri-ye beh Maghāmāt-e Mas'ul*, 14–15.

48. See, for example, A. Badakhshān, S. Shaibāni, and M. Olfat, "Techniques and Experiences of Measurement and Observation of Major Atmospheric

Pollutants in Iran," *The Bulletin of the Iranian Petroleum Institute* 54 (1974): 13–20. Badakhshan also presented their findings at WMO-WHO Technical Conference on Observation and Measurement of Atmospheric Pollution in Helsinki, July 30 – August 4, 1973.

49. "Paykār bā Āludegi-ye Havā," *Nāmeh-ye San'at-e Naft-e Iran* (Farvardin 1351), 34; Zerbonia, R. and B. Soraya, "Air Pollution Control in Iran," 335.

50. Manuchehr Olfat, "Āludehkonandeh-ha-ye Mohem-i Havā – Tashkhis va Sonjesh-e Ānhā," *Nashriyeh-ye Anjoman-e Naft-e Iran* 47 (1351), 9–15.

51. 'Alirezā Moshref Razavi, "Tasir-e Inverzhun dar Tarākam-e Āludegi-ye Havā va Vaz'-e Inverzhun-e Tehran," *Nashriyeh-ye Anjoman-e Naft-e Iran* 51 (1351), 32–37.

52. Ibid., 32–37.

53. Ibid., 20–21, 31.

54. Hozeh-ye Shomāreh-ye 9, *Hoshdāri-yi beh Maghāmāt-e Mas'ul*, 12.

55. "Sukht-i Tehran Bāyad Taghir Konad." See also Sāleh as reported in "Dud-e Tehran Tahdid Mikonad," 13.

56. "Simpozium-i Āludegi-ye Havā," *Nāmeh-ye San'at-e Naft-e Iran* (Esfand 1350), 13–14.

57. Hozeh-ye Shomāreh-ye 9, *Hoshdāri-ye beh Maghāmāt-e Mas'ul*, 15; "Eghdāmāt-e Sherkat-e Melli-ye Naft-e Iran barāyeh Jelugiri az Āludegi-ye Havā," 31–32.

58. "Dud-e Tehran Tahdid Mikonad," *Ettelā'āt*, 13.

59. For the printed version of the speech, see Sa'id Shaybani, *Mas'aleh-yi Āludegi-ye Havā va Barnāmeh-ha-ye Sherkat-e Melli-ye Naft-e Iran dar in Zamineh* (Tehran[?]: Ravābat-e Omumi-ye San'at-e Naft-e Iran, 1350).

60. "Eghdāmāt-e Sherkat-e Melli-ye Naft-e Iran barāyeh Jelugiri az Āludegi-ye Havā," 12–30.

61. "Naqsh-e Gāz dar Taqlil-e Āludegi-ye Mohit-e Zist," *Nāmeh-ye San'at-e Naft-e Iran* (1353), 18–20.

62. Mohsen Shirāzi, *San'at-e Gāz-e Iran: Az Aghaz ta Astaneh-ye Enqelab*, Part 2, interview by Gholāmrezā Afkhami (Bethesda, MD: Foundation for Iranian Studies, 1999), available online at https://fis-iran.org/fa/ebook/gas/ (accessed September 2, 2025).

63. Zoé Chateau et al., "Integrating Sociotechnical and Spatial Imaginaries in Researching Energy Futures," *Energy Research & Social Science* 80 (2021), 102207.

64. Paul Warde, Libby Robin, and Sverker Sörlin, *The Environment: A History of the Idea* (Baltimore, MD: Johns Hopkins University Press, 2018), 25–72, 122–150.

65. Frank Uekoetter, *The Age of Smoke: Environmental Policy in Germany and the United States, 1880–1970* (Pittsburgh, PA: University of Pittsburgh Press, 2009).

66. Letter from Ahmad Naser to heads of Felesh-Melavi, Senā, and Chakosh asphalt companies, no document number, undated; Letter from Mayor of Tehran to Agha-yi M'aynian, no document number, 28 Aban 1352; and Letter from Mayor of Tehran to Agha-yi M'aynian, no document number, 29 Aban 1352; *Tabdil-e No'-e Sukht*, NLAI.

67. Sherkat-e Melli-ye Gāz-e Iran, *San'at-e Gāz-e Iran* (Tehran[?]: Enteshārāt-e Ravābat-e Omumi-ye San'at-e Naft-e Iran, 1352/1973–74), 29–44.

68. Shirāzi, *San'at-e Gāz-e Iran*, Part 2.

69. Letter from Ahmad Nāser to heads of Felesh-Melāvi, Senā, and Chakosh asphalt companies, no document number, undated; *Tabdil-e No'-e Sukht*, NLAI.

70. Letter from Mayor of Tehran to Āghā-ye M'aynian, no document number, 28 Ābān 1352; *Tabdil-e No'-e Sukht*, NLAI.

71. Letter from Mayor of Tehran to Āghā-ye M'aynian, no document number, 29 Ābān 1352; *Tabdil-e No'-e Sukht*, NLAI.

72. Letter from Sherkat-e Sahāmi-ye Ajorfeshāri-ye Shiraz to the Provincial Government of Fārs, 24, 17 Khordād 1351; *Mekātebāt-e Sherkat-e Melli-ye Gāz*, NLAI.

73. Ibid.; Letter from Sherkat-e Sahāmi-ye Ajorfeshāri-ye Shiraz to Sherkat-e Melli-ye Gāz-e Iran Mantaqeh-ye Shiraz, 32, 28 Khordād 1351; *Mekātebāt-e Sherkat-e Melli-ye Gāz*, NLAI.

74. Letter from Manuchehr Piruz to Manuchehr Eqbāl, no document number, no document date; *Mekātebāt-e Sherkat-e Melli-ye Gāz*, NLAI.

75. Letter from Iraj 'Atārad to Manuchehr Piruz, 11967, 5 Mordād 1351; *Mekātebāt-e Sherkat-e Melli-ye Gāz*, NLAI.

76. Transcript of Committee Meeting held 5 Mordād 1351, no document number, no document date; attached to Document 11967; *Mekātebāt-e Sherkat-e Melli-ye Gāz*, NLAI.

77. See Transcript of Committee Meeting held 5 Mordad 1351, no document number, no document date; attached to Document 11967; Letter from Manuchehr Piruz to NIGC, 26570, 7 Bahman 1351; and Letter from Taqi Mosaddeqi to Manuchehr Piruz, 1744, 18 Aban 1351, attached to 19902, 10 Dey 1351; *Mekātebāt-e Sherkat-e Melli-ye Gāz*, NLAI.

78. Letter from Muhammad Sām to Nikpey, m/13974, 18 Esfand 1352; Letter from Nāser of Region 11, no document number, 7 Āzar 1352; *Tabdil-e No'-e Sukht*, NLAI.

79. Letter from Ahmad Nāser to Dr. Nikpey, mm11/49, 11 Āzar 1352, pg. 1; *Tabdil-e No'-e Sukht*, NLAI.

80. Letter from head of the Special Office of the Shah to Qolāmrezā Nikpey, m/400–45, 18 Āzar 1352; Letter from the Municipality of Tehran to Āghā-ye Mosaddeqi, m/2108, 20 Āzar 1352; *Tabdil-e No'-e Sukht*, NLAI.

81. Letter from Mayor of Tehran to Āghā-ye M'aynian, m/2348, 1 Dey 1352; *Tabdil-e No'-e Sukht*, NLAI.

82. Shirāzi, *San'at-e Gāz-e Iran*, Part 2.

83. Letter from Muhammad Sām to Nikpey, m/13974, 18 Esfand 1352; *Tabdil-e No'-e Sukht*, NLAI.

84. For the quote, see Letter from Mayor of Tehran to Ministry of the Interior, document h/52, 13 Farvardin 1353; *Tabdil-e No'-e Sukht*, NLAI; for the list of consumers, see "Naqsh-e Gāz dar Tabdil-e Sukht-e Kureh-hā-ye Ājorsāzi," *Nāmeh-ye San'at-e Naft-e Iran* 14, 3 (1354), 11–19.

85. Shirāzi, *San'at-e Gāz-e Iran*, Part 3.

86. "Naqsh-e Gāz dar Tabdil-e Sukht-e Kureh-hā-ye Ājorsāzi," 11.

87. Letter from Ahmad Naser to Dr. Nikpey, mm11/49, 11 Āzar 1352, pg. 1, 19; *Tabdil-e No'-e Sukht*, NLAI.

88. Muhammad Sām to Manuchehr Piruz,, m/13974, 18 Esfand 1352; *Tarh-e Gāzresāni beh Kureh-ha va Kucheh-ha va Khiābān-ha va Hazineh-ha-ye Tarh-e Mazbur* (98–293–4335), Ostāndāri-ye Fārs, National Archives of Iran, Fārs Document Center, Shiraz. [hereafter *Tarh-e Gāzresāni beh Kureh-ha*, NLAI].

89. Mostafā Qahremāni to the Management of the Economic Administration of Fars, 350, 23 Esfand 1352, attached to document 7995; *Tarh-e Gāzresāni beh Kureh-ha*, NLAI.

90. Letter from Mahmud Nāji to Fārs Provincial Government, pf/shz/2391, 27 Āzar 1353; Letter from the Minister of the Interior to the Provincial Government of Fārs, m/9417, 25 Ābān 1354; *Tarh-e Gāzresāni beh Kureh-ha*, NLAI.

91. Proceedings of Meeting, no document number; 31 Farvardin 1353, pg. 1; Letter from Javād Haqān to Managers of Brickmaking Factories, 1934, 4 Ordibehesht 1353, attached to document 6/6321; *Tarh-e Gāzresāni beh Kureh-ha*, NLAI.

92. Letter from Shirāzi to Feruzān, Āryā, Kāzerun, Hāfez, and Golriz brickmakers, document 6060, 24 Tir 1353; Letter from the Syndicate of Ceramic Employers of Shiraz to the Economic Administration of Fārs, 18, 25 Tir 1353, attached to 12648; *Tarh-e Gāzresāni beh Kureh-ha*, NLAI.

93. Letter from Ahmad Behzād to Fārs Provincial Government, pf/shz/1354, 24 Tir 1353, attached to 11310; *Tarh-e Gāzresāni beh Kureh-ha*, NLAI.

94. Letter from NIGC to Abbās Golshen owner of the Reza Brickmaking Factory, 125, 20 Esfand 1352, pg. 1, attached to letter sent 14 Mehr 2535; *Tarh-e Gāzresāni beh Kureh-ha*, NLAI.

95. Letter from the Pressure Brick Corporation of Shiraz to NIGC, Shiraz Region, 120, 1 Dey 1354; *Tarh-e Gāzresāni beh Kureh-ha*, NLAI.

96. Letter from Ahmad Behzād to the Governor of Fārs, pf/shz/1801, 2 Mordād 2535; *Tarh-e Gāzresāni beh Kureh-ha*, NLAI Shiraz.

97. Letter from Muhammad Qahremāni to NIGC, 774418, 15 Mehr 2535; *Tarh-e Gāzresāni beh Kureh-ha*, NLAI.

98. Letter from Nāvi, the mayor of Region 4, to Kamālzādeh, document 1653, 1 Farvardin 2535 and Letter from Muhammad Zāyefi to Kamālzādeh, document 1570, 21 Farvardin 2535; Letter from Muhammad Zāyefi to Kamālzādeh, 1570, 21 Farvardin 2535; *Tabdil-e No'-e Sukht*, NLAI.

99. Letter from Ahmad Behzād to Fārs Provincial Government, 1755, 25 Shahrivar 1353; *Tarh-e Gāzresāni beh Kureh-ha*, NLAI.

100. Letter on behalf of the Minister of the Interior to Manuchehr Piruz, m/309, 18 Farvardin 1354; and Letter on behalf of the Minister of the Interior to Manuchehr Piruz, m/3236, 2 Tir 1354; Letter from Manuchehr Piruz to NIGC, 8024, 10 Tir 1354; Letter from Minister of the Interior to Manuchehr Piruz, p/4365, 4 Mordād 1354; Letter on behalf of Morteză Sālehi to Manuchehr Piruz, m/12311, 5 Bahman 1354; *Tarh-e Gāzresāni beh Kureh-ha*, NLAI.

101. Letter from the Minister of the Interior to the Provincial Government of Fārs, m/9417, 25 Ābān 1354; Proceedings of meeting held at 11:00am on 30 Tir 1354, no document number, pg. 1–2; attached to Letter from Ahmad Behzād to Governor Piruz, document pf/shz/843, 20 Mordād 1354; The estimated cost for the system was 300,000 rial, either payable in cash at the time of signing or in 15,000 rial installments spread across eighteen months with a 30,000 rial down payment. See Letter from NIGC to Abbās Golshen owner of the Reza Brick-making Factory, 125, 20 Esfand 1352, attached to letter sent 14 Mehr 2535; *Tarh-e Gāzresāni beh Kureh-ha*, NLAI.

102. Letter from the Municipality of the Capital to Morteză Sālehi, 91344, 7 Shahrivar 2535; See Letter from Mayor of Tehran to Deputy for Development Affairs of the Ministry of the Interior, h/378/362, 8 Ordibehesht 2535; Letter from Kolāhi to Kamālzādeh, 1239, 25 Esfand 2535; and Letter from Morteză Sālehi to Qolāmrezā Nikpey, m/622, 21 Farvardin 2535; *Tabdil-e No'-e Sukht*, NLAI.

103. "10 Kureh-ye Ājorpazi Bemanzur-e Tabdil-e Sukht T'atil Shod," *Ettelā'āt* (10 Esfand 1356).

104. Cyrus Schayegh, "'Seeing Like a State': An Essay on the Historiography of Modern Iran," *International Journal of Middle East Studies* 42, no. 1 (2010), 37–61.

105. "Naqsh-e Gāz dar Taqlil-e Āludegi-ye Mohit-e Zist," *Nāmeh-ye San'at-e Naft Iran* (Winter 1353), 21.

106. Letter from Husayn Razm-Ārā to Feyli, p. 1, z,h-655, 12 Bahman 2536; attached to letter 22717; *Kharid-e Otobus-hā-ye Gāzsuz* (340–460), Shahrdāri-ye Tehran, National Archives of Iran, Tehran [hereafter *Kharid-e Otobus-hā-ye Gāzsuz*].

107. Letter from Taqi Mosaddeqi to Fāzeli, document 1-56/3384/g, 10 Esfand 2535; Memorandum from Mayor of Tehran to Mossadeqi 33184/19353, 21 Esfand 2535; attached to 2601, 31 Mordād 2536; *Kharid-e Otobus-hā-ye Gāzsuz.*

108. Razm-Ārā, to Feyli, p. 1.

109. Letter from ʿAbulnabizādeh to Governor Piruz, document 120, 8 Dey 1354; attached to letter 24772; *Tarh-e Gāzresāni.*

110. Letter from from Ir.P. Tiedema and J. van der Weide to the governor of Fārs, p. 1-4, Attachment 2, Appendix 4, document 76-1073, 26 January 1976; attached to document 25-10974; *Tarh-e Gāzresāni.*

111. Razm-Ārā, to Feyli, p. 1–2; Letter from Mayor of Tehran to Hedāyati, p. 2, document h/4016, 18 Bahman 2535; Attachment "LPG Car" to Letter from Mayor of Tehran to Hedāyati, document h/4016, 18 Bahman 2535; *Kharid-e Otobus-hā-ye Gāzsuz.*

112. Letter from Ibrāhim Zahadi to the Special Office of the Shah, document t/873, 11 Esfand 1352; *Kharid-e Otobus-hā-ye Gāzsuz.* Butane-Propane News Incorporated, *Propane in America: The First 100 Years, 1912–2012* (Arcadia, CA: Butane-Propane News Incorporated, 2012), 59.

113. Arthur D. Little, Inc. and Chicago Transit Authority, *Comparative Economics of Propane and Diesel Buses: Report to Chicago Transit Authority* (Chicago: Little, 1960); Letter from Husayn Razm-Ārā to Khādemi, document 22717, 17 Esfand 2536; *Kharid-e Otobus-hā-ye Gāzsuz.*

114. Mosaddeqi to Fāzeli, p. 2, *Kharid-e Otobus-hā-ye Gāzsuz.*

115. Memorandum from Taqi Mosaddeqi to Nikpur, Mayor of Tehran, document 3-56/3387/g, 112 Esfand 2535; attached to report 2601; *Kharid-e Otobus-hā-ye Gāzsuz.*

116. Memorandum from Hamid Khazāri and Abolfatah Sʿaidi to Chief Executive of the United Bus Company of Tehran, no document number, no date; attached to report 2601; *Kharid-e Otobus-hā-ye Gāzsuz.*

117. Letter from Mayor of Tehran to Hedāyati, p. 1–2, document h/4016, 18 Bahman 2535; Letter from Mayor of Tehran to Lt. Gen. Fāzeli, document h/5792, 21 Esfand 2535; Letter from Mayor of Tehran to Mosaddeqi, document 33184/19353, 21 Esfand 2535; Record of meeting held 10 Ordibehesht 2536, p. 2, document 3927, 10 Ordibehesht 2536; *Kharid-e Otobus-hā-ye Gāzsuz.*

118. Letter from Ibrāhim Zahadi to the Mayor of Tehran, document th/167, 24 Khordād 1353; Letter from Mostafā Sultāni to Shahrestāni document h/403, 14 Shahrivar 2536; *Kharid-e Otobus-hā-ye Gāzsuz.*

119. Memorandum from Hamid Kharāzi and Abolfatah Saʿidi, no document number, undated; attached to report 2601; *Kharid-e Otobus-hā-ye Gāzsuz.*

120. Report prepared by the Technical Office of the United Bus Company of Tehran, p. 1–4, document 2601, 31 Mordād 2536; *Kharid-e Otobus-hā-ye Gāzsuz.*

121. Record of Meeting held 10 Ordibehesht 2536, p. 2, document 3927, 10 Ordibehesht 2536; *Kharid-e Otobus-hā-ye Gāzsuz.*

122. Memorandum of meeting between Leyland Motors and the United Bus Company of Tehran, no document number, 12 Ābān 2536; Letter to Shahrestāni, document 34345, 10 Esfand 2536; Report on meeting between the United Bus Company, Mercedes-Benz Germany, and the Iran National Company, no document number, 23 Ābān 2536; Razm-Ārā, to Feyli, 2, *Kharid-e Otobus-hā-ye Gāzsuz.*

123. Letter from Husayn Razm-Ārā to Khādemi, p. 1–2, document 22717, 17 Esfand 2536; Razm-Ārā to Feyli, 1–2; *Kharid-e Otobus-hā-ye Gāzsuz.*

124. McNeill and Engelke, *The Great Acceleration*, 120–122.

125. Ervand Abrahamian, *Iran: Between Two Revolutions* (Princeton, NJ: Princeton University Press, 1982), 430–431.

126. Muhammad 'Ali Mostofizādeh, "Tabdil-e Sukht-e Vasāyat-e Naqlieh be Gāz," *Nashriyeh-ye Anjoman-e Naft-e Iran* (4th Quarter 2534), 9–15, 19.

127. Ibid., 13.

128. Ibid., 13–18.

129. Ibid., 18–19; Vāhed-e Mohandesi-ye Iran Gāz (Melli Shodeh), *Matn-e Sokhanrāni-ye Āghā-ye Mohandes Ashrāqi dar Avalin Seminār-e Barrasi-ye Prozheh-ye Gāzsuzkardan-e Khodro-hā dar Iran* (Tehran: Vāhed-e Mohandesi-ye Iran Gāz (Melli Shodeh), 1362), 10.

130. Letter from Manuchehr Piruz to Mosaddeqi, document m/626, 28 Mehr 1354; Letter from Manuchehr Piruz to Mosaddeqi, no document number, undated but probably late 1354; *Tarh-e Gāzresāni beh Kureh-hā va Kucheh-hā va Khiābān-hā va Hazineh-hā-ye Tarh-e Mazbur* (98-293-4335), Ostāndāri-ye Fārs, National Archives of Iran, Fārs Document Center, Shiraz [hereafter *Tarh-e Gāzresāni*].

131. Robert Steele, *The Shah's Imperial Celebrations of 1971: Nationalism, Culture and Politics in Late Pahlavi Iran* (London: I.B. Tauris, 2021).

132. Report on negotiations held 23 Ābān 1354, no document number, undated; attached to letter 61-6798; Letter from Iskandar Firuz to Nasr Isfahāni, document 61-6798, 30 Farvardin 2535; Letter from the provincial government of Fārs to the mayor of Shiraz, document 18-157, 8 Ordibehesht 2535; Letter from Abdul Husayn Shamszādeh to the governor of Fārs, document 609, 04 Mordād 2535; *Tarh-e Gāzresāni.*

133. Taghi Ebtekar, "The Prospect for Alternative Fuel Vehicles in Iran: An Environmental Assessment," *Proceedings of the 28th Intersociety Energy Conversion Engineering Conference*, vol. 2, 893–897, (Washington, D.C.: American Chemical Society, 1993), 895.

134. Taghi Ebtekar, "Reconstruction of Shiraz CNG Project," in *International Congress Exhibition on the Use of Methane in the Transport Sector* (Bologna, Italy: Fedormentano, 1986), 212–213.

135. Mosaddeqi to Fāzeli, *Kharid-e Otobus-hā-ye Gāzsuz.*

136. "Simpozium-e Āludegi-ye Havā," *Nāmeh-ye San'at-e Naft-e Iran* (Esfand 1350), 13.

137. "Naft va Hefāzat-e Mohit-e Zist," *Nāmeh-ye San'at-e Naft-e Iran* (Spring 1354), 7.

138. "Naqsh-e Gāz dar Taqlil-e Āludegi-ye Mohit-e Zist," *Nāmeh-ye San'at-e Naft-e Iran* 13, no. 4 (1353), 18–20.

139. Jalal Al-e Ahmad, *Gharbzadegi* (Tehran: Nashr-e Jāmeh-ye Darān, 1384). See Liora Hendelman-Baavur, "The Odyssey of Jalal Al-e Ahmad's *Gharbzadegi* – Five Decades After," in *Persian Language, Literature and Culture: New Leaves, Fresh Looks*, ed. Kamran Talatoff, 258–286 (London: Routledge, 2015). See also Hamid Dabashi, *Theology of Discontent: The Ideological Foundations of the Islamic Revolution in Iran* (New York: New York University Press, 1993. Reprint, New Brunswick, NJ: Transaction Publishers, 2008), 39–101.

140. For more on Muhammad Reza Shah's "Great Civilization," see chapter 7 of Ali M. Ansari's *Modern Iran: The Pahlavis and After*, 2nd ed. (New York: Routledge, 2011) and Cyrus Schayegh's "Iran's Global Long 1970s: An Empire Project, Civilisational Developmentalism, and the Crisis of the Global North," in *The Age of Aryamehr: Late Pahlavi Iran and Its Global Entanglements*, ed. Roham Alvandi (London: Gingko, 2018), 260–289. Kathleen John-Alder has also touched upon such themes in her study of the 1970s-era Pardisan park project; see chapter 6 of *Ian McHarg and the Search for Ideal Order* (London: Routledge Press, 2019).

141. Much of this chapter previously appeared as Ciruce Movahedi-Lankarani, "A Ghoul at the Gates: Natural Gas Energy and the Environment in Pahlavi Iran, 1960–1979," *International Journal of Middle East Studies* 54, no. 1 (2022), 80–99. © The Author(s), 2022. Published by Cambridge University Press. Reprinted with permission.

Chapter 5

1. Residents of Shishehgari, no document number, 5 Mehr 2536, attached to pf/gf/3686, *Residegi beh Darkhāst-hā-ye Ahāli-ye Ostān-e Fārs dar Khosus: Sodur-e Gozarnāmeh, Komak-hā-ye Māli, Estekhdām va . . .* (98-293-3326), Ostāndāri-ye Fārs, National Archives of Iran, Fars Document Center [hereafter *Residegi beh Darkhāst*].

2. Brian Larkin, *Signal and Noise: Media, Infrastructure, and Urban Culture in Nigeria* (Durham, NC: Duke University Press, 2008).

3. Charlotte Lemanski, "Infrastructural Citizenship: The Everyday Citizenships of Adapting and/or Destroying Public Infrastructure in Cape Town, South Africa," *Transactions of the Institute of British Geographers* 45, no. 3 (2019), 2; Antina von Schnitzler, *Democracy's Infrastructure: Techno-Politics and Protest after Apartheid* (Princeton, NJ: Princeton University Press, 2016); Nikhil Anand, *Hydraulic City: Water and the Infrastructures of Citizenship in Mumbai* (Durham, NC: Duke University Press, 2017); Rosalind Fredericks, *Garbage Citizenship: Vital Infrastructures of Labor in Dakar, Senegal* (Durham, NC: Duke University Press, 2018).

4. Setrag Manoukian, *City of Knowledge in Twentieth Century Iran: Shiraz, History and Poetry* (New York: Routledge, 2012).

5. Letter from Donald W. Gilfillan to H.E. Mehdi Farrokh, no document number, August 24, 1955, p. 1; Letter from 'Abāssqali Balāli to NIOC, Document 22387, 9 Dey 1334, *Parvandeh-ye Ahdās-e Luleh-ye Gāz va Kārkhāneh-ye Kudshimiāi* (98-293-12524), Ostāndāri-ye Fārs, National Archives of Iran, Fars Document Center [hereafter *Parvandeh-ye Ahdās-e Luleh-ye Gāz*].

6. Memorandum from NIOC to Plan Organization, Document 422/19531, 16 Dey 1334, *Parvandeh-ye Ahdās-e Luleh-ye Gāz.*

7. Memorandum from NIOC to Prime Minister, Document 422/19936, 19 Esfand 1344, *Parvandeh-ye Ahdās-e Luleh-ye Gāz.*

8. Memorandum from Plan Organization to Governor of Fārs, Document 34139, 21 Bahman 1335; Memorandum from Governor of Fars to Ministry of the Interior, Document 699, 19 Farvardin 1336, *Parvandeh-ye Ahdās-e Luleh-ye Gāz.*

9. Meeting Minutes, 27 Farvardin 1336, p. 1–2, *Parvandeh-ye Ahdās-e Luleh-ye Gāz.*

10. Memorandum from Governor of Fārs to Minister of Industry and Mines, Document 15624, 23 Shahrivar 1337; Memorandum from Ministry of Industry and Mines to Governor of Fārs, Document 650/90/26063, 23 Mehr 1337; Memorandum from Governor of Fārs to Office of the Shah, Document 24937, 12 Bahman 1337, *Parvandeh-ye Ahdās-e Luleh-ye Gāz.*

11. Letter from the Chairman of the Board of NIOC to the Governor of Fārs, Document 5/1/7-66144, 3 Bahman 1338, *Parvandeh-ye Ahdās-e Luleh-ye Gāz.*

12. Letter from Prime Minister to NIOC, Document 21186, 16 Bahman 1338; *Parvandeh-ye Ahdās-e Luleh-ye Gāz.*

13. Letter from Governor of Fārs to Chairman of the Board of NIOC, no document number, undated, *Parvandeh-ye Ahdās-e Luleh-ye Gāz.*

14. Sherkat-e Melli-ye Gāz-e Iran, *San'at-e Gāz-e Iran*, 36.

15. Memorandum from Minister of Industry and Mines to Prime Minister, Document 1/14/7/232299, 2 Dey 1340, *Makātebāt dar Movarrad-e Prozheh-ye*

Gāzresāni beh Shirāz va Anjām Omur Marbut be Ān (98-293-7311), Ostāndāri-ye Fārs, National Archives of Iran, Fārs Document Center [hereafter *Makātebāt dar Movarrad-e Prozheh-ye Gāzresāni*].

16. Memorandum from Governor of Fārs to Prime Minister, Document 1982, 2 Ordibehesht 1341, *Makātebāt dar Movarrad-e Prozheh-ye Gāzresāni.*

17. Memorandum from Governor of Fārs to Plan Organization, Urban Development Section of Area 3, Social Affairs, Document 1981, 3 Ordibehesht 1341; Memorandum from Plan Organization to Governor of Fārs, Document 39/1839, 13 Ordibehesht 1341, *Makātebāt dar Movarrad-e Prozheh-ye Gāzresāni.*

18. Letter from Agricultural Union of Fārs, no document number, undated; Letter from Head of the Chamber of Commerce of Bandar Bushehr to Governor of Fārs, Document 1461, 29 Bahman 1337, *Parvandeh-ye Ahdās-e Luleh-ye Gāz.*

19. Letter from Shiraz Electrical Company to Mayor of Shiraz, Document 10492, 1 Shahrivar 1341, p. 1–2, *Makātebāt dar Movarrad-e Prozheh-ye Gāzresāni.*

20. Letter from Muhammad Bāqar-e Mantaqi to Governor of Fārs, Document 136, 2 Mordād 1341, *Makātebāt dar Movarrad-e Prozheh-ye Gāzresāni.*

21. Memorandum from NIOC to Governor of Fārs, Document 6553005, 5 Shahrivar 1341, *Makātebāt dar Movarrad-e Prozheh-ye Gāzresāni.*

22. Memorandum from Mayor of Shiraz to Ministry of Industry and Mines, no document number, 2 Ordibehesth 1341 with additional notes added 18 and 21 Shahrivar 1341, *Makātebāt dar Movarrad-e Prozheh-ye Gāzresāni.*

23. Meeting Minutes, no document number, 6 Shahrivar 1341, *Makātebāt dar Movarrad-e Prozheh-ye Gāzresāni.*

24. Letter from Manuchehr Farmānfarmiān, Supervisor of Distribution Organization of the NIOC to the Governor of Fārs, Document 425/288307, 21 Ābān 1341, *Makātebāt dar Movarrad-e Prozheh-ye Gāzresāni.*

25. Meeting Minutes, no document number, 27 Ābān 1341, p. 1, *Makātebāt dar Movarrad-e Prozheh-ye Gāzresāni.*

26. Sherkat-e Melli-ye Gāz-e Iran, *San'at-e Gāz-e Iran*, 36.

27. Ibid., 36.

28. Mohsen Shirāzi, *San'at-e Gāz-e Iran: Az Āghāz tā Āstāneh-ye Enqelāb*, Part 2, interview by Gholāmrezā Afkhami (Bethesda, MD: Foundation for Iranian Studies, 1999), available online at https://fis-iran.org/fa/ebook/gas/ (accessed September 2, 2025).

29. Robert Steele, *The Shah's Imperial Celebrations of 1971: Nationalism, Culture and Politics in Late Pahlavi Iran* (London: I.B. Tauris, 2021), 41.

30. Ibid., 29–30, 41, 73–91, 126–128.

31. Sherkat-e Melli-ye Naft-e Iran, *Naft va Zendegi* (Tehran: Enteshārāt-e Ravābat-e Omumi-ye San'at-e Naft-e Iran, 1352), 58–62.

32. Ibid., 63–64, attachments 3–8.

33. Ibid., 65–69, attachments 10–14.

34. Sherkat-e Melli-ye Gaz-e Iran, *San'at-e Gāz-e Iran* ([Tehran?]: Enteshārāt-e Ravābat-e Omumi-ye San'at-e Naft-e Iran, 1352), 34.

35. Sherkat-e Melli-ye Naft-e Iran, *Naft va Zendegi*, 72–74.

36. Ibid., 55.

37. Shirāzi, *San'at-e Gāz-e Iran*, Part 2.

38. Gholam-Reza Afkhami, "Interview with Mosaddeqi, Taqi," November 30, 1987, Alexandria, VA, available online at https://fis-iran.org/oral-history/mosaddeqi-taqi/ (accessed January 14, 2023).

39. Shirāzi, *San'at-e Gāz-e Iran*, Part 2.

40. Sherkat-e Melli-ye Gaz-e Iran, *San'at-e Gāz-e Iran*, 29–31.

41. Sherkat-e Melli-ye Gāz-e Iran, "F'āliat-hā va 'Amalkard-e Sherkat-e Melli-ye Gāz-e Iran tay Durān-e Barnāmeh-hā-ye 'Omrāni-ye Chahārom va Panjom," 16 Bahman 2535, p. 7; Documents and Publication Center of the Plan and Budget Organization of the Islamic Republic of Iran.

42. Sherkat-e Melli-ye Gaz-e Iran, *San'at-e Gāz-e Iran*, 31.

43. Shirāzi, *San'at-e Gāz-e Iran*, Part 2.

44. Azadeh Mashayekhi, "The 1968 Tehran Master Plan and the Politics of Planning Development in Iran (1945–1979)," *Planning Perspectives* 34, no. 5 (2019), 863–866.

45. Shirāzi, *San'at-e Gāz-e Iran*, Part 2.

46. For more, see Hāmed Tālebi, 'Aisi Hojjat, and Muhammad Farziān, "Kuy-hā-ye Maskuni-ye Kutāh-ye Moratabeh-ye Tehran dar Dureh-ye Pahlavi-ye Dovvom," *Contemporary Architecture of Iran* (Spring 1393), available online at http://www.caoi.ir/en/study/976-کوی-های-مسکونی-تهران-در-دوران-پهلوی.html (accessed on July 2, 2019).

47. Though this author has been thus far unable to document the origin of this rumor or the length of its circulation within Iranian society, it is a common belief that is repeated with regularity by contemporary Iranians across many social and professional classes whenever the subject of natural gas is brought up.

48. Mohsen Shirāzi, *San'at-e Gāz-e Iran*, Part 2.

49. Ibid.

50. Sherkat-e Melli-ye Gaz-e Iran, *San'at-e Gāz-e Iran*, 31.

51. Sherkat-e Melli-ye Gāz-e Iran, "F'āliat-hā va 'Amalkard-e Sherkat-e Melli-ye Gāz-e Iran," 8.

52. Mohsen Shirāzi, *San'at-e Gāz-e Iran*, Part 2.

53. Sherkat-e Melli-ye Gāz-e Iran, "F'āliat-hā va 'Amalkard-e Sherkat-e Melli-ye Gāz-e Iran," table 2–2.

54. Ibid., table 2–1.

55. Calculated from Sherkat-e Melli-ye Gāz-e Iran, "F'āliat-hā va 'Amalkard-e Sherkat-e Melli-ye Gāz-e Iran," table 2–9.

56. Sherkat-e Melli-ye Gāz-e Iran, "F'āliat-hā va 'Amalkard-e Sherkat-e Melli-ye Gāz-e Iran," 14–18; Hossein Razavi and Firouz Vakil, *The Political Environment of Economic Planning in Iran, 1971–1983: From Monarchy to Islamic Republic* (Boulder, CO: Westview Press, 1984), 36.

57. Calculated from Sherkat-e Melli-ye Gāz-e Iran, "F'āliat-hā va 'Amalkard-e Sherkat-e Melli-ye Gāz-e Iran," table 2–6.

58. Afkhami, "Interview with Mosaddeqi, Taqi," 21–22.

59. Sherkat-e Melli-ye Naft-e Iran, *Naft va Zendegi*, 75–76.

60. Pirooz Ashraf, "Natural Gas Industry in Iran," Table 3, *Encyclopaedia Iranica*, online edition, 2016, available at http://www.iranicaonline.org/articles/natural-gas-industry-in-iran (accessed on May 6, 2020).

61. Pooya Azadi, Mohsen B. Mesgaran, and Matin Mirramezani, *The Struggle for Development in Iran: The Evolution of Governance, Economy, and Society* (Stanford, CA: Stanford University Press, 2022), 98–102.

62. Nikki R. Keddie, *Roots of Revolution: An Interpretive History of Modern Iran* (New Haven, CT: Yale University Press, 1981), 142–182; Eric J. Hooglund, *Land and Revolution in Iran, 1960–1980* (Austin, TX: University of Texas Press, 1982); Afsaneh Najmabadi, *Land Reform and Social Change in Iran* (Salt Lake City, UT: University of Utah Press, 1987); Mohammad Gholi Majd, *Resistance to the Shah: Landowners and Ulama in Iran* (Gainesville, FL: University of Florida Press, 2000).

63. For more on the economic, social, and political history of Iran in the 1960s and 1970s, see chapters 9, 10, and 11 of Ervand Abrahamian's *Iran: Between Two Revolutions* (Princeton, NJ: Princeton University Press, 1982) as well as chapter 7 of Nikki R. Keddie's *Modern Iran: Roots and Results of Revolution*. See also chapter 3 of Pamela Karimi's *Domesticity and Consumer Culture in Iran: Interior Revolutions of the Modern Era* (New York: Routledge, 2013). For more on the ideological content and orientation of Khomeini's views, see Ervand Abrahamian's *Khomeinism: Essays on the Islamic Republic* (Berkeley, CA: University of California Press, 1993), particularly chapters 1 and 2. For more on Khomieni's thought and its contextualization within the work of seven other major revolutionary thinkers, see Hamid Dabashi's essential *Theology of Discontent: The Ideological Foundations of the Islamic Revolution in Iran* (New York: New York University Press, 1993. Reprint, New Brunswick, NJ: Transaction Publishers, 2008).

64. Dr. Esfandiyār Amirshāhi to Governor of Fārs, 6/29046, 25 Dey 1352; Memo from Ostāndāri-ye Fārs to NIGC, 24328, 8 Ābān 1352; Memo from

Ostāndāri-ye Fārs to NIGC Shiraz, 23328, 1 Dey 1351; Residents of 10-meter Alley Kholdbarin to Ostāndāri-ye Fārs, undated but 1352 based on archival stamp, *Residegi beh Shakāyāt va Darkhāst-hā-ye Ahāli-ye Fārs (1352–1353)* (98-293-1262), Ostāndāri-ye Fārs, National Archives of Iran, Fārs Document Center.

65. Proceedings of the Administrative Council of the County of Sepidān, 23 Khordād 2535, attached to Memo 464, 31 Khordād 2535, attached to Memo, Political and Law Enforcement Office of Fārs Province to Shiraz Regional Gas, 16 Tir 2535, 1–2, *Omur Marbut be Rāh va Tarābari-ye Ostān Fārs az Jomleh-ye Rāhsāzi va Marmat Rāh-ha va Lulehkeshi-ye Gāz-e Sepidān* (293-12649), Ostāndāri-ye Fārs, National Archive of Iran, Tehran [hereafter *Omur Marbut be Anjoman-e Shahrestān-e Kāzerun*]. See also Saghar Sadeghian, "The Caspian Forests of Northern Iran during the Qajar and Pahlavi Periods: Deforestation, Regulation, and Reforestation," *Iranian Studies* 49, no. 6 (2016), 973–996 and Farshad Amiraslani and Deirdre Dragovich, "Forest Management Policies and Oil Wealth in Iran over the Last Century: A Review," 167–176.

66. Amiraslani and Dragovich, "Forest Management Policies and Oil Wealth in Iran over the Last Century," 167–176.

67. Proceedings of the County Board of Kāzerun, no. 58, 30 Ordibehesht 2537, attached to Memo 1087, 17 Khordād 2537, p. 1, *Omur Marbut be Anjoman-e Shahrestān-e Kāzerun.*

68. Letter from Abdullāh Hushdārān to Manuchehr Āzemun, Document 364, 2 Tir 2537, *Omur Marbut be Anjoman-e Shahrestān-e Kāzerun.*

69. Proceedings of the County Board of Kāzerun, Document 66, 2 Shahrivar 2537, attached to Letter 2698, 20 Shahrivar 2537, attached to Letter 19238/22, 26 Shahrivar 1357, p. 1, *Omur Marbut be Anjoman-e Shahrestān-e Kāzerun.*

70. Proceedings of the County Board of Kāzerun, Document 68, 22 Shahrivar 2537, attached to Letter 21022/2, 4 Mehr 2537, p. 2, *Omur Marbut be Anjoman-e Shahrestān-e Kāzerun.*

71. Ibid.

72. Meeting minutes of the Committee to Fight Price Gouging and Hoarding, no document number, undated but the meeting was held 30 Dey 1359, *Suratjalaseh Marbut beh Tozie'-ye Gāz (1309–1348)* (96-293-20426), Ostāndāri-ye Yazd, National Archives of Iran, Yazd Document Center.

73. Letter from Ahmad Behzād to Javād Imāmi, Document 4094, 23 Mehr 1357, attached to Letter 5195, 26 Dey 1357, attached to letter 29839/22, 15 Farvardin 1358, *Omur Marbut be Anjoman-e Shahrestān-e Kāzerun.*

74. See Letter from Javād Imāmi to NIGC-Fārs Region, Document 21022/22, 16 Mehr 1357; Letter from Javād Imāmi to NIGC-Fārs Region, Document 23300/22, 8 Ābān 1357; Ahmad Behzād to NIGC-Fārs Region, Document

26820/22, 6 Āzar 1357; Letter from Javād Imāmi to NIGC-Fārs Region, Document 26494/22, 7 Āzar 1357, *Omur Marbut be Anjoman-e Shahrestān-e Kāzerun.*

75. Proceedings of the County Board of Kāzerun, Document 79, 20 Dey 1357, attached to Memo 4480, 9 Bahman 1357, p. 1, *Omur Marbut be Anjoman-e Shahrestān-e Kāzerun.*

76. Abrahamian, *Khomeinism*, 13–38.

77. Proceedings of the County Board of Kāzerun, Document 79, 20 Dey 1357, p. 1; Letter from Abdullāh Hushdārān to the Governor of Kāzerun, 2, 19 Farvardin 1358, attached 1945/2, 2 Ordibehesht 1358, *Omur Marbut be Anjoman-e Shahrestān-e Kāzerun.*

78. Sherkat-e Melli-ye Gāz-e Iran, *Sherkat-e Melli-ye Gāz-e Iran: Shesh Sāl B'ad az Piruzi-ye Enqelāb-e Islāmi* (Tehran[?]: Ravābat-e 'Omumi va Ershād-e Vezārat-e Naft, 1364), 7–8.

79. Ravābat-e Omumi va Ershād-e Islami-ye Vezārat-e Naft, *Sherkat-e Melli-ye Gāz dar Khedmat-e Mostaz'afin* (Tehran: Enteshārāt-e Ravābat-e Omumi va Ershād-e Islam-ye Vezārat-e Naft, 1363), 3–4.

80. Ashraf, "Natural Gas Industry in Iran," Table 3.

81. Sherkat-e Melli-ye Gāz-e Iran, *Sherkat-e Melli-ye Gāz-e Iran*, 33–35.

82. Vezārat-e Naft, *Sherkat-e Melli-ye Gāz dar Khedmat-e Mostaz'afin*, 4–7.

83. Sherkat-e Melli-ye Gāz-e Iran, *Sherkat-e Melli-ye Gāz-e Iran*, 33–35.

84. Sherkat-e Melli-ye Gāz-e Iran, *Sherkat-e Melli-ye Gāz-e Iran*, cover, 18, 24–25, 31, 39; Ravābat-e Omumi va Ershād-e Islami-ye Vezārat-e Naft, *Sherkat-e Melli-ye Gāz dar Khedmat-e Mostaz'afin*, 4, 6. Such depictions were particularly common in *Nedā-ye Gāz*, the NIGC monthly magazine that began publication in March 1991. See, for example, *Nedā-ye Gāz* 1, no. 2 (1370), inside cover; *Nedā-ye Gāz* 1, no. 3 (1370), back cover; *Nedā-ye Gāz* 1, no. 4 (1370), inside back cover, back cover; *Nedā-ye Gāz* no. 5 (1370), inside cover, 5–9; and more.

85. Sherkat-e Melli-ye Gāz-e Iran, *Sherkat-e Melli-ye Gāz-e Iran*, 12.

86. *Energy Institute Statistical Review of World Energy 2024*, "Gas: Consumption – Bcf (from 1965)," available at https://www.energyinst.org/statistical-review.

87. Elham Hassanzadeh, *Iran's Natural Gas Industry in the Post-Revolutionary Period: Optimism, Scepticism, and Potential* (Oxford, UK: Oxford University Press, 2014), 12–47.

88. Taghi Ebtekar, "Environmental Impact of Alternative Fuel on Tehran Air Pollution," in *Proceedings of the 30th Intersociety Energy Conversion Engineering Conference*, vol. 2, 31–36, (New York: The American Society of Mechanical Engineers, 1995), 32.

89. Muhammad Imran Khan, Tabassum Yasmin, and Abdul Shakoor, "Technical Overview of Compressed Natural Gas (CNG) as a Transportation Fuel," *Renewable and Sustainable Energy Review* 51 (2015), 786.

90. "Tarh-e Āzarbāyijān, Jedāl bā Tabiy'at-e Sakht-e Kuhestān," *Nedā-ye Gāz* 1, no. 2 (1370), 8.

91. See, for example, Sherkat-e Melli-ye Gāz-e Iran, *Sherkat-e Melli-ye Gāz-e Iran: Shesh Sāl B'ad az Piruzi-ye Enqelāb-e Islāmi*, 24–25; Sherkat-e Melli-ye Gāz-e Iran, *Rāhnamā-ye Imani barāye Masrafkonnandegān-e Gāz-e Tabi'i* (Tehran[?]: Ravābat-e 'Omumi va Ershād-e Sherkat-e Melli-ye Gāz-e Iran, 1368), 3; Majid Bujārzādeh and Ali Bahādar, *Gāz, Enerzhi-ye Pāk bā Nim-e Qarn-e Talāsh* (Tehran: Sherkat-e Melli-ye Gāz-e Iran – Ravābat-e 'Omumi, 1395), 190, 272, 319, 379, 457, 518, 639, 665, 683, 701, 791, 877, 928–929, 991, 1002.

Conclusion

1. Caterina Scaramelli, *How to Make a Wetland: Water and Moral Ecology in Turkey* (Stanford, CA: Stanford University Press, 2021).

2. Victor Seow, *Carbon Technocracy: Energy Regimes in Modern East Asia* (Chicago: University of Chicago Press, 2021), 7.

3. Christopher R.W. Dietrich, *Oil Revolution: Anticolonial Elites, Sovereign Rights, and the Economic Culture of Decolonization* (Cambridge, UK: Cambridge University Press, 2107), 129.

4. "Indonesian President: Ice Boxes on Screen Sabotaged Colonialism," *Variety*, June 6, 1956, 61, Entertainment Industry Magazine Archive. See also Marshall McLuhan and Quentin Fiore, *The Medium is the Massage: An Inventory of Effects* (Berkeley, CA: Gingko Press, 1967), 131.

5. Sherkat-e Melli-ye Gaz-e Iran, *San'at-e Gāz-e Iran* ([Tehran?]: Enteshārāt-e Ravābat-e Omumi-ye San'at-e Naft-e Iran, 1352), 34–36; Sherkat-e Melli-ye Gāz-e Iran, "F'āliat-hā va 'Amalkard-e Sherkat-e Melli-ye Gāz-e Iran tay Durān-e Barnāmeh-hā-ye 'Omrāni-ye Chahārom va Panjom," 16 Bahman 2535, 4–8, Documents and Publication Center of the Plan and Budget Organization of the Islamic Republic of Iran; Mohsen Shirāzi, *San'at-e Gāz-e Iran: Az Āghāz tā Āstāneh-ye Enqelāb*, Part 2, interview by Gholāmrezā Afkhami (Bethesda, MD: Foundation for Iranian Studies, 1999), available online at https://fis-iran.org/fa/ebook/gas/ (accessed September 2, 2025).

6. Sherkat-e Melli-ye Gaz-e Iran, *San'at-e Gāz-e Iran*, 41–47; Sherkat-e Melli-ye Naft-e Iran, *Naft va Zendegi* (Tehran: Enteshārāt-e Ravābat-e Omumi-ye San'at-e Naft-e Iran, 1352), 78–85; Sherkat-e Melli-ye Gāz-e Iran, "F'āliat-hā va 'Amalkard-e Sherkat-e Melli-ye Gāz-e Iran, 4–10; Pirooz Ashraf, "Natural Gas Industry in Iran," Table 3, *Encyclopaedia Iranica*, online edition, 2016, available at http://www.iranicaonline.org/articles/natural-gas-industry-in-iran (accessed on May 18, 2020).

7. Eskandar Firouz and Daniel Balland, "Environmental Protection," *Encyclopedia Iranica*, online edition, 2011, available at http://www.iranicaonline.org/articles/environmental-protection (accessed online on July 22, 2019).

8. Letter from Sayyid Ahmad Nazari to Office of Environmental Protection on Zanjān, 13926, 11 Shahrivar 1359, *Ehdās-e Kureh-hā-ye Ājorpazi*, NLAI; "Beh 'Illat-e Āludeh Kardan-e Mohit-e Zist 3 Kureh Ājorpazi va 2 Vāhed-e Tolid Movād-e Shimiāi dar Varāmin T'atil Shod," *Ettelā'āt* (23 Khordād 1373); 8 Kureh-ye Ājorpazi-ye Gheyr-e Mojāz dar Ostān Hamadān T'atil Shod," *Ettelā'āt* (21 Tir 1375); "Tamāmi Kureh-hā-ye Ājorpazi-ye Āludehkonnandeh Tuyserkān T'atil Shod," *Ettelā'āt* (18 Āzar 1375); "Beh Dalil-e Āludeh Kardan-e Havā Kureh-hā-ye Ājorpazi-ye Sonati dar Ostān Hamadān T'atil Shod," *Ettelā'āt* (10 Khordād 1376); "Mas'ul-e Setād-e Sālemsāzi-ye Mohit-e Zist-e Ostān-e Qom: Kureh-hā-ye Ājorpazi-ye Qom har Sāl 7 ton Gāz-hā-ye Ālāyandeh dar Havā Parākandeh Mikonand," *Ettelā'āt* (6 Shahrivar 1376) and "Beh Dalil-e Āludehsāzi-ye Mohit-e Zist 53 Kureh-ye Ājorpazi dar Qom T'atil Shod," *Ettelā'āt* (19 Mordād 1376).

9. "Sāzmān-e Hefāzat-e Mohit-e Zist Dalā'el-e T'atil-e Kureh-ha-ye Ājorpazi-ye Shahrak-e Qarchak Varāmin rā A'lām Kard," *Ettelā'āt* (3 Āzār 1363).

10. Global Environment Facility, *Tehran Transport Emissions Reduction Project* (Washington, D.C.: The World Bank, 1993), 6.

11. Farhad Atash, "The Deterioration of Urban Environments in Developing Countries: Mitigating the Air Pollution Crisis in Tehran, Iran," *Cities* 24, no. 6 (2007), 399–409.

12. Vahid Hosseini and Hossein Shahbazi, "Urban Air Pollution in Iran," *Iranian Studies* 49, no. 6 (2016), 1035–1037; Hamed Vafa-Arani, Salman Jahani, Hossein Dashti, Jafar Heydari, and Saeed Moazen, "A System Dynamics Modeling for Urban Air Pollution: A Case Study of Tehran, Iran," *Transportation Research* 31, Part D (2014), 22; Nastaran Ansari and Abbas Seifi, "A System Dynamics Model for Analyzing Energy Consumption and CO_2 Emission in Iranian Cement Industry under Various Production and Export Scenarios," *Energy Policy* 58 (2013), 75; A.R. Karbassi, M. Abbasspour, M.S. Sekhavatjou, F. Ziviyar, and M. Saeedi, "Potential for Reducing Air Pollution from Oil Refineries," *Environmental Monitoring and Assessment* 145, no. 1–3 (2008): 159–166; Global Environment Facility, *Tehran Transport Emissions Reduction Project*, 2.

13. For a detailed discussion of that failure, see S.A. Hashemian, N. Mansouri, and M.A. Morady, "Investigating the Impacts of Retrofitted CNG Vehicles on Air Pollutant Emissions in Tehran," *International Journal of Environmental Research* 7, no. 3 (2013): 669–678.

14. Paul J. Crutzen and Eugene F. Stoermer, "The 'Anthropocene,'" *Global Climate Newsletter* 41 (2000), 17–18.

15. Kenneth Pomeranz, *The Great Divergence: China, Europe, and the Making of the Modern World Economy* (Princeton, NJ: Princeton University Press, 2000).

16. Andreas Malm, *Fossil Capital: The Rise of Steam Power and the Roots of Global Warming* (London: Verso, 2016).

17. Simon L. Lewis and Mark A. Maslin, "Defining the Anthropocene," *Nature* 519 (2015), 171–180.

18. J.R. McNeill and Peter Engelke, *The Great Acceleration: An Environmental History of the Anthropocene Since 1945* (Cambridge, MA: Harvard University Press, 2014), 2–4, 9–10, 132–133, 151.

19. Seow, *Carbon Technocracy*, 7.

20. Cara New Daggett, *The Birth of Energy: Fossil Fuels, Thermodynamics, and the Politics of Work* (Durham, NC: Duke University Press, 2019).

21. Dipesh Chakrabarty, *Provincializing Europe: Postcolonial Thought and Historical Difference* (Princeton, NJ: Princeton University Press, 2000, 2007); On Barak, *Powering Empire: How Coal Made the Middle East and Sparked Global Carbonization* (Oakland, CA: University of California Press, 2020).

22. Barak, *Powering Empire*, 8.

23. Elizabeth Chatterjee, "The Asian Anthropocene: Electricity and Fossil Developmentalism," *The Journal of Asian Studies* 79, no. 1 (2020), 3–24.

24. Sara Lorenzini, *Global Development: A Cold War History* (Princeton, NJ: Princeton University Press, 2019).

25. Paul Warde, Libby Robin, and Sverker Sörlin, *The Environment: A History of the Idea* (Baltimore, MD: Johns Hopkins University Press, 2018), 122–150.

26. Jason W. Moore, *Capitalism in the Web of Life: Ecology and the Accumulation of Capital* (London: Verso, 2015); Jason W. Moore, "The Capitalocene, Part I: On the Nature and Origins of our Ecological Crisis," *The Journal of Peasant Studies* 44, no. 3 (2015), 594–630; Donna Haraway et al., "Anthropologists are Talking – About the Anthropocene," *Ethnos* 81, no. 3 (2015), 535–564.

27. Donna Haraway, *Staying with the Trouble: Making Kin in the Chthulucene* (Durham, NC: Duke University Press, 2016).

28. Anna L. Tsing, *The Mushroom at the End of the World: On the Possibility of Life in Capitalist Ruins* (Princeton, NJ: Princeton University Press, 2015), 20.

29. International Energy Agency, *The Role of Gas in Today's Energy Transitions* (Paris: IEA, 2019).

30. *Energy Institute Statistical Review of World Energy 2024*, "Carbon: Carbon Dioxide Emissions from Energy (from 1965)," available at https://www.energyinst.org/statistical-review.

31. On Barak, "Three Watersheds in the History of Energy," *Comparative Studies of South Asia, Africa and the Middle East* 34, no. 3 (2014), 440–453.

32. Kaveh Madani et al., "Iran's Socio-economic Drought: Challenges of a Water-Bankrupt Nation," *Iranian Studies* 49, no. 6 (2016), 997–1016.

Bibliography

Archives, Document Collections, and Libraries

British Petroleum Archive, University of Warwick, Coventry, United Kingdom.
Central Library of the Ministry of Energy, Tehran, Iran.
Development and Resources Corporation Records, Public Policy Papers, Department of Special Collections, Princeton University Library.
Documents and Publications Center of the Plan and Budget Organization of the Islamic Republic of Iran, Tehran, Iran.
Institute for Iranian Contemporary Historical Studies, Tehran, Iran.
Iran Petroleum Museum and Document Center, Tehran, Iran.
Iran Social Science Data Portal. Available at https://irandataportal.syr.edu/.
Majlis Library, Museum, and Document Center, Tehran, Iran.
National Iranian Oil Company Information Center and Central Library, Tehran, Iran.
National Library and Archives of the Islamic Republic of Iran.
  Fārs Document Center, Shiraz, Iran
    Provincial Government of Fārs
  National Archive of Iran, Tehran, Iran
    Governorate of Zanjān
    Municipality of Tehran
    Presidential Files
    Provincial Government of Fārs

Provincial Government of Khuzestān
National Library of Iran, Tehran, Iran
U.S. National Archives at College Park, Department of State, RG 59.

Periodicals in Persian

Ettelā'āt
Nāmeh-ye San'at-e Naft-e Iran
Nashriyeh-ye Anjoman-e Naft-e Iran
Nedā-ye Gāz

Published Sources in Persian

Afkhami, Gholāmrezā. "Interview with Mosaddeqi, Taqi," November 30, 1987. Alexandria, VA. Foundation for Iranian Studies. https://fis-iran.org/oral-history/mosaddeqi-taqi/ (accessed January 14, 2023).

Al-e Ahmad, Jalal. *Gharbzadegi*. Tehran: Nashr-e Jāmeh-ye Darān, 1384.

Hozeh-ye Shomāreh-ye 9. *Hoshdāri beh Maghāmāt-e Mas'ul dar Movarrad-e Āludegi-ye Havā-ye Tehran*. Tehran: n.p., 1346.

Modiriat-e Mohandesi va Tarh-ha, Sherkat-e Melli-ye Gāz-e Māy'a-ye Iran. *Moqarrāt va Zavābet-e Sākht va Nasb-e Sistem-ha-ye Gāz-e Māye'-ye Vasāt-e Naqlieh-e Motori*. n.p.: Sherkat-e Melli-ye Gāz-e Māye'-ye Iran, 1372.

Moshref Razavi, 'Alirezā. "Tasir-e Inverzhun dar Tarākam-e Āludegi-ye Havā va Vaz'-e Inverzhun-e Tehran." *Nashriyeh-ye Anjoman-e Naft-e Iran*, no. 47 (Khordād 1351): 16–37.

Mostofizādeh, Muhammad 'Ali. "Tabdil-e Sukht-e Vasāyat-e Naqlieh be Gāz." *Nashriyeh-ye Anjoman-e Naft-e Iran*, no. 62 (2534): 9–19.

Olfat, Manuchehr. "Āludehkonandeh-ha-ye Mohem-e Havā – Tashkhis va Sonjesh-e Anhā." *Nashriyeh-ye Anjoman-e Naft-e Iran*, no. 47 (Khordād 1351): 9–15.

Ravābat-e 'Omumi va Ershād-e Islami-ye Vezārat-e Naft. *Sherkat-e Melli-ye Gāz dar Khedmat-e Mostaz'afin*. Tehran[?]: Enteshārāt-e Ravābet-e 'Omumi va Ershād-e Islami-ye Vezārat-e Naft, 1363.

Rouhani, Fuad. *San'at-e Naft-e Iran: Bist Sāl pas az Melli Shodan*. Tehran: Sherkat-e Sahāmi-ye Ketāb-hā-ye Jibi, 2536.

Sāzmān-e Hefāzat-e Mohit-e Zist. *Āmār-e Āludegi-ye Havā-ye Tehran, Seh Māheh-ye Sevvom-e Sāl-e 1357*. Tehran[?]: Enteshārāt-e Sāzmān-e Hefāzat-e Mohit-e Zist, 1357[?].

Sāzmān-e Hefāzat-e Mohit-e Zist. *Barnāmeh-ye Hefāzat-e Mohit-e Zist dar Durān-e Barnāmeh-ye 'Omrāni-ye Panjom-e Keshvar.* n.p.: Sāzmān-e Hefāzat-e Mohit-e Zist, n.d. [early 1970s?].

Shaybāni, Sa'id. *Mas'aleh-ye Āludegi-ye Havā va Barnāmeh-hā-ye Sherkat-e Melli-ye Naft-e Iran dar in Zamineh.* Tehran[?]: Ravābat-e 'Omumi-ye San'at-e Naft-e Iran, 1350.

Sherkat-e Melli-ye Gāz-e Iran. *Rāhnamā-ye Imani barāye Masrafkonnandegān-e Gāz-e Tabi'i.* Tehran[?]: Ravābat-e 'Omumi va Ershād-e Sherkat-e Melli-ye Gāz-e Iran, 1368.

Sherkat-e Melli-ye Gāz-e Iran. *San'at-e Gāz-e Iran.* Tehran[?]: Enteshārāt-e Ravābat-e 'Omumi-ye San'at-e Naft-e Iran, 1352.

Sherkat-e Melli-ye Gāz-e Iran. *Sherkat-e Melli-ye Gāz-e Iran: Shesh Sāl B'ad az Piruzi-ye Enqelāb-e Islāmi.* Tehran[?]: Ravābat-e 'Omumi va Ershād-e Vezārat-e Naft, 1364.

Sherkat-e Melli-ye Naft-e Iran. *Naft va Zendegi.* Tehran: Enteshārāt-e Ravābat-e 'Omumi-ye San'at-e Naft-e Iran, 1352.

Sherkat-e Melli-ye Naft-e Iran. *Shāhluleh-ye Gāz-e Iran.* Tehran[?]: Enteshārāt-e Ravābat-e 'Omumi-ye San'at-e Naft-e Iran, n.d.

Sherkat-e Sahāmi-ye Kudshimiāi-ye Iran. *Kārkhāneh-ye Shimiāi-ye Shiraz.* Tehran: Enteshārāt-e Omur-e Ravābat-e 'Omumi-ye San'at-e Naft-e Iran, 1353.

Shirāzi, Mohsen. *San'at-e Gāz-e Iran: Az Āghāz tā Āstāneh-ye Enqelāb.* Interview by Gholāmrezā Afkhami. Bethesda, MD: Foundation for Iranian Studies, 1999. Available online https://fis-iran.org/fa/ebook/gas/ (accessed September 2, 2025).

Vāhed-e Mohandesi-ye Sherkat-e Iran Gāz (Melli Shodeh). *Matn-e Sokhanrāni-ye Āghā-ye Mohandes Ashrāqi dar Avalin Seminār-e Barrasi-ye Prozheh-ye Gāzsuzkardan-e Khodro-ha dar Iran.* Tehran: Vāhed-e Mohandesi-ye Sherkat-e Iran Gāz (Melli Shodeh), 1362.

Vezārat-e Āmuzesh va Parvaresh. *Gāz.* Kitāb-hā-ye Khāndani barāye Dāneshāmuzān-e Dabestān. Tehran[?]: Vezārat-e Āmuzesh va Parvaresh, 1351.

Vezārat-e Niru, Daftar-e Barnāmehrizi-e Enerzhi. *Tarāznāmeh-ye Enerzhi-ye Keshvar, 1346–1365.* Tehran[?]: Vezārat-e Niru, 1366.

Published Sources in English

Abaie, I., H.J. Ansari, A. Badakhshan, and A. Jaafari. "History and Development of the Alborz and Sarajeh Fields of Central Iran." Paper presented at the *6th World Petroleum Congress, Frankfurt am Mein, Germany, June 1963.* OnePetro.

Arthur D. Little, Inc., and Chicago Transit Authority. *Comparative Economics of Propane and Diesel Buses: Report to Chicago Transit Authority*. Chicago, IL: Little, 1960.

Badakhshan, A., S. Shaibāni, and M. Olfat. "Techniques and Experiences of Measurement and Observation of Major Atmospheric Pollutants in Iran." *The Bulletin of the Iranian Petroleum Institute* 54 (1974): 13–20.

Bechtel International Corporation. *Natural Gas Supply for Europe: Preliminary Study for Natural Gas Pipe Line*. San Francisco: Bechtel International Corporation, 1951.

Ebtekar, Taghi. "Environmental Impact of Alternative Fuel on Tehran Air Pollution." In *Proceedings of the 30th Intersociety Energy Conversion Conference*, vol. 2, 31–36. New York: The American Society of Mechanical Engineers, 1995.

———. "The Prospect for Alternative Fuel Vehicles in Iran: An Environmental Assessment." In *Proceedings of the 28th Intersociety Energy Conversion Engineering Conference*, vol. 2, 893–897. Washington, D.C.: American Chemical Society, 1993.

———. "Reconstruction of Shiraz CNG Project." In *International Congress Exhibition on the Use of Methane in the Transportation Sector*, 211–213. Bologna, Italy: Fedormentano, 1986.

Energy Institute Statistical Review of World Energy 2024. Available online at https://www.energyinst.org/statistical-review.

Firouz, Eskandar. "Environmental and Nature Conservation in Iran." *Environmental Conservation* 3, no. 1 (Spring 1976): 33–42.

Gibson, H.S. "Multistage Stabilization of Crude." *Transactions of the AIME* 136, no. 1 (December 1940): 25–36.

Global Environment Facility. *Tehran Transport Emissions Reduction Project*. Washington, D.C.: The World Bank, 1993.

"Indonesian President: Ice Boxes on Screen Sabotaged Colonialism." *Variety*, June 6, 1956.

International Engineering Company, Morrison-Knudsen International Company, Inc. *Report on Program for the Development of Iran*. San Francisco: n.p., 1947.

McLeod, Thos H. *National Planning in Iran: A Report Based on the Experiences of the Harvard Advisory Group in Iran*. Regina, SK: 1964.

Naghavi, S., and N. Manoochehri. "Iranian Gas Trunkline Project." Paper presented at *7th World Petroleum Conference, Mexico City, Mexico, April 1967*. OnePetro.

Pahlavi, Muhammad Reza. *Mission for My Country*. New York: McGraw-Hill Book Company, Inc., 1961.

United Nations. Economic Commission for Asia and the Far East. *Proceedings of the Second Symposium on the Development of Petroleum Resources of Asia and the Far East.* Vol. 1 & 2. New York: United Nations, 1963.

United Nations. Economic Commission for Asia and the Far East, United Nations, and Bureau of Technical Assistance Operations. *Proceedings of the Seminar on the Development and Utilization of Natural Gas Resources.* New York: United Nations, 1965.

Zerbonia, R. and B. Soraya. "Air Pollution Control in Iran." *Journal of the Air Pollution Control Association* 28, no. 4 (1978): 334–337.

Secondary Literature in Persian

Ardekāni, Husayn Mahbubi. *Tārikh-e Moassesāt-e Tamaddoni-ye Jadid dar Iran.* Vol. 3. 2nd ed. Tehran: Enteshārāt-e Dāneshgāh-e Tehran, 1376.

Bujārzādeh, Majid, and 'Ali Bahādar. *Gāz, Enerzhi-ye Pāk bā Nim-e Qarn-e Talāsh: Panjāhomin Sāl-e Tasis-e Sherkat-e Melli-ye Gāz-e Iran.* Tehran: Sherkat-e Melli-ye Gāz-e Iran – Ravābat-e 'Omumi, 1395.

Hasantāsh, Seyyed Gholāmhusayn, and Mikāyil 'Azimi. *Tārikh-e San'at-e Gāz-e Māy'a-ye Iran.* Tehran: Kavir, 1394.

Tālebi, Hāmed, 'Aisi Hojjat, and Muhammad Farziān. "Kuy-hā-ye Maskuni-ye Kutāh-ye Moratabeh-ye Tehran dar Dureh-ye Pahlavi-ye Dovvom." *Contemporary Architecture of Iran* (Spring 1393), available online at http://www.caoi.ir/en/study/976-کوی-های-مسکونی-تهران-در-دوران-پهلوی.html (accessed on July 2, 2019).

Razāqi, Ibrāhim. *Eqtesād-e Iran.* Tehran: Nashr-e Ney, 1367.

Secondary Literature in English

Abrahamian, Ervand. *A History of Modern Iran.* Cambridge, UK: Cambridge University Press, 2008.

———. *Iran: Between Two Revolutions.* Princeton, NJ: Princeton University Press, 1982.

———. *Khomeinism: Essays on the Islamic Republic.* Berkeley, CA: University of California Press, 1993.

———. *Oil Crisis in Iran: From Nationalism to Coup d'Etat.* Cambridge, UK: Cambridge University Press, 2021.

Adas, Michael. "Modernization Theory and the American Revival of the Scientific and Technological Standards of Social Achievement and Human Worth."

In *Staging Growth: Modernization, Development, and the Global Cold War*, edited by David C. Engerman, Nils Gilman, Mark H. Haefele, and Michael E. Latham, 25–45. Amherst, MA: University of Massachusetts Press, 2003.

Afary, Janet. *Sexual Politics in Modern Iran*. Cambridge, UK: Cambridge University Press, 2009.

Afrasiabi, Kaveh L. "The Environmental Movement in Iran: Perspectives from Below and Above." *The Middle East Journal* 57, no. 3 (2003): 432–448.

Aftalion, Fred. *A History of the International Chemical Industry*. Translated by Otto Theodor Benfey. Philadelphia, PA: University of Pennsylvania Press, 1991.

Alsharhan, A.S., and A.E.M. Nairn. *Sedimentary Basins and Petroleum Geology of the Middle East*. Amsterdam, Netherlands: Elsevier, 1997.

Alvandi, Roham. "Flirting with Neutrality: The Shah, Khrushchev, and the Failed 1959 Soviet-Iranian Negotiations." *Iranian Studies* 47, no. 3 (2014): 419–440.

———. "The Shah's Détente with Khrushchev: Iran's 1962 Missile Base Pledge to the Soviet Union." *Cold War History* 14, no. 3 (2014): 423–444.

Amin, Camron Michael. *The Making of the Modern Iranian Woman: Gender, State Policy, and Popular Culture, 1865–1946*. Gainesville, FL: University Press of Florida, 2002.

Amirahmadi, Hooshang, and Farhad Atash. "Dynamics of Provincial Development and Disparity in Iran, 1956–1984." *Third World Planning Review* 9, no. 2 (1987): 155–185.

Amiraslani, Farshad, and Deirdre Dragovich. "Forest Management Policies and Oil Wealth in Iran over the Last Century: A Review." *Natural Resources Forum* 37, no. 3 (2013): 167–176.

Anand, Nikhil. *Hydraulic City: Water and the Infrastructures of Citizenship in Mumbai*. Durham, NC: Duke University Press, 2017.

Ansari, Ali M. *Modern Iran: The Pahlavis and After*, 2nd ed. New York: Routledge, 2011.

———. "The Myth of the White Revolution: Mohammad Reza Shah, 'Modernization' and the Consolidation of Power." *Middle Eastern Studies* 37, no. 3 (July 2001): 1–24.

———. *The Politics of Nationalism in Modern Iran*. Cambridge, UK: Cambridge University Press, 2012.

Ansari, Nastaran, and Abbas Seifi. "A System Dynamics Model for Analyzing Energy Consumption and CO_2 Emission in Iranian Cement Industry Under Various Production and Export Scenarios." *Energy Policy* 58 (2013): 75–89.

Appadurai, Arjun. *The Future as Cultural Fact: Essays on the Global Condition*. London: Verso, 2013.

Apter, David E. *The Politics of Modernization*. Chicago: The University of Chicago Press, 1965.

Atabaki, Touraj. "Far From Home, But at Home: Indian Migrant Workers in the Iranian Oil Industry." *Studies in History* 31, no. 1 (2015): 85–114.

———. "From *'Amaleh* (Labor) to *Kargar* (Worker): Recruitment, Work Discipline and Making of the Working Class in the Persian/Iranian Oil Industry." *International Labor and Working-Class History* 84 (2013): 159–175.

———. "Indian Migrant Workers in the Iranian Oil Industry 1908–1951." In *Working for Oil: Comparative Social Histories of Labor in the Global Oil Industry*, edited by Touraj Atabaki, Elisabetta Bini, and Kaveh Ehsani, 189–226. Cham, Switzerland: Palgrave Macmillan, 2018.

Atabaki, Touraj, and Erik J. Zürcher. *Men of Order: Authoritarian Modernization Under Atatürk and Reza Shah*. London: I.B. Tauris & Co. Ltd, 2004.

Atash, Farhad. "The Deterioration of Urban Environments in Developing Countries: Mitigating the Air Pollution Crisis in Tehran, Iran." *Cities* 24, no. 6 (2007): 399–409.

Azadi, Pooya, Mohsen B. Mesgaran, and Matin Mirramezani. *The Struggle for Development in Iran: The Evolution of Governance, Economy, and Society*. Stanford, CA: Stanford University Press, 2022.

Babb, Sarah. *Managing Mexico: Economists from Nationalism to Neoliberalism*. Princeton, NJ: Princeton University Press, 2001.

Bakhtin, M.M. *The Dialogic Imagination: Four Essays*, translated by Caryl Emerson and Michael Holquist, edited by Michael Holquist. Austin, TX: University of Texas Press, 1981, 2014.

Bamberg, J.H. *The History of the British Petroleum Company, Vol. 2: The Anglo-Iranian Years, 1928–1954*. Cambridge, UK: Cambridge University Press, 1994.

Banani, Amin. *The Modernization of Iran, 1921–1941*. Stanford, CA: Stanford University Press, 1961.

Barak, On. *On Time: Technology and Temporality in Modern Egypt*. Berkeley, CA: University of California Press, 2013.

———. *Powering Empire: How Coal Made the Middle East and Sparked Global Carbonization*. Oakland, CA: University of California Press, 2020.

———. "Three Watersheds in the History of Energy." *Comparative Studies of South Asia, Africa and the Middle East* 34, no. 3 (2014): 440–453.

Baran, Paul A. *The Political Economy of Growth*. New York: Monthly Review Press, 1957.

Barnes, Jessica. *Cultivating the Nile: The Everyday Politics of Water in Egypt*. Durham, NC: Duke University Press, 2014.

Barnes, Joe, Mark H. Hayes, Amy M. Jaffe, and David G. Victor. "Introduction to the Study." In *Natural Gas and Geopolitics: From 1970 to 2040*, edited by David G. Victor, Amy M. Jaffe, and Mark H. Hayes, 3–24. Cambridge, UK: Cambridge University Press, 2006.

Barthes, Roland. *Camera Lucida: Reflections on Photography*. Translated by Richard Howard. New York: Hill and Wang, 1981.

Bell, Wendell. "What Do We Mean by Futures Studies?" In *New Thinking for A New Millennium*, edited by Richard A. Slaughter, 3–25. London: Routledge, 1996.

Bennett, Jane. "The Agency of Assemblages and the North American Blackout." *Public Culture* 17, no. 3 (2005): 445–465.

———. *Vibrant Matter: A Political Ecology of Things*. Durham, NC: Duke University Press, 2010.

Bet-Shlimon, Arbella. *City of Black Gold: Oil, Ethnicity, and the Making of Modern Kirkuk*. Stanford, CA: Stanford University Press, 2019.

Biglari, Mattin. "Iranian Oil Nationalisation as Decolonisation: Historiographical Reflections, Global History, and Postcolonial Theory." In *Iran and Global Decolonisation: Politics and Resistance After Empire*, edited by Firoozeh Kashani-Sabet and Robert Steele, 101–133. London: Gingko, 2023.

———. "Toxic Standards: Pollution, 'Slow Violence', and the Environmental History of the Abadan Oil Refinery, Iran." *Journal of Energy History* 12, no. 1 (2024): 1–17.

Blackbourn, David. *The Conquest of Nature: Water, Landscape, and the Making of Modern Germany*. New York: W.W. Norton & Company, 2006.

Blaszczyk, Regina Lee. "Synthetics for the Shah: DuPont and the Challenges to Multinationals in 1970s Iran." *Enterprise & Society* 9, no. 4 (2008): 670–723.

Bostock, Frances, and Geoffrey Jones. *Planning and Power in Iran: Ebtehaj and Economic Development Under the Shah*. London: Frank Cass and Company Limited, 1989.

Bowker, Geoffrey C. *Science on the Run: Information Management and Industrial Geophysics at Schlumberger, 1920–1940*. Cambridge, MA: MIT Press, 1994.

Brew, Gregory. *Petroleum and Progress in Iran: Oil, Development, and the Cold War*. Cambridge, UK: Cambridge University Press, 2022.

Brownell, Emily. *Gone to Ground: A History of Environment and Infrastructure in Dar es Salaam*. Pittsburgh, PA: University of Pittsburgh Press, 2020.

Butane-Propane News Incorporated. *Propane in America: The First 100 Years, 1912–2012*. Arcadia, CA: Butane-Propane News Incorporated, 2012.

Chakrabarty, Dipesh. *Provincializing Europe: Postcolonial Thought and Historical Difference*. Princeton, NJ: Princeton University Press, 2000, 2007.

Chateau, Zoé, Patrick Devine-Wright, and Jane Wills. "Integrating Sociotechnical and Spatial Imaginaries in Researching Energy Futures." *Energy Research & Social Science* 80 (2021): 102–207.

Chatterjee, Elizabeth. "The Asian Anthropocene: Electricity and Fossil Developmentalism." *The Journal of Asian Studies* 79, no. 1 (2020): 3–24.

Clawson, Patrick. "Knitting Iran Together: The Land Transport Revolution, 1920–1940." *Iranian Studies* 26, no. 3–4 (Summer/Fall 1993): 235–250.

Cronin, Stephanie. *The Making of Modern Iran: State and Society Under Riza Shah, 1921–1941*. New York: Routledge, 2003.

Cronin, Stephanie. *Soldiers, Shahs and Subalterns in Iran: Opposition, Protest and Revolt, 1921–1941*. New York: Palgrave Macmillan, 2010.

Crutzen, Paul J., and Eugene F. Stoermer. "The 'Anthropocene.'" *Global Climate Newsletter* 41 (2000): 17–18.

Cullather, Nick. *The Hungry World: America's Cold War Battle Against Poverty in Asia*. Cambridge, MA: Harvard University Press, 2010.

Dabashi, Hamid. *Iran: A People Interrupted*. New York: The New Press, 2007.

———. *Theology of Discontent: The Ideological Foundations of the Islamic Revolution in Iran*. New York: New York University Press, 1993. Reprint, New Brunswick, NJ: Transaction Publishers, 2008.

Daftary, Farhad. "Development Planning in Iran: A Historical Survey." *Iranian Studies* 6, no. 4 (1973): 176–228.

Daggett, Cara New. *The Birth of Energy: Fossil Fuels, Thermodynamics, and the Politics of Work*. Durham, NC: Duke University Press, 2019.

Davoudi, Leonardo. *Persian Petroleum: Oil, Empire and Revolution in Late Qajar Iran*. London: I.B. Tauris, 2021.

De Grazia, Victoria. *Irresistible Empire: America's Advance Through Twentieth-Century Europe*. Cambridge, MA: The Belknap Press of Harvard University Press, 2005.

Derr, Jennifer L. *The Lived Nile: Environment, Disease, and Material Colonial Economy in Egypt*. Stanford, CA: Stanford University Press, 2019.

Deuten, J. Jasper, and Arie Rip. "The Narrative Shaping of a Production Creation Process." In *Contested Futures: A Sociology of Prospective Techno-Science*, edited by Nik Brown, Brian Rappert, and Andrew Webster, 65–86. London: Routledge, 2000.

Dewey, Matías, and Kedron Thomas. "Futurity Beyond the State: Illegal Markets and Imagined Futures in Latin America." *Latin American Politics and Society* 64, no. 4 (2022): 1–23.

Di Palma, Vittoria. *Wasteland: A History*. New Haven, CT: Yale University Press, 2014.

Dietrich, Christopher R.W. *Oil Revolution: Anticolonial Elites, Sovereign Rights, and the Economic Culture of Decolonization*. Cambridge, UK: Cambridge University Press, 2017.

Douglas, Mary. *Purity and Danger: An Analysis of Concepts of Pollution and Taboo.* London: Routledge, 1966, 2002.

Ehsani, Kaveh. "Disappearing the Workers: How Labor in the Oil Complex Has Been Made Invisible." In *Working for Oil: Comparative Social Histories of Labor in the Global Oil Industry*, edited by Touraj Atabaki, Elisabetta Bini, and Kaveh Ehsani, 11–34. Cham, Switzerland: Palgrave Macmillan, 2018.

———. "Social Engineering and the Contradictions of Modernization in Khuzestan's Company Towns: A Look at Abadan and Masjed-Soleyman." *International Review of Social History* 48, no. 3 (2003): 361–399.

Ekbladh, David. *The Great American Mission: Modernization and the Construction of an American World Order*. Princeton, NJ: Princeton University Press, 2010.

Elm, Mostafa. *Oil, Power, and Principle: Iran's Oil Nationalization and Its Aftermath*. Syracuse, NY: Syracuse University Press, 1992.

Escobar, Arturo. *Encountering Development: The Making and Unmaking of the Third World*. Princeton, NJ: Princeton University Press, 1995, 2012.

Esfahani, Hadi Salehi, and M. Hashem Pesaran. "The Iranian Economy in the Twentieth Century: A Global Perspective." *Iranian Studies* 42, no. 2 (2009): 177–211.

Fadaee, Simin. *Social Movements in Iran: Environmentalism and Civil Society.* New York: Routledge, 2012.

Ferguson, James. *The Anti-Politics Machine: "Development," Depoliticization, and Bureaucratic Power in Lesotho*. Minneapolis, MN: University of Minnesota Press, 1990, 1994.

Ferrier, R.W. *The History of the British Petroleum Company, Vol. 1: The Developing Years, 1901–1932*. Cambridge, UK: Cambridge University Press, 1982.

Fischer, Edward F. *The Good Life: Aspiration, Dignity, and the Anthropology of Wellbeing*. Stanford, CA: Stanford University Press, 2014.

Forbes, R.J., and D.R. O'Beirne. *The Technical Development of the Royal Dutch/Shell, 1890–1940*. Leiden, Netherlands: E.J. Brill, 1957.

Fredericks, Rosalind. *Garbage Citizenship: Vital Infrastructures of Labor in Dakar, Senegal*. Durham, NC: Duke University Press, 2018.

Galambos, Louis, Takashi Hikino, and Vera Zamagni, eds. *The Global Chemical Industry in the Age of the Petrochemical Revolution*. Cambridge, UK: Cambridge University Press, 2007.

Gamson, William A., and Andre Modigliani. "Media Discourse and Public Opinion on Nuclear Power: A Constructionist Approach." *American Journal of Sociology* 95, no. 1 (July 1989): 1–37.

Garavini, Guiliano. *The Rise and Fall of OPEC in the Twentieth Century*. Oxford, UK: Oxford University Press, 2019.

Gheissari, Ali. *Iranian Intellectuals in the 20th Century*. Austin, TX: University of Texas Press, 1998.

Ghosh, Amitav. "Petrofiction: The Oil Encounter and the Novel." *The New Republic* (March 2, 1992): 29–34.

Gille, Zsuzsa. *From the Cult of Waste to the Trash Heap of History: The Politics of Waste in Socialist and Postsocialist Hungary*. Bloomington, IN: Indiana University Press, 2007.

Gilman, Nils. *Mandarins of the Future: Modernization Theory in Cold War America*. Baltimore, MD: Johns Hopkins University Press, 2003.

Grigas, Agnia. *The New Geopolitics of Natural Gas*. Cambridge, MA: Harvard University Press, 2017.

Gustafson, James M. *The Lion and the Sun: Environmental History and the Formation of Modern Iran*. London: I.B. Tauris, 2025.

Hajer, Maarten A. *The Politics of Environmental Discourse: Ecological Modernization and the Policy Process*. Oxford, UK: Oxford University Press, 1995.

Halliday, Fred. *Iran: Dictatorship and Development*. New York: Penguin Books, Ltd., 1979.

Haraway, Donna. "A Cyborg Manifesto: Science, Technology, and Socialist-Feminism in the Late Twentieth Century." *Socialist Review* 15 (1985): 65–107.

———. *Staying with the Trouble: Making Kin in the Chthulucene*. Durham, NC: Duke University Press, 2016.

Haraway, Donna, Noboru Ishikawa, Scott F. Gilbert, Kenneth Olwig, Anna L. Tsing, and Nils Bubandt. "Anthropologists are Talking – About the Anthropocene." *Ethnos* 81, no. 3 (2015): 535–564.

Harvey, Penny, and Hannah Knox. *Roads: An Anthropology of Infrastructure and Expertise*. Ithaca, NY: Cornell University Press, 2015.

Hashemian, S.A., N. Mansouri, and M.A. Morady. "Investigating the Impacts of Retrofitted CNG Vehicles on Air Pollutant Emissions in Tehran." *International Journal of Environmental Research* 7, no. 3 (2013): 669–678.

Hassanzadeh, Elham. *Iran's Natural Gas Industry in the Post-Revolutionary Period: Optimism, Scepticism, and Potential*. Oxford, UK: Oxford University Press, 2014.

Hecht, Gabrielle. *The Radiance of France: Nuclear Power and National Identity After World War II*. Cambridge, MA: MIT Press, 1998.

Hein, Carola, and Mohamad Sedighi. "Iran's Global Petroleumscape: The Role of Oil in Shaping Khuzestan and Tehran." *Architectural Theory Review* 21, no. 3 (2017): 349–374.

Helleiner, Eric. *Forgotten Foundations of Bretton Woods: International Development and the Making of the Postwar Order.* Ithaca, NY: Cornell University Press, 2014.

Hendelman-Baavur, Liora. "The Odyssey of Jalal Al-e Ahmad's *Gharbzadegi* – Five Decades After." In *Persian Language, Literature and Culture: New Leaves, Fresh Looks*, edited by Kamran Talatoff, 258–286. London: Routledge, 2015.

Högselius, Per. *Red Gas: Russia and the Origins of European Energy Dependence.* New York: Palgrave Macmillan, 2013.

Hooglund, Eric J. *Land and Revolution in Iran, 1960–1980.* Austin, TX: University of Texas Press, 1982.

Hosseini, Vahid, and Hossein Shahbazi. "Urban Air Pollution in Iran." *Iranian Studies* 49, no. 6 (2016): 1029–1046.

Huber, Matthew T. *Lifeblood: Oil, Freedom, and the Forces of Capital.* Minnesota, MN: University of Minnesota Press, 2013.

International Energy Agency. *The Role of Gas in Today's Energy Transitions.* Paris: IEA, 2019. Available at https://www.iea.org/reports/the-role-of-gas-in-todays-energy-transitions.

Jasanoff, Sheila. "Future Imperfect: Science, Technology, and the Imaginations of Modernity." In *Dreamscapes of Modernity: Sociotechnical Imaginaries and the Fabrication of Power*, edited by Sheila Jasanoff and Sang-Hyun Kim, 1–33. Chicago: University of Chicago Press, 2015.

Jasanoff, Sheila, and Sang-Hyun Kim. "Containing the Atom: Sociotechnical Imaginaries and Nuclear Power in the United States and South Korea." *Minerva* 47, no. 2 (June 2009): 119–146.

Jefroudi, Maral. "Revisiting 'the Long Night' of Iranian Workers: Labor Activism in the Iranian Oil Industry in the 1960s." *International Labor and Working-Class History* 84 (2013): 176–194.

John-Alder, Kathleen. *Ian McHarg and the Search for Ideal Order.* London: Routledge Press, 2019.

Johnson, Bob. *Carbon Nation: Fossil Fuels in the Making of American Culture.* Lawrence, KA: University Press of Kansas, 2014.

Jones, Christopher F. *Routes of Power: Energy and Modern America.* Cambridge, MA: Harvard University Press, 2014.

Jones, Toby Craig. *Desert Kingdom: How Oil and Water Forged Modern Saudi Arabia.* Cambridge, MA: Harvard University Press, 2010.

Kander, Astrid, Paolo Malanima, and Paul Warde. *Power to the People: Energy in Europe Over the Last Five Centuries.* Princeton, NJ: Princeton University Press, 2013.

Karbassi, A.R., M. Abbasspour, M.S. Sekhavatjou, F. Ziviyar, and M. Saeedi. "Potential for Reducing Air Pollution from Oil Refineries." *Environmental Monitoring and Assessment* 145, no. 1–3 (2008): 159–166.

Karimi, Pamela. *Domesticity and Consumer Culture in Iran: Interior Revolutions of the Modern Era.* New York: Routledge, 2013.

Kashani-Sabet, Firoozeh. "American Crosses, Persian Crescents: Religion and the Diplomacy of US-Iranian Relations, 1834–1911." *Iranian Studies* 44, no. 5 (2011): 607–625.

———. *Conceiving Citizens: Women and the Politics of Motherhood in Iran.* Oxford, UK: Oxford University Press, 2011.

———. "Dressing Up (or Down): Veils, Hats, and Consumer Fashions in Interwar Iran." In *Anti-Veiling Campaigns in the Muslim World: Gender, Modernism and the Politics of Dress*, edited by Stephanie Cronin, 149–162. New York: Routledge, 2014.

———. *Frontier Fictions: Shaping the Iranian Nation, 1804–1946.* Princeton, NJ: Princeton University Press, 1999.

———. *Heroes to Hostages: America and Iran, 1800–1988.* Cambridge, UK: Cambridge University Press, 2023.

Katouzian, Homa. *The Political Economy of Modern Iran: Despotism and Pseudo-Modernism, 1926–1979.* New York: New York University Press, 1981.

———. "Riza Shah's Political Legitimacy and Social Base, 1921–1941." In *The Making of Modern Iran: State and Society Under Riza Shah, 1921–1941*, edited by Stephanie Cronin, 15–36. London: Routledge, 2003.

Keddie, Nikki R. *Roots of Revolution: An Interpretive History of Modern Iran.* New Haven, CT: Yale University Press, 1981.

Khalili, Laleh. "Crude Knowledge: Petro-Periodicals and Resource Sovereignty." *International Journal of Middle East Studies* 56, no. 3 (2024): 446–464.

Khan, Muhammad Imran, Tabassum Yasmin, and Abdul Shakoor. "Technical Overview of Compressed Natural Gas (CNG) as a Transportation Fuel." *Renewable and Sustainable Energy Review* 51 (2015): 785–797.

Khazeni, Arash. *Tribes and Empire on the Margins of Nineteenth-Century Iran.* Seattle, WA: University of Washington Press, 2009.

Kia, Mana. *Persianate Selves: Memories of Place and Origin Before Nationalism.* Stanford, CA: Stanford University Press, 2020.

Koyagi, Mikiya. *Iran in Motion: Mobility, Space, and the Trans-Iranian Railway.* Stanford, CA: Stanford University Press, 2021.

Kuniholm, Bruce R. *The Origins of the Cold War in the Near East: Great Power Conflict and Diplomacy in Iran, Turkey, and Greece.* Princeton, NJ: Princeton University Press, 1980, 1994.

Larkin, Brian. "The Politics and Poetics of Infrastructure." *Annual Review of Anthropology* 42 (2013): 327–343.

———. *Signal and Noise: Media, Infrastructure, and Urban Culture in Nigeria.* Durham, NC: Duke University Press, 2008.

Latham, Michael E. *Modernization as Ideology: American Social Science and "Nation Building" in the Kennedy Era.* Chapel Hill, NC: University of North Carolina Press, 2000.

Latour, Bruno. *Aramis, or the Love of Technology.* Translated by Catherine Porter. Cambridge, MA: Harvard University Press, 1996.

———. *The Pasteurization of France.* Translated by Alan Sheridan and John Law. Cambridge, MA: Harvard University Press, 1988.

———. *We Have Never Been Modern.* Translated by Catherine Porter. Cambridge, MA: Harvard University Press, 1993.

Lees, G.M. "Persia." In *The Science of Petroleum: A Comprehensive Treatise of the Principles and Practice of the Production, Refining, Transport and Distribution of Mineral Oil, Vol. 6, Part 1, The World's Oil Fields: The Eastern Hemisphere,* edited by V.C. Illing, 73–82. London: Oxford University Press, 1953.

Lemanski, Charlotte. "Infrastructural Citizenship: The Everyday Citizenships of Adapting and/or Destroying Public Infrastructure in Cape Town, South Africa." *Transactions of the Institute of British Geographers* 45, no. 3 (2019): 1–17.

Lente, Harro van. "Forceful Futures: From Promise to Requirement." In *Contested Futures: A Sociology of Prospective Techno-Science,* edited by Nik Brown, Brian Rappert, and Andrew Webster, 43–63. London: Routledge, 2000.

Lerner, Daniel. *The Passing of Traditional Society: Modernizing the Middle East.* New York: The Free Press, 1958.

Levy, Eugene. "The Aesthetics of Power: High-Voltage Transmission Systems and the American Landscape." *Technology and Culture* 38, no. 3 (1997): 575–607.

Lewis, Simon L., and Mark A. Maslin. "Defining the Anthropocene." *Nature* 519 (2015): 171–180.

Limbert, Mondana E. *In the Time of Oil: Piety, Memory, and Social Life in an Omani Town.* Stanford, CA: Stanford University Press, 2010.

Lockhart, Laurence. "Iranian Petroleum in Ancient and Medieval Times." *Journal of the Institute of Petroleum* 25, no. 183 (1939): 1–18.

Lorenzini, Sara. *Global Development: A Cold War History.* Princeton, NJ: Princeton University Press, 2019.

Macdonald, Graeme. "Containing Oil: The Pipeline in Petroculture." In *Petrocultures: Oil, Politics, Culture*, edited by Sheena Wilson, Adam Carlson, and Imre Szeman, 36–77. Montreal, Canada: McGill-Queen's University Press, 2017.

Madani, Kaveh, Amir AghaKouchak, and Ali Mirchi. "Iran's Socio-Economic Drought: Challenges of a Water-Bankrupt Nation." *Iranian Studies* 49, no. 6 (2016): 997–1016.

Madanipour, Ali. *Tehran: The Making of a Metropolis*. Chichester, UK: John Wiley and Sons, Ltd. 1998.

Mahdavy, H. "The Patterns and Problems of Economic Development in Rentier States: The Case of Iran." In *Studies in the Economic History of the Middle East*, edited by M.A. Cook, 428–467. London: Routledge, 1970.

Majd, Mohammad Gholi. *Resistance to the Shah: Landowners and Ulama in Iran*. Gainesville, FL: University of Florida Press, 2000.

Malm, Andreas. *Fossil Capital: The Rise of Steam Power and the Roots of Global Warming*. London: Verso, 2016.

———. *The Progress of This Storm: Nature and Society in a Warming World*. London: Verso, 2018.

Manoukian, Setrag. *City of Knowledge in Twentieth Century Iran: Shiraz, History and Poetry*. New York: Routledge, 2012.

Marashi, Afshin. *Nationalizing Iran: Culture, Power, and the State, 1870–1940*. Seattle, WA: University of Washington Press, 2008.

Marcel, Valérie, and John V. Mitchell. *Oil Titans: National Oil Companies in the Middle East*. Baltimore, MD: Brookings Institution Press, 2006.

Mashayekhi, Azadeh. "The 1968 Tehran Master Plan and the Politics of Planning Development in Iran (1945–1979)." *Planning Perspectives* 34, no. 5 (2019): 849–876.

McLuhan, Marshall, and Quentin Fiore. *The Medium is the Massage: An Inventory of Effects*. Berkeley, CA: Gingko Press, 1967.

McNeill, J.R., and Peter Engelke. *The Great Acceleration: An Environmental History of the Anthropocene Since 1945*. Cambridge, MA: Harvard University Press, 2014.

Melamid, Alexander. "The Geographical Pattern of Iranian Oil Development." *Economic Geography* 35, no. 3 (1959): 199–218.

Mirsepassi, Ali. *Intellectual Discourse and the Politics of Modernization: Negotiating Modernity in Iran*. Cambridge, UK: Cambridge University Press, 2000.

———. *Iran's Troubled Modernity: Debating Ahmad Fardid's Legacy*. Cambridge, UK: Cambridge University Press, 2019.

———. *Political Islam, Iran, and the Enlightenment: Philosophies of Hope and Despair.* Cambridge, UK: Cambridge University Press, 2011.

Mitchell, Timothy. *Carbon Democracy: Political Power in the Age of Oil.* London: Verso, 2011.

Mitchell, Timothy. *Rule of Experts: Egypt, Techno-Politics, Modernity.* Berkeley, CA: University of California Press, 2002.

Mofid, Kamran. *Development Planning in Iran: From Monarchy to Islamic Republic.* Cambridgeshire, UK: Middle East and North Africa Press, Ltd., 1987.

Moore, Jason W. *Capitalism in the Web of Life: Ecology and the Accumulation of Capital.* London: Verso, 2015.

———. "The Capitalocene, Part I: On the Nature and Origins of our Ecological Crisis." *The Journal of Peasant Studies* 44, no. 3 (2015): 594–630.

Mossavar-Rahmani, Bijan. *Energy Policy in Iran: Domestic Choices and International Implications.* New York: Pergamon Press, 1981.

Movahedi-Lankarani, Ciruce. "A Firm Policy Decision: Infrastructural Form and Pahlavi Developmentalism at the Ahvāz Pipe Mill." *Iranian Studies* 58, no. 2 (2025): 1–17.

———. "A Ghoul at the Gates: Natural Gas Energy and the Environment in Pahlavi Iran, 1960–1979." *International Journal of Middle East Studies* 54, no. 1 (2022): 80–99.

———. "Precarious Petroleum: Volatile Reservoirs, Varied Natural Gas Compositions, and Development in 1960s Iran." *Comparative Studies of South Asia, Africa and the Middle East* 44, no. 1 (2024): 3–17.

Najmabadi, Afsaneh. *Land Reform and Social Change in Iran.* Salt Lake City, UT: University of Utah Press, 1987.

———. *Women with Mustaches and Men Without Beards: Gender and Sexual Anxieties of Iranian Modernity.* Berkeley, CA: University of California Press, 2005.

Nassehi, Ramin. "Domesticating Cold War Economic Ideas: The Rise of Iranian Developmentalism in the 1950s and 1960s." In *The Age of Aryamehr: Late Pahlavi Iran and Its Global Entanglements*, edited by Roham Alvandi, 35–69. London: Gingko Library, 2018.

Nekrasov, Viacheslav. "Decision-Making in the Soviet Energy Sector in Post-Stalinist Times: The Failure of Khrushchev's Economic Modernization Strategy." In *Cold War Energy: A Transnational History of Soviet Oil and Gas*, edited by Jeronim Perović, 165–200. Cham, Switzerland: Palgrave Macmillan, 2016.

Nemchenok, Victor V. "'That So Fair a Thing Should Be So Frail:' The Ford Foundation and the Failure of Rural Development in Iran, 1953–1964." *Middle East Journal* 63, no. 2 (2009): 261–284.

Nye, David E. *American Technological Sublime*. Cambridge, MA: MIT Press, 1994.

Painter, David S., and Gregory Brew. *The Struggle for Iran: Oil, Autocracy, and the Cold War, 1951–1954*. Chapel Hill, NC: The University of North Carolina Press, 2022.

Perović, Jeronim, ed. *Cold War Energy: A Transnational History of Soviet Oil and Gas*. Cham, Switzerland: Palgrave Macmillan, 2016.

Pomeranz, Kenneth. *The Great Divergence: China, Europe, and the Making of the Modern World Economy*. Princeton, NJ: Princeton University Press, 2000.

Porter, Theodore M. *Trust in Numbers: The Pursuit of Objectivity in Science and Public Life*. Princeton, NJ: Princeton University Press, 1995.

Rajaee, Farhang. *Islamism and Modernism: The Changing Discourse in Iran*. Austin, TX: University of Texas Press, 2007.

Razavi, Hossein, and Firouz Vakil. *The Political Environment of Economic Planning in Iran, 1971–1983: From Monarchy to Islamic Republic*. Boulder, CO: Westview Press, 1984.

Ricks, Thomas M. "U.S. Military Missions to Iran, 1943–1978: The Political Economy of Military Assistance." *Iranian Studies* 12, no. 3–4 (1979): 163–193.

Rosenberg, Daniel, and Susan Harding. "Introduction: Histories of the Future." In *Histories of the Future*, edited by Daniel Rosenberg and Susan Harding, 3–18. Durham, NC: Duke University Press, 2005.

Ross, Michael L. *The Oil Curse: How Petroleum Wealth Shapes the Development of Nations*. Princeton, NJ: Princeton University Press, 2012.

Rostam-Kolayi, Jasamin. "The New Frontier Meets the White Revolution: The Peace Corps in Iran, 1962–76." *Iranian Studies* 51, no. 4 (2018): 587–612.

Rostow, W.W. *The Stages of Economic Growth: A Non-Communist Manifesto*. Cambridge, UK: Cambridge University Press, 1960.

Sadan, Amit. "Masters of the Garden: An Environmental History of Nation-Building in Pahlavi Iran." PhD diss. University of Michigan, 2023.

Sadeghian, Saghar. "The Caspian Forests of Northern Iran During the Qajar and Pahlavi Periods: Deforestation, Regulation, and Reforestation." *Iranian Studies* 49, no. 6 (2016): 973–996.

Saheb, Tahereh. "Air Pollution Governance in Iran: Inhibiting Factors." PhD diss. Rensselaer Polytechnic Institute, 2015.

Scaramelli, Caterina. *How to Make a Wetland: Water and Moral Ecology in Turkey*. Stanford, CA: Stanford University Press, 2021.

Schayegh, Cyrus. "Iran's Global Long 1970s: An Empire Project, Civilisational Developmentalism, and the Crisis of the Global North." In *The Age of*

Aryamehr: Late Pahlavi Iran and Its Global Entanglements, edited by Roham Alvandi, 260–289. London: Gingko, 2018.
———. "Iran's Karaj Dam Affair: Emerging Mass Consumerism, the Politics of Promise, and the Cold War in the Third World." *Comparative Studies in Society and History* 54, no. 3 (2012): 612–643.
———. "'Seeing Like a State': An Essay on the Historiography of Modern Iran." *International Journal of Middle East Studies* 42, no. 1 (2010): 37–61.
———. *Who Is Knowledgeable Is Strong: Science, Class, and the Formation of Modern Iranian Society*. Berkeley, CA: University of California Press, 2009.
Scott, James C. *Seeing Like a State: How Certain Schemes to Improve the Human Condition Have Failed*. New Haven, CT: Yale University Press, 1998.
Selcer, Perrin. *The Postwar Origins of the Global Environment: How the United Nations Built Spaceship Earth*. New York: Columbia University Press, 2018.
Seow, Victor. *Carbon Technocracy: Energy Regimes in Modern East Asia*. Chicago: University of Chicago Press, 2021.
Shafiee, Katayoun. *Machineries of Oil: An Infrastructural History of BP in Iran*. Cambridge, MA: MIT Press, 2018.
Shahnavaz, Shahbaz. *Britain and the Opening Up of South-West Persia, 1880–1914: A Study in Imperialism and Economic Dependence*. London: RoutledgeCurzon, 2005.
Shamir, Ronen. *Current Flow: The Electrification of Palestine*. Stanford, CA: Stanford University Press, 2013.
Shively, Jacob. "'Good Deeds Aren't Enough': Point Four in Iran, 1949–1953." *Diplomacy & Statecraft* 29, no. 3 (2018): 413–431.
Sitzes, Bryan Campbell. "Wildlife and Beyond: The Evolution of Modern Iranian Environmentalism." PhD diss. University of Texas at Austin, 2024.
Sontag, Susan. *On Photography*. New York: Farrar, Straus and Giroux, 1973. Reprint, New York: Picador, 1977.
Spitz, Peter H. *Petrochemicals: The Rise of an Industry*. New York: John Wiley & Sons, 1988.
Sreberny-Mohammadi, Annabelle, and Ali Mohammadi. *Small Media, Big Revolution: Communication, Culture, and the Iranian Revolution*. Minneapolis, MN: University of Minnesota Press, 1994.
Staples, Amy L.S. *The Birth of Development: How the World Bank, Food and Agriculture Organization, and World Health Organization Changed the World, 1945–1965*. Kent, OH: The Kent State University Press, 2006.
Star, Susan Leigh. "The Ethnography of Infrastructure." *American Behavioral Scientist* 43, no. 3 (1999): 377–391.
Steele, Robert. *The Shah's Imperial Celebrations of 1971: Nationalism, Culture and Politics in Late Pahlavi Iran*. London: I.B. Tauris, 2021.

Summit, April R. "For a White Revolution: John F. Kennedy and the Shah of Iran." *Middle East Journal* 58, no. 4 (2004): 560–575.

Swyngedouw, Erik. *Liquid Power: Contested Hydro-Modernities in Twentieth-Century Spain*. Cambridge, MA: Harvard University Press, 2015.

Tageldin, Shaden M. *Disarming Words: Empire and the Seductions of Translation in Egypt*. Berkeley, CA: University of California Press, 2011.

Thompson, T. Jack. *Light on Darkness?: Missionary Photography of Africa in the Nineteenth and Early Twentieth Centuries*. Grand Rapids, MI: William B. Eerdmans Publishing Company, 2012.

Todd, Zoe. "An Indigenous Feminist's Take on the Ontological Turn: 'Ontology' Is Just Another Word for Colonialism." *Journal of Historical Sociology* 29, no. 1 (2016): 4–22.

Tsing, Anna L. *The Mushroom at the End of the World: On the Possibility of Life in Capitalist Ruins*. Princeton, NJ: Princeton University Press, 2015.

Uekoetter, Frank. *The Age of Smoke: Environmental Policy in Germany and the United States, 1880–1970*. Pittsburgh, PA: University of Pittsburgh Press, 2009.

Vafa-Arani, Hamed, Salman Jahani, Hossein Dashti, Jafar Heydari, and Saeed Moazen. "A System Dynamics Modeling for Urban Air Pollution: A Case Study of Tehran, Iran." *Transportation Research* 31, Part D (2014): 21–36.

Vahdat, Farzin. *God and Juggernaut: Iran's Intellectual Encounter with Modernity*. Syracuse, NY: Syracuse University Press, 2002.

Verhoeven, Harry. *Water, Civilisation and Power in Sudan: The Political Economy of Military-Islamist State Building*. Cambridge, UK: Cambridge University Press, 2015.

Victor, David G., Amy M. Jaffe, and Mark H. Hayes, eds. *Natural Gas and Geopolitics: From 1970 to 2040*. Cambridge, UK: Cambridge University Press, 2006.

Vitalis, Robert. *America's Kingdom: Mythmaking on the Saudi Oil Frontier*. Stanford, CA: Stanford University Press, 2007.

von Schnitzler, Antina. *Democracy's Infrastructure: Techno-Politics and Protest After Apartheid*. Princeton, NJ: Princeton University Press, 2016.

Walker, Brett L. *Toxic Archipelago: A History of Industrial Disease in Japan*. Seattle, WA: University of Washington Press, 2010.

Wallerstein, Immanuel. *The Modern World-System I: Capitalist Agriculture and the Origins of the European World-Economy in the Sixteenth Century*. New York: Academic Press, 1974.

Ward, Barbara, and René Dubos. *Only One Earth: The Care and Maintenance of a Small Planet*. New York: W.W. Norton & Company Inc., 1972.

Warde, Paul, Libby Robin, and Sverker Sörlin. *The Environment: A History of the Idea*. Baltimore, MD: Johns Hopkins University Press, 2018.

Wilson, Sheena, Adam Carlson, and Imre Szeman, eds. *Petrocultures: Oil, Politics, Culture.* Montreal, Canada: McGill-Queen's University Press, 2017.

Winner, Langdon. "Do Artifacts Have Politics?" *Daedalus* 109, no. 1 (1980): 121–136.

———. "Upon Opening the Black Box and Finding It Empty: Social Constructivism and the Philosophy of Technology." *Science, Technology, & Human Values* 18, no. 3 (1993): 362–378.

Yeh, Sonia. "An Empirical Analysis on the Adoption of Alternative Fuel Vehicles: The Case of Natural Gas Vehicles." *Energy Policy* 35, no. 11 (2007): 5865–5875.

Yergin, Daniel. *The Prize: The Epic Quest for Oil, Money & Power.* New York: Free Press, 1991.

———. *The Quest: Energy, Security, and the Remaking of the Modern World.* New York: The Penguin Press, 2011.

Young, Oran R. *Resource Regimes: Natural Resources and Social Institutions.* Berkeley, CA: University of California Press, 1982.

Zia-Ebrahimi, Reza. "'Arab Invasion' and Decline, Or the Import of European Racial Thought by Iranian Nationalists." *Ethnic and Racial Studies* 37, no. 6 (2014): 1043–1061.

Index

The authorized representative in the EU for product safety and compliance is:
Mare Nostrum Group
B.V Doelen 72
4831 GR Breda
The Netherlands

www.ingramcontent.com/pod-product-compliance
Lightning Source LLC
Jackson TN
JSHW021433090226
97521JS00004B/5

* 9 7 8 1 5 0 3 6 4 6 2 2 3 *